Managing Forest Ecosystems

Volume 45

Series Editors

Margarida Tomé, Instituto Superior de Agronomía, Lisboa, Portugal

Thomas Seifert, Faculty of Environment and Natural Resources, University of Freiburg, Freiburg, Germany

Mikko Kurttila, Natural Resources Institute, Helsinki, Finland

The aim of the book series Managing Forest Ecosystems is to present state-of-the-art research results relating to the practice of forest management. Contributions are solicited from prominent authors. Each reference book, monograph or proceedings volume will be focused to deal with a specific context. Typical issues of the series are: resource assessment techniques, evaluating sustainability for even-aged and uneven-aged forests, multi-objective management, predicting forest development, optimizing forest management, biodiversity management and monitoring, risk assessment and economic analysis.

Pasi Rautio • Johanna Routa •
Saija Huuskonen • Emma Holmström •
Jonas Cedergren • Christian Kuehne
Editors

Continuous Cover Forestry in Boreal Nordic Countries

 Springer

Editors

Pasi Rautio
Natural Resources Unit
Natural Resources Institute Finland (Luke)
Rovaniemi, Finland

Saija Huuskonen
Natural Resources Unit
Natural Resources Institute Finland (Luke)
Helsinki, Finland

Jonas Cedergren
Skogforsk
Uppsala, Sweden

Johanna Routa
Production Systems Unit
Natural Resources Institute Finland (Luke)
Joensuu, Finland

Emma Holmström
S. S. Forest Research Centre
Swedish University of Agricultural Sciences
Lomma, Sweden

Christian Kuehne
Division of Forest and Forest Resources
Norwegian Institute of Bioeconomy
Research (NIBIO)
Ås, Norway

ISSN 1568-1319   ISSN 2352-3956   (electronic)
Managing Forest Ecosystems
ISBN 978-3-031-70483-3   ISBN 978-3-031-70484-0   (eBook)
https://doi.org/10.1007/978-3-031-70484-0

Natural Resources Institute Finland (LUKE), Swedish University of Agricultural Sciences (SLU),
Norwegian Institute of Bioeconomy Research (Nibio), Skogforsk

Acknowledgements

This book has received funding from Nordic Forest Research (SNS no. N2023-07), Natural Resources Institute Finland (Luke), the Swedish University of Agricultural Sciences (SLU) and Future Forests, Skogforsk (the Forestry Research Institute of Sweden), Norwegian Institute of Bioeconomy Research (NIBIO), and REBOUND project (funded by Strategic Research Council within the Research Council of Finland, decisions No 358482 + 358497). All the chapters were reviewed by programme director Mikko Kurttila (Luke), Professor Tomas Lundmark (SLU), and at least two of the editors. Language was edited by Carl Salk and Leslie Walke.

Contents

About the Editors

Pasi Rautio currently works as a research professor in silviculture at Natural Resources Institute Finland (Luke). His expertise is in silviculture as well as in forest and environmental ecology. His roles involve forest monitoring and studies on forest regeneration and forest management methods in areas that are facing pressures of other land uses, such as tourism, recreation, and reindeer herding. Pasi is the coordinator for ArcticHubs, a H2020 project investigating solutions to Arctic land use conflicts and long-term sustainability of the region. His previous work at the University of Oulu in Finland and Stockholm University in Sweden has concerned issues like biodiversity and restoration of seminatural pastures; evolutionary ecology of plant–animal interactions; and air pollution effects on forest ecosystems. He also worked as a forest policy officer in the European Commission (DG Environment). He is a member of UNECE Team of Specialists on Boreal Forests and Programme Co-ordinating Group of UNECE ICP Forests.

Johanna Routa works as a research manager in Profitable and responsible primary production research programme at Luke. She is a docent of Sustainable forest biomass production and quality management at the University of Eastern Finland. Her expertise is in sustainable forest biomass production, biomass supply chains, costs of different management practices, and the quality management of forest biomass. She has experience in research leadership and bioenergy R&D with companies and other stakeholders. Johanna is the coordinator of H2020 project BRANCHES (Boosting Rural Bioeconomy Networks following multi-actors approaches) which aims to increase the implementation of new technologies in bioeconomy. She is a member of the IEA bioenergy group.

Saija Huuskonen is a senior scientist at Luke and a docent of silviculture at the University of Helsinki. Her core expertise lies in forest management and silvicultural practices, focusing on their impact on wood production and sustainable forest use. She is involved in the developing of the forest management decision support system, Motti. Saija leads mixed forests projects, which evaluate mixed forests from various perspectives, including ecosystem services, forest management, forest resources, risk management, and growth dynamics. She also established new long-term mixed forest field experiments to create new research infrastructure. She is a member of national forest management committees and deputy coordinator at IUFRO 1.01.01 Boreal Forest Silviculture and Management division. Dr. Huuskonen has contributed to dissemination efforts through peer-reviewed articles, textbooks, reports, and speaking engagements. Her diverse knowledge and experience in silviculture in the Nordic boreal region spans research, teaching, and practical applications.

Emma Holmström is an associate professor in silviculture at SLU. Her research spans over forest management for production, forest conservation, and multiuse purposes. She studies forest experiments and monitoring data from the Swedish national forest inventory, aiming for results useful for both the applied forestry and the theoretical framework of modelling growth and yield. She has a special interest in silviculture of mixed forest and of the broadleaved tree species in northern Europe. In recent years, she has established several long-term experiments and projects investigating how climate change-induced droughts might be mitigated by forest management in pine and spruce. Emma is the coordinator of the interdisciplinary platform Future Forests since 2021, which is a research infrastructure used for developing knowledge on CCF both in past programme periods and in the current.

Jonas Cedergren is a research officer working for the Swedish Forest Research Institute. His area of expertise is the zone between silviculture and harvesting system and methods. He has conducted practical studies in Sweden and Malaysia. Focus has been on response of the residual stand to partial harvests and the implications of logging damage. Mr. Cedergren has worked for more than a decade with selective logging operations in Malaysia. There he conducted experiments on felling damage, established a large-scale and long-term growth and yield study, and developed a prototype logging system for peat swamp forest. He coordinated a major Swedish project on closed canopy forest management methods in Sweden. He has worked extensively with the FAO Global Forest Resource Assessment. Finally, he has worked a decade at the FAO as a specialist officer in forest technology, with a focus on occupational safety and health.

Christian Kuehne currently works as a researcher in forest management at the Norwegian Institute for Bioeconomy Research (NIBIO) in Ås, Norway. He lectured on silviculture and forest restoration at the University of Freiburg, Germany, and forest mensuration at the University of Maine, USA. His studies examine the effects of manipulating forest structure through silvicultural interventions on stand dynamics such as regeneration establishment and growth response in residual trees. He has expertise in forest growth and yield modelling and also studied challenges and limitations of the close-to-nature forest management paradigm in Central Europe.

Chapter 1
Introduction

Jonas Cedergren, Emma Holmström, Johanna Routa, Saija Huuskonen, Christian Kuehne, and Pasi Rautio

Abstract

- In this book we summarize peer-reviewed scientific articles and research reports from Finland, Sweden, and Norway on continuous cover forestry (CCF), i.e. forestry without clearcutting
- This book originates from growing interest in CCF among various stakeholders, and aims to promote discussion, further research, and inform decision-makers
- The book targets those interested in boreal forests, forest management, and ecosystem services
- In this chapter we review the background to the use of CCF and the reasons that led to its prohibition and subsequent resurgence in the Nordic countries

J. Cedergren
Skogforsk, Uppsala Science Park, Uppsala, Sweden
e-mail: jonas.cedergren@skogforsk.se

E. Holmström
Swedish University of Agricultural Sciences, S. S. Forest Research Centre, Inst för sydsvensk skogsvetenskap, Lomma, Sweden
e-mail: emma.holmstrom@slu.se

J. Routa
Natural Resources Institute Finland, Production Systems Unit, Joensuu, Finland
e-mail: johanna.routa@luke.fi

S. Huuskonen
Natural Resources Institute Finland, Natural Resources Unit, Helsinki, Finland
e-mail: saija.huuskonen@luke.fi

C. Kuehne
Norwegian Institute of Bioeconomy Research, Ås, Norway
e-mail: christian.kuehne@nibio.no

P. Rautio (✉)
Natural Resources Institute Finland, Natural Resources Unit, Rovaniemi, Finland
e-mail: pasi.rautio@luke.fi

© The Author(s) 2025
P. Rautio et al. (eds.), *Continuous Cover Forestry in Boreal Nordic Countries*,
Managing Forest Ecosystems 45, https://doi.org/10.1007/978-3-031-70484-0_1

1

Keywords Continuous cover forestry · Rotation forestry · Forest management · Forestry without clearcutting · Nordic countries

1.1 Why This Book

In this book we have compiled research results, mainly from Finland, Sweden and Norway, as a basis for describing the current state of knowledge about, and experience of, forestry using methods without clearcutting.

"Without clearcutting" opens up for a multitude of definitions and ambiguities. Clearcutting is the prevailing method of wood procurement in the three countries, involving a final felling of forests, creating clearcuts that are then regenerated to form even-aged stands. In this book, we will be addressing forest management methods that involve no clearcutting, and throughout the text we call this continuous cover forestry (CCF).

Our focus is on forestry and silviculture in the three Nordic countries, Norway, Sweden and Finland, now and in the near future, with the assumption that needs and demands relating to the ecosystem services will remain the same as today. The idea of writing this book originates from the increasing interest in CCF among practitioners, politicians and society in general. We saw an interest and a need to compile and summarise our knowledge as researchers.

Our geographical framework is the boreal forest of the Fennoscandian Peninsula, a region dominated by coniferous forests of primarily Norway spruce (*Picea abies*) and Scots pine (*Pinus sylvestris*). The most common broadleaves in the region are silver and downy birch (*Betula pendula*, *Betula pubescens*), aspen (*Populus tremula*), and alder (*Alnus incana*, *Alnus glutinosa*). The climate is shaped by the northerly latitude, with large contrasts between winter and summer light and temperatures, but is also influenced by proximity to the Atlantic current, the Gulf Stream, which moderates cold winter temperatures.

The potential reader of the book (you) is someone who has, in some way, invested in learning more about the boreal forest and/or the use of forest products and ecosystem services. You may be a forest owner, forest worker, timber buyer, farmer, or reindeer herder, or you may simply be interested in forests and how they are managed. The text offers valuable information for researchers and students of natural resources, and for any students who want to learn more about CCF. We hope this book can promote discussion and further research, and believe it could be useful for decision makers at local, national, or international levels.

Our aim is to give an overview of the complexity and scope of the subject in the region, and to provide pathways to further reading and inspiration.

1.2 How to Use This Book, and a Brief Summary of the Chapters

We have summarised the published knowledge, mainly from peer-reviewed scientific articles and research reports. Experts across a diverse range of scientific disciplines have contributed their insights and experience (Fig. 1.1), and each chapter is written by of a team of researchers from Norway, Sweden and Finland. The authors have validated the sources and synthesised the most relevant knowledge from their field. They have also identified current knowledge gaps regarding conversion to CCF and CCF methods.

Each chapter can be read more or less as a standalone review of a certain aspect of CCF, but we recommend the reader to begin with Chap. 2, where the authors clarify definitions and methods described in the remainder of the book. We define what we understand as CCF but also what we mean by conversion to CCF from other forest management types. We also introduce terminology and methods used in the three Nordic countries, and national legislation pertaining to the issue. Throughout the book, CCF methods are compared to forest management methods

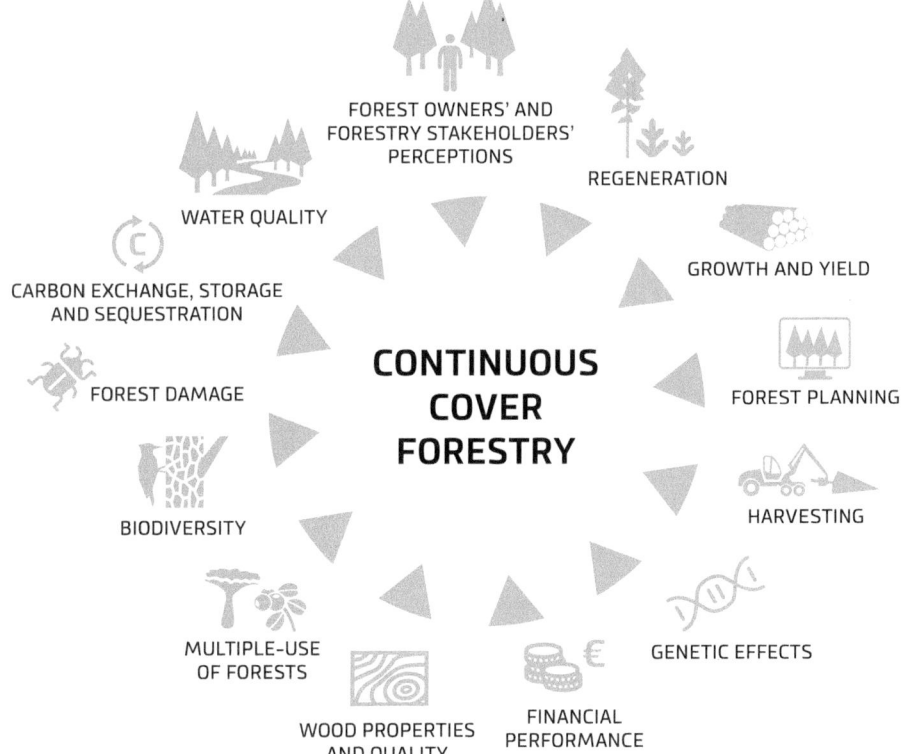

Fig. 1.1 Continuous cover forestry issues covered in this book

that do not aim to maintain a continuous forest cover. In practical forestry and media, these methods are commonly referred to as rotation forest management or rotation forestry (RF). Strictly speaking, this is not entirely accurate. As explained in Chap. 2, some cutting regimes under RF can also be considered CCF. However, the term RF, used as an alternative to CCF, has become established in the language of forestry professionals, media and even science. We therefore use the term RF also in this book in this context.

In Chaps. 3 and 4 we review cultivation of the boreal tree species under CCF, in terms of regeneration methods, growth and yield. Regeneration refers to active methods to achieve recruitment after one or more trees have been harvested, and includes both site preparation and the recruitment source (e.g., seeds or seedlings). In addressing growth and yield we highlight experiences gained from previous experiments, observations, and scenario analyses relating to CCF methods. In Chap. 5 we examine forest planning and how the national decision support systems can be applied to CCF systems.

Related to forest management are logging operations and possible similarities and differences between CCF and other management methods. These are addressed in Chap. 6, acknowledging that, even though the forest owner makes the decisions on which management system is to be used, the harvester operator is a crucial factor in the quality of the work.

Interest in the impacts of CCF methods on long-term growth and population genetics of trees will increase in view of a changing climate. In Chap. 7 the authors elaborate on what are important factors for breeding and how management can affect population genetics.

Chapter 8 addresses financial implications of CCF, together with a compilation of how financial aspects are analysed in many forest management systems. The financial outcome is also impacted by timber prices and management and harvesting costs, and how wood quality and characteristics are affected by the CCF method; the latter are outlined in Chap. 9. However, the forest produces more than wood, so in Chap. 10 we show how other provisioning services might be affected by CCF, in an overview of the multiple use of forests.

The effects of CCF on forest conservation and biodiversity are reviewed in Chap. 11, with emphasis on comparisons with clearcutting methods. Common to all the harvesting systems is the reduction in amount of large, old trees and dead wood. We consider biodiversity and forest conservation measures that can be important for CCF.

Forest health is important for the future provisioning of forest ecosystem services, so in Chap. 12 we assess how biotic and abiotic damage agents can interact with the choice of forestry system. Carbon exchange, storage and sequestration are in focus in Chap. 13, where we elaborate on how CCF methods might change the magnitude of carbon pool fluxes. In Chap. 14, the impact of CCF on water quality and riparian zones is discussed, and compared with corresponding existing knowledge regarding clearcutting systems.

Finally, in Chap. 15, we conclude by considering barriers to applying, and/or converting to, CCF, by reviewing the opinions and beliefs of stakeholders. We describe three main stakeholder groups: forest owners, the forest industry, and forest professionals, and discuss their attitudes towards CCF and potential implications.

1.3   Background to CCF in the Nordic Countries

1.3.1   Selection Cutting and Selective Logging

Discussions between advocates of CCF and RF have a long history in Fennoscandia. For Sweden, see e.g. Ström (1830), Kempe (1894) and Amilon (1930); for Norway, see Opsahl (1923), Eide (1936) and Barth (1937); for Finland, see Cajander (1910), Heikinheimo (1924), Hertz (1930) and Sarvas (1944).

Both systems can be traced back to the Middle Ages. There is evidence that regulated selective logging was applied in French broadleaved forests in the 1200s. Early development of selective logging included minimum periods between cuts (cutting cycles) and limits to harvesting rates (Hawkins 1962; Lundqvist 2005).

Policies and land tenure have changed considerably in Europe over the centuries. These often resulted in widespread use of exploitive and destructive forms of selection harvesting, today often referred to as selective logging (O'Hara 2014). Valuable trees were harvested, and stand recovery was left to chance. As a result, selective logging was restricted and banned in state- and company-owned forests in the mid-1800s. RF began to dominate in these forests instead.

In Fennoscandia, CCF was generally practised in the form of selective logging with pre-mature re-entries. One result was depleted stands with unstocked gaps (Lundqvist 2017), which led to general concern about future timber supplies (Leikola 1987). In Finland, for example, CCF was generally considered unsuitable for Finnish forests in the early 1900s (Cajander 1910).

During the 1940s in Fennoscandia, it was concluded that selection harvesting had not produced well-regenerated forests (e.g. Sarvas 1944; Opsahl 1953). This was partly a result of misuse of the selection system, as emphasised by its advocates (Barth 1937; Bøhmer 1957), and partly because the system was applied in unsuitable forests. Consequently, selective cutting and associated systematic forms of the selection system were effectively banned in Sweden and Finland (Söderström 1971). Research on selective alternatives to RF stopped, and experimental plots were abandoned in these two countries. In Norway, selective cutting was more gradually replaced by RF, because forest administrations endorsed and eventually promoted a shift towards RF advocated by influential actors in forest research (Andreassen 1994; Nygaard and Øyen 2020).

1.3.2 Shelterwood and Group Selection

There is evidence to suggest that a shelterwood approach was used in Central Europe in the 1600s (Hånell and Holgén 1997). The system in its present form seems to have first emerged during the 1800s in Prussia. Uniform shelterwood dominates present approaches, but there are other variants.

Although shelterwood cutting has only played a minor role in the management of spruce and pine stands in Norway, the system has inspired various research initiatives, summarised in Lexerød (2001). Irregularity of seed production of Norway spruce in northern Sweden and susceptibility of both pine and spruce to wind damage has limited the use of shelterwood in Sweden (Hagner 1958, 1962; Karlsson et al. 2017). In Finland, shelterwood cutting has generally been used in RF as a method of natural regeneration for spruce (Metsänhoidon suositukset – Metsien kestävän hoidon ja käytön perusteet 2022).

Group selection has been the subject of few studies in Fennoscandia. Group selection in combination with thinning from above was advocated as early as the study by Wallmo (1897). Group selection was also recommended for Norwegian spruce and pine stands by Barth (1937), Bøhmer (1957) and Børset (1986). In recent years, gap cutting has attracted more interest in Finland, and has been studied in pine (Hallikainen et al. 2019) and spruce forests (Valkonen et al. 2011) on both mineral soils and on peatland (Hökkä et al. 2011).

1.3.3 Renaissance of CCF

Interest in conservation started to grow in the mid-1970s, but did not initially impact timber production. Social concerns, archaeological sites, reindeer husbandry, and other public interests also started to leave their mark. It took until the early 1980s for research on the subject to start afresh. Interest in alternatives to RF has increased over time, and alternatives are now practised, although still on a modest scale. Forest laws are more permissive, and forest policies can be said to encourage diversification of forest management. The same period has also witnessed changes in RF, with conservation and social aspects playing a bigger role than before.

Much attention is currently paid to a number of concepts, for example close(r)-to-nature forestry or Pro Silva, new forestry, Liberich, and the Lubeck method. However, these concepts are approaches for individual operations rather than silvicultural systems. They could be called forest management philosophies within the realm of CCF (Albrektson et al. 2008).

Another important message from history is the relatively long timescale in boreal forest systems, from a tiny seed to an old giant tree. The forest we see today is the result of decisions made in a previous society, and the changes we decide on today will take another hundred years to manifest fully.

1.4 Looking Ahead, Forest Land Use and Needs for Adaptation to CCF, Sustainability, and Societal Needs

Forest management must meet many different objectives, and discussion is ongoing about how to reconcile these objectives. The production of raw materials for various purposes, as well as carbon sequestration, biodiversity conservation, and adaptation to climate change, all create new challenges. Aiming simultaneously at different goals can also result in conflicts. Geographical location, different types of growing conditions, and the history of forest management shape our forests and affect management options.

There is an urgent need for information on the various effects of different CCF methods. To fulfil the multiple goals of forest utilisation, we need not only diversified forests but also versatile and appropriate management methods. However, the mere choice of forest management method does not guarantee the realisation of all goals. For example, securing forest biodiversity requires active conservation strategies, both in RF and CCF.

Unlike much of central Europe, CCF has not been practiced on a large scale in the past 30–40 years in Fennoscandia. Whether greater implementation of CCF in the region is realistic, and suitable for a more balanced provision of forest ecosystem services, is currently the subject of heated debate. This book aims to provide an overview of the current state of knowledge about CCF in the Nordic region, so that the discussion can be based on scientific facts, while acknowledging knowledge gaps and associated research needs.

In recent years, several new projects and research focusing on CCF have been funded in the Nordic countries, reflecting the renewed interest among forest owners and society. Current research focuses on conversion and how a broader implementation of CCF might affect the outcomes of forest management at regional and national levels. Interest in CCF and its implementation is growing rapidly across the three countries, in part because of the new EU forest strategy for 2030. Implementation particularly applies to Norway, where CCF has already been adopted in the new PEFC standards. A key question is whether calls for more CCF in the Nordic countries are appropriate and realistic. This book does not offer a definitive answer to that question, but does provide an up-to-date review of scientific knowledge relevant to the Nordic region.

References

Amilon JA (1930) Wallmoblädningen å Högsjö. Sveriges skogsvårdsförbunds tidskrift:343–425
Andreassen K (1994) Bledning og bledningsskogen - en litteraturstudie. Aktuelt fra Skogforsk 2-94, 23 p
Barth A (1937) Norsk Skogbruk under utvikling. Tidskrift for Skogbruk 45
Bøhmer JG (1957) Bledningsskog II. Tidskrift for skogbruk 65:203-247
Børset O (1986) Skogskjøtsel II - Skogskjøtselens teknikk. Landbruksforlaget, Oslo

Cajander AK (1910) Metsiemme uudistushakkuut toisiinsa verrattuina. Ylipainos "Maahengestä". Otava. https://helda.helsinki.fi/bitstream/handle/10138/160522/1910_Cajander_Metsiemme. pdf?sequence=1. Accessed 5 Feb 2024

Eide E (1936) Norsk skogbruk under utvikling. Tidsskr Skogbr 44(2):1–16

Hagner S (1958) Om kott och fröproduktionen i svenska barrskogar. Meddelanden från Statens skogsforskningsinstitut 47(8):1–120. https://pub.epsilon.slu.se/10015/1/medd_statens_skogs-forskningsinst_047_08.pdf. Accessed 5 Feb 2024

Hagner S (1962) Naturlig föryngring under skärm. En analys av föryngringsmetoden, dess möjligheter och begränsningar i mellannorrländskt skogsbruk. Meddelanden från Statens skogsfor-skningsinstitut 52(4):1–263. https://publications.slu.se/?file=publ/show&id=125204. Accessed 5 Feb 2024

Hallikainen V, Hökkä H, Hyppönen M et al (2019) Natural regeneration after gap cutting in scots pine stands in northern Finland. Scand J For Res 34(2):115–125. https://doi.org/10.108 0/02827581.2018.1557248

Hånell B, Holgén, P (1997) Skärmskogsbruk i Sverige. Fakta Skog Nr 5

Hawkins PJ (1962) European selection forests with special reference to methods of yield determina-tion in comparison with *Callitris glauca* (Cypress pine) of southern Queensland. Dissertation. University of Oxford

Heikinheimo O (1924) Suomen metsien metsänhoidollinen tila. Commun Inst For Fenn 9:1–12

Hertz M (1930) Kuusi ja sen merkitys maamme metsätaloudessa. Yksityismetsänhoitajayhdistyksen julkaisuja, Helsinki

Hökkä H, Repola J, Moilanen M, Saarinen M (2011) Seedling survival and establishment in small canopy openings in drained spruce mires in northern Finland. Silva Fenn 45(4):633–645. https://doi.org/10.14214/sf.97

Karlsson C, Sikström U, Örlander G et al (2017) Naturlig föryngring av gran och tall. In Skogsstyrelsen (Swedish Forest Agency), Skogsskötselserien (Fries C ed) https://www.skogs-styrelsen.se/globalassets/mer-om-skog/skogsskotselserien/skogsskotselserien-04-naturlig-foryngring-av-tall-och-gran-2017.pdf. Accessed 5 Feb 2024

Kempe F (1894) I den norrländska skogsfrågan. Uppsatser i den norrländska skogsfrågan samt Landshövding Curry Treffenbergs motion on revision af skogslagstiftningen. Stockholm

Leikola M (1987) Metsien hoidon aatehistoriaa. Summary: leading ideas in Finnish silviculture. Silva Fenn 21(4):332–341. https://doi.org/10.14214/sf.a15480

Lexerød N (2001) Alternative skogbehandlinger - produksjon, virkeskvalitet, driftsteknikk & økonomi. Aktuelt fra skogforskningen 4/01. https://www.nb.no/items/URN:NBN:no-nb_digi-bok_2008110500072. Accessed 5 Feb 2024

Lundqvist L (2005) Blädning. Report 61. Department of Silviculture, SLU, Umeå

Lundqvist L (2017) Tamm review: selection system reduces long-term volume growth in Fennoscandic uneven-aged Norway spruce forests. Forest Ecol Manag 391:362–375. https://doi.org/10.1016/j.foreco.2017.02.011

Metsänhoidon suositukset – Metsien kestävän hoidon ja käytön perusteet (2022) Tapion julkaisuja. https://metsanhoidonsuositukset.fi/fi/metsanhoidon-suositukset-metsien-kestavan-hoidon-ja-kayton-perusteet. Accessed 5 Feb 2024

Nygaard PH, Øyen B (2020) Skoghistorisk tilbakeblikk med vekt på utviklingen av bestandssk-ogbruket i Norge. NIBIO rapport 6/45/2020. https://nibio.brage.unit.no/nibio-xmlui/han-dle/11250/2654343. Accessed 5 Feb 2024

O'Hara KL (2014) Multiaged Silviculture: managing for complex Forest stand structures. Oxford University Press, Oxford

Opsahl W (1923) Hugger vi vore skoger rationelt? Skogeigeren 10:109–113, 171–174

Opsahl W (1953) Trekk fra skogbrukets utvikling 1928–1953. Rich. Andvord, Oslo

Sarvas R (1944) Tukkipuun harsintojen vaikutus Etelä-Suomen yksityismetsiin. Referat: Einwirkung den Sägestamm plenterungen auf die Privatwälder Sudfinnlands. Commun Inst For Fenn 33:1–268. https://jukuri.luke.fi/handle/10024/522442. Accessed 5 Feb 2024

Söderström V (1971) Varför hyggen? In: Ymer 1971: Tätort och glesbygd, pp 161–175

Ström IA (1830) Handbok för skogshushållare, 2nd edn. Nordströms förlag, Stockholm
Valkonen S, Koskinen K, Mäkinen J, Vanha-Majamaa I (2011) Natural regeneration in patch clear-cutting in Picea abies stands in southern Finland. Scand J For Res 26:530–542. https://doi.org/1 0.1080/02827581.2011.611818
Wallmo U (1897) Rationell skogsafverkning, Stockholm
Albrektson A, Elfving B,Lundqvist L, Valinger E (2008) Skogsskötselns grunder och samband. Skogsskötselserien del 1

Chapter 2
Definitions and Terminology: What Is Continuous Cover Forestry in Fennoscandia?

Andreas Brunner, Sauli Valkonen, Martin Goude, Kjersti Holt Hanssen, and Charlotta Erefur

Abstract

- Definitions of continuous cover forestry (CCF) vary among countries, and are often a political compromise.
- We offer a common definition of CCF for this book, which can be found in a text box below.
- The three silvicultural systems included in CCF are described briefly.
- Conversion to CCF will be an important activity in the near future, but approaches to and experiences of conversion to CCF are largely lacking in Fennoscandia.
- Methods need to be developed for how to assess the suitability of forest stands for CCF or conversion to CCF.
- Bad practices and experiences with selective cutting in Fennoscandia before 1950 have led to a loss of experience and willingness to apply CCF.
- Climate adaptation will make it necessary to modify CCF approaches in the region, especially given the limited number of shade-tolerant species.

A. Brunner (✉)
Norwegian University of Life Sciences (NMBU), Ås, Norway
e-mail: andreas.brunner@nmbu.no

S. Valkonen
National Resources Institute Finland (LUKE), Helsinki, Finland
e-mail: sauli.valkonen@luke.fi

M. Goude
Swedish University of Agricultural Sciences (SLU), Tönnersjöheden försökspark, Simlångdalen, Sweden
e-mail: martin.goude@slu.se

K. H. Hanssen
Norwegian Institute of Bioeconomy Research (NIBIO), Ås, Norway
e-mail: kjersti.hanssen@nibio.no

C. Erefur
Swedish University of Agricultural Sciences (SLU), Umeå, Sweden
e-mail: charlotta.erefur@slu.se

© The Author(s) 2025
P. Rautio et al. (eds.), *Continuous Cover Forestry in Boreal Nordic Countries*,
Managing Forest Ecosystems 45, https://doi.org/10.1007/978-3-031-70484-0_2

Keywords Silvicultural systems · Selection system · Shelterwood system · Group system · Conversion

2.1 Definitions and Terminology

In this chapter, we define the terms essential for this book, i.e., continuous cover forestry (CCF), conversion to CCF, and other related terms. We also give a brief overview of the silvicultural systems included in CCF in our region. The context for these definitions and terms will be presented in Sect. 2.2. Clear definitions of terms are essential for communication, especially when using technical terms, which are often abbreviations of complex concepts. Terminology also varies between countries, and the English terminology is often not well known.

2.1.1 Continuous Cover Forestry

CCF has been defined in many different contexts, and definitions vary. It has never been a scientific concept, and does not have its origin in research or academic textbooks. When CCF is defined in the context of forest policy at different levels, these definitions are typically not very detailed or operational (e.g., Vitkova et al. 2014; Hertog et al. 2022; Egan et al. 1999). Silvicultural textbooks and scientific literature rarely offer a clear definition of CCF. Frequently, equivalent terms in national languages are used, which sometimes overlap with CCF as it is defined here. Some of these terms are reviewed below.

Based on the terms and definitions currently used in Fennoscandia, *we define CCF* as the group of silvicultural systems that maintain a continuous forest cover. CCF is not defined by the occurrence of specific stand structures over a fixed period, nor by individual silvicultural interventions, e.g., regeneration cuts. However, the silvicultural systems included lead to variations in stand structures within certain limits. The terminology of silvicultural systems and definition of the term silvicultural system draws upon Matthews (1989).

Silvicultural systems that are, or can be, used in Fennoscandia to maintain a continuous cover are the *selection system,* the *shelterwood system*, and the *group system*. The shelterwood and group systems also belong to the *rotation forest management* paradigm, while the selection system does not (Table 2.1). The rotation forest management includes silvicultural systems designed for more or less even-aged stands, where all trees are harvested more or less at the same time, i.e., at the end of the rotation. The use of shelterwood and group systems can only be considered as CCF if a continuous minimum forest cover is maintained and is an explicit aim. In that sense, a fast complete release of an established regeneration in the shelterwood system or large, many, or frequent gaps in the group system do not

Table 2.1 Summary of terms used for silvicultural systems

Management paradigm	Silvicultural system	Individual cuttings	Continuous cover forestry
Selection management	Selection system	Selection cutting	Yes
Rotation forest management/rotation forestry	Group system	Gap cutting, gap extension	When continuous cover is the aim and maintained
	Shelterwood system	Preparatory cutting, shelterwood cutting(s), shelter removal	When continuous cover is the aim and maintained
	Clearcutting system	Clearcutting, seed tree cutting, seed tree removal	No

maintain a forest cover. These practices of the shelterwood and group system are therefore outside what we define as CCF.

What is CCF in Fennoscandia?

- We define *CCF* as the group of silvicultural systems that maintain a continuous forest cover.
- We refrain from including a detailed definition of a *minimum forest cover* in terms of density, height, homogeneity, and continuity of the cover in our definition of CCF. An overview of definitions for a minimum forest cover in the three countries is given.
- *Silvicultural systems* that are, or can be, used in Fennoscandia to maintain a continuous cover are the *selection system*, the *shelterwood system*, and the *group system*.
- Only the long-term application of one of these silvicultural systems qualifies as CCF.
- *Conversion to CCF* is not included as CCF in our definition. However, conversion will be an important activity in the near future for forest owners who wish to apply CCF.

CCF is defined here as an application of a silvicultural system over at least one rotation, for the silvicultural systems where the term rotation still applies, or over an equivalent period for the selection system. Individual cuts only qualify as CCF if they are part of a silvicultural system and its long-term application. For example, an individual selective cut is only CCF if it is part of the selection system; the same applies for gap cuttings and the group system, and for shelter cuts or late thinnings and the shelterwood system.

Our definition of CCF only includes the practice of the silvicultural systems in stands that possess and maintain these systems' typical structures. Conversion to

CCF is not included in this definition. However, in Finland and many other political contexts, conversion to CCF is also included under this umbrella (Äijälä et al. 2019; Valkonen 2020).

CCF is most often thought of as being connected to *natural regeneration*. However, it can also be practised in combination with or purely based on artificial regeneration methods.

None of the CCF methods are strictly limited to *shade-tolerant species*, but the methods were originally developed for them (Schütz 2001; Burschel and Huss 1997). The current call in Fennoscandia to apply CCF to intolerant species such as Scots pine or mixed forests containing shade-intolerant species makes modifications to the original prescriptions necessary.

When discussing CCF, it is frequently necessary to specify the opposite, i.e., summarise silvicultural systems, or certain practices of silvicultural systems, which do not meet our definition of CCF, under a common term. Such terms exist in our national languages and for the national definitions of CCF, e.g., *åpne hogster* in Norwegian and *hyggesskogsbruk* in Swedish. These terms summarise the traditional practice during the last 70 years and include the clearcutting system and the seed tree system. We do not introduce a common definition in English for non-CCF methods in this book, and instead refer to them as *traditional management* or give details about which silvicultural methods have been addressed in the studies.

2.1.1.1   Minimum Density and Continuity of Forest Cover

We refrain from giving a detailed definition of a *minimum forest cover* in terms of density, height, homogeneity, and continuity of the cover in our definition of CCF. However, such definitions are understandably necessary to implement forest policies, legislation, and management prescriptions. In this context, the first attempts to define CCF more specifically, including minimum forest cover specifications, have been made in Fennoscandia.

In Norway, *lukkede hogster* (closed cuts) means that at least 150 trees/ha are left after a regeneration cut or that gap size is less than 0.2 ha. In addition to the selection system, the shelterwood system, and the group system, it is also customary to include *fjellskoghogst* (mountain forest selection cutting), a method combining the group and selection systems, in this term. The term *lukkede hogster* has been in use in silvicultural textbooks for decades. This definition is not legally binding, except for the Norwegian PEFC standard (revision 2023), which uses the limits shown above for stand density and gap size.

However, the number of trees alone does not specify a forest cover, nor the duration of the cover. Assuming tree dimensions in a shelterwood system of DBH = 32 cm, 150 trees/ha would translate to 12 m^2/ha basal area, which likely comprises a forest cover in that situation. The current practice of *mountain forest selection cutting* often leaves much more open stands (Granhus et al. 2020).

In Sweden, the Forestry Act (§5 and §10) defines a minimum stand density (standing volume as a function of basal-area-weighted mean height) that does not trigger an obligation for active regeneration efforts. These density thresholds define a minimum forest cover for CCF. Current definitions of CCF-related terms provide detailed definitions (Appelqvist et al. 2021). The Swedish definition of *hyggesfritt skogsbruk* (clearcut-free forestry) has its baseline in the statutory restriction in harvesting level at a given height, which distinguishes a thinning from a cutting that requires regeneration, the "§5 curve" (illustrated in Appelqvist et al. 2021, page 7, https://www.skogsstyrelsen.se/globalassets/om-oss/rapporter/rapporter-20222021202020192018/rapport-2021-8-hyggesfritt-skogsbruk%2D%2D-skogsstyrelsens-definition.pdf). However, this only applies until regeneration is established. In shelterwood cutting, the stand volume can be reduced to half the standing volume defined in the §5 curve after a satisfying regeneration is established. The number of seedlings needed for this is defined in the Forestry Act, and ranges from 1000 to 1500 /ha depending on site quality. When the regeneration reaches 2.5 m, the remaining shelterwood can be reduced to 25 trees/ha above 10 m height in coniferous stands and 5 trees/ha above 10 m in noble broadleaf stands (e.g., beech). In gap cutting, gap size is limited to 0.25 ha, and additional gaps that reduce the standing volume below the §5 curve can be created after a satisfying regeneration is established and reaches 2.5 m height.

In Finnish legislation, CCF is defined by cutting methods, with the explicit inclusion of only selection and gap cutting. Minimum post-harvest stand basal area limits are defined for harvest entries in the Forest Act. They vary by latitude (lower limits in the north) and soil fertility (lower limits where fertility is lower). The values range from 5 m^2/ha on xeric and subxeric sites in northern Lapland to 10 m^2/ha on mesic sites in southern Finland. In the widely recognised Finnish best practices (Äijälä et al. 2019), the recommended minimum basal areas are generally 2–3 m^2/ha greater than the statutory limits. The motivation is to provide a buffer in the event of post-harvest damage and mortality, and also to safeguard a high level of volume production.

2.1.1.2 Conversion to CCF

Conversion is a distinct change in silvicultural system, stand structure, and/or tree species composition. *Transformation* indicates a smaller or more gradual change in silvicultural system, stand structure, and/or tree species composition. This definition has been used in Central Europe for many decades and is also reflected in Hasenauer (2004). Shifting from the clearcutting and seed tree system to CCF, as currently intended in many places in Fennoscandia, falls under the term conversion.

Given the current rarity of CCF in Fennoscandia, conversion to CCF will be an important activity in the coming decades. A time horizon of many decades is the compound consequence of only individual properties or stands being converted, only a part of the forest stands being readily suitable for conversion, and the time horizon of many decades for a conversion of specific stands. Knowledge about conversion is therefore currently much more relevant than detailed prescriptions for the silvicultural systems included in CCF in conditions after conversion.

Special silvicultural treatments to convert to CCF are only necessary for some silvicultural systems and stand conditions (see Sect. 2.1.2.2 for further details). In some cases, silvicultural systems under CCF can be applied without conversion, for example the shelterwood system in previously thinned stands or the group system. In these cases, the first regeneration cut might mark the shift to CCF.

2.1.1.3 Other Terms

Uneven-aged management is used as a more general term in many contexts. Some of the silvicultural systems included in our definition of CCF, e.g., the shelterwood system, might only generate a limited range of age variation for longer periods of stand development.

The term *multi-storeyed (flersjiktet)* is frequently used in the context of CCF and related terms. Our definition of CCF is wider in that it does not set up a certain stand structure as an objective, but accommodates variation in stand structures according to the dynamics of the silvicultural systems involved.

Mountain forest selection cutting (*fjellskoghogst*) is practiced in Norway and Sweden, often reducing standing volume intensively at large cutting intervals (Granhus et al. 2020). Even though the current practice might not fulfil criteria for a minimum forest cover, a modified practice that maintains a larger standing volume and acknowledges the very long regeneration times on these sites might be classified as CCF.

National terms only partly overlap with our definition of CCF.

In Norway, *lukkede hogster* (closed cuts) only provides a very simple and hardly operational classification of silvicultural systems.

In Sweden, *hyggesfritt skogsbruk* (clearcut-free forestry) is mainly a negative definition that excludes clearcutting (Appelqvist et al. 2021). However, the detailed definitions supply a coherent definition of CCF. *Naturkultur* (Liberich) is a proposal for an individual tree-based management system (Hagner et al. 2001), where performance at stand level is uncertain and not documented, but currently being tested in a series of field trials (Goude et al. 2022).

In Finland, *jatkuvapeitteinen metsänkasvatus* (continuous cover forest management), or *jatkuva kasvatus* (continuous cover silviculture), is wider than our definition of CCF, because it also includes methods for conversion to CCF and some overlap with two-storey stands (Äijälä et al. 2019).

In the same way that CCF is most often a political concept, many other terms have been coined to describe forest management systems that follow a certain philosophy or political agreement. Most prominently, *close-to-nature forestry* has been promoted in Europe for decades (Brang et al. 2014). More recently, the EU has introduced the term *closeR-to-nature forest management* and asked researchers to define it (Larsen et al. 2022; European_Commission 2023). In North America, *forest ecosystem management* and *ecological forestry* have been promoted as alternatives to traditional forest management in recent decades. Egan et al. (1999) discuss the lack of clarity of these terms and their relationship to other basic concepts of forest management, i.e., sustainability and multiple use.

Retention forestry (Gustafsson et al. 2020) is a concept closely linked to the clearcutting system. However, in leaving single trees and groups of trees, retention forestry can generate stand structures that might facilitate conversion to CCF in the future. Retention will also be necessary in most CCF methods.

Natural disturbance-based forest management (Kuuluvainen et al. 2021) promotes a philosophy that asks forest management to mimic natural disturbances. Harvesting with single-tree selection and small gaps would reflect small-scale natural disturbances, while larger gaps would represent intermediate disturbances, and larger clearcuts large-scale disturbances, respectively. A much greater volume of retained trees, snags, and coarse woody debris would be characteristics for such management (Koivula et al. 2014). In contrast to this philosophy, the silvicultural systems included in our definition of CCF do not explicitly intend to mimic natural disturbances. They might briefly generate stand structures that also can be found in forests under natural dynamics (Rouvinen and Kuuluvainen 2005). However, systematic target diameter harvest in the selection system or a homogeneous shelter in the shelterwood system do not have equivalents in natural forest dynamics.

2.1.2   Methods of Continuous Cover Forestry

In this section, to give a more detailed definition of CCF likely to be practised in Fennoscandia, we briefly describe the silvicultural systems and conversion methods involved. Due to the current lack of practice of these methods, the description will largely rely on silviculture textbooks from Central Europe and experiences from a limited number of experiments in Fennoscandia. We also address necessary modifications to current forest management in Fennoscandia due to adaptation to climate change.

2.1.2.1   Silvicultural Systems under CCF

The following descriptions of the three silvicultural systems are only intended as a definition of these CCF methods. Detailed research results from the region that describe regeneration, genetic effects, growth and yield, pathogens and pests, and wood quality will be reviewed in later chapters. In contrast to the simple, standardised instructions currently practised in the context of the dominant clearcutting and seed tree system, silvicultural prescriptions for CCF methods will be much more complex and detailed, and require more intensive status descriptions and monitoring by professional managers and forest owners.

Other silvicultural systems relevant in this context include combined systems, such as irregular shelterwood. They are not described in detail here.

2.1.2.1.1   The Selection System

The selection system is based on the harvest of individual trees and the continuous maintenance of a multi-layered stand structure. It has been practised for centuries, mostly by farmers in different regions of Europe. Around 150 years ago it was developed into a more systematic management system, often accompanied by detailed inventories controlling stand structure and stand density (Schütz 2001; Lundqvist et al. 2014). Given the continuously high stand density, the selection system is best suited for shade-tolerant species, e.g., Norway spruce in Fennoscandia. However, trials with less shade-tolerant species, e.g., Scots pine, have been set up. To guarantee continuous ingrowth of less shade-tolerant species, stand density has to be reduced, which limits productivity.

An alternative approach to maintaining presence of these species is to create larger openings in a group selection system. The group selection system systematically opens small gaps, either in addition to or combined with single tree harvesting, to allow regeneration of groups of trees. In addition to allowing regeneration of less shade-tolerant species, groups of young trees are an advantage for species that require conspecific competition to form a straight trunk, such as most broadleaved species.

The multi-layered structure in stands managed by the selection system is characterised by a falling diameter distribution. This diameter distribution indicates that trees of all size classes are to be found in a stand, and is often thought of as indicating a vertical structure with multiple canopy layers. However, these stands are also richly structured horizontally, with small openings and dense groups (Figs. 2.1 and 2.2). Quantifying a 3D stand structure or diameter distributions in practical forest management probably exceeds the resources available in forest management in Fennoscandia. Simplifications for these stand structure requirements have therefore been proposed (Lundqvist et al. 2014).

To maintain the multi-storeyed stand structure, a stand density that guarantees high productivity and sufficient ingrowth must be preserved over time. Cutting intervals and amounts need to be adapted to this. Individual tree selection is most often guided by principles of target diameter harvest, but might also include elements of stand tending.

The selection system should not be misunderstood as directly copying natural disturbance dynamics. It is an intensive management system that maintains a certain phase of stand dynamics. Given the harvesting of mostly the largest trees and the low target diameters in Fennoscandia, large living and dead trees will not automatically be present in stands managed according to this system. To provide habitats for organisms adapted to these structures, the selection system must also apply retention principles and leave large trees of various species. This also applies to the shelterwood and group systems.

Fig. 2.1 A stand managed with the selection system at two points in time. Even though the structure of the stand has changed due to tree growth and removals, the visual impression is near constant

2.1.2.1.2 The Shelterwood System

The shelterwood system has been practised on small areas in Fennoscandia in recent decades, both in Norway spruce and in Scots pine (Fig. 2.3). Starting with stands that have been regularly thinned during their lifetime, and therefore consisting of individual trees with long crowns and high individual-tree stability, the first preparatory cuts initiate a series of cuts that maintain a shelter over many decades. This initiates regeneration at rather high density, distributed homogeneously over the entire area (Figs. 2.4 and 2.5). The last shelter trees should not be removed before

Fig. 2.2 Stand structure and cut instructions for a sample plot in Norway managed with the selection system. Dark green = spruce, light green = birch. Spruce trees to be removed during the next selection cut are shown in light blue. The planned cut will remove approximately 100 m³/ha of the standing volume of 255 m³/ha

the entire area is regenerated and covered by small trees with a minimum height to be established, i.e., at least 50 cm (Fig. 2.5). This variant of the system produces a stand in the next generation with an age variation of up to 30 years, which most often develops into a rather homogeneous stand structure despite the large age range. In a contemporary Finnish variant for pine (Fig. 2.4), some of the overstorey

Fig. 2.3 Advance regeneration of Scots pine under a shelter in Løiten Almenning, Norway. Photo: Andreas Brunner

trees are retained throughout the rotation to promote ecosystem services and production of high-quality timber (Äijälä et al. 2019; Valkonen 2020). Wind damage is common in this system during the shelter phase, but varies greatly according to stand and meteorological conditions (Hånell and Ottosson-Löfvenius 1994).

2.1.2.1.3   The Group System

In the group system, stands are opened up by creating a series of gaps. Gap size, number, and placement, chronological gap sequence, and gap extension can vary widely in this system (Figs. 2.6 and 2.7). Gaps might also be initiated as shelter groups. After initiating the regeneration process with gaps, the remaining parts of the stand might also be harvested at once or with other patterns. However, gap extension is a common prescription in this system. The irregular shelterwood system also creates gaps as part of the regeneration process. Gap size and sequence have to be adapted to the shade tolerance of the species targeted in the next generation. There are no long-term examples of the practical application of group systems in Fennoscandia, and in Central Europe a description of the full regeneration period is also lacking (Brunner et al. 2004).

In Fennoscandia, the group system is often proposed for shade-intolerant species such as Scots pine. The minimum gap size for natural regeneration of Scots pine and birch species is not known, but might vary greatly depending on site conditions and along latitudinal gradients.

TIME

Fig. 2.4 A Scots pine stand managed with the shelterwood system at three points in time. This variant of the shelterwood system keeps a dense shelter over an extended period, until the next generation has reached heights that approach the heights of the shelter trees

2.1.2.2 Methods for Conversion to CCF

Conversion to CCF is most challenging in the case of conversion to the selection system, because it requires developing multi-layered stand structures. This can only be accomplished over a period of many decades, involving natural regeneration processes and requiring sufficient individual tree stability and remaining lifetime of the adult trees (Schütz 2001). This section therefore focuses mainly on this task.

Application of the shelterwood and group systems often only requires stands that are sufficiently thinned during earlier stand development to consist of trees that are individually highly stable. No further conversion is necessary, as regeneration is initiated during the regeneration cuts in these systems.

Conversion to the selection system will be an important topic in the context of CCF for many forest managers in Fennoscandia in the coming decades. Forest management practices in Fennoscandia since 1950 have created stand structures that are difficult to convert to CCF. These are even-aged monospecific stands, often dense and unthinned (e.g., in Norway) and therefore with very low individual tree stability and basically no other option than clearcutting in older stands. However, in younger stands, lower risk of wind damage permits creation of openings, initiating recruitment, and generating multi-layered stands over the subsequent decades.

Approaches for conversion to the selection system have only been developed and tried for a few decades in other regions, e.g., Central Europe and western North America. The first empirical results are available, but development and testing are still taking place. For boreal forests and Fennoscandia, approaches and trials are

Fig. 2.5 Time series of the shelterwood phase in a shelterwood system. After the preparatory cut at age 40 (subfigure 2), two shelterwood cuts at ages 50 and 60 (subfigs. 3 and 4) create an open shelter that allows advance regeneration. The shelter is removed at age 70 (subfig. 5) when the advance regeneration has reached a height of at least 50 cm. By age 70 the regeneration established after the first shelterwood cut has reached a height of about 2 m

very scarce and still need substantial work for development or adaptation. Practices from other regions that have been tried in Fennoscandia are *target diameter harvest* and *variable-density thinning*, which are described in more detail below.

Conversion scenarios have been studied with growth models, which often show lower growth during the conversion period (Reventlow et al. 2021; Hilmers et al. 2020; Hanewinkel and Pretzsch 2000; Brunner et al. 2006). Reduced growth is caused by reducing the stand density to initiate recruitment. The exact reduction in stand growth is uncertain, because growth models are often not designed for these stand structures (Drössler et al. 2014; Brunner et al. 2006), or because they use rough assumptions for the establishment and growth of seedlings and saplings (Drössler et al. 2014; Reventlow et al. 2021; Hanewinkel and Pretzsch 2000; Hilmers et al. 2020). In some scenario studies, the conversion method is described in insufficient detail, so cannot be adapted to other forest types (Reventlow et al. 2021; Hanewinkel and Pretzsch 2000; Drössler et al. 2014; Hilmers et al. 2020). None of the simulators include risks for windthrow or other risks that reduce

Fig. 2.6 A stand managed with the group system at two points in time. Here, the group system is applied in a stand with large variation in tree heights. Gap cuttings remove only a few trees per gap, creating small openings in the beginning, which are slightly extended in later cuts. This example is therefore close to the group selection system

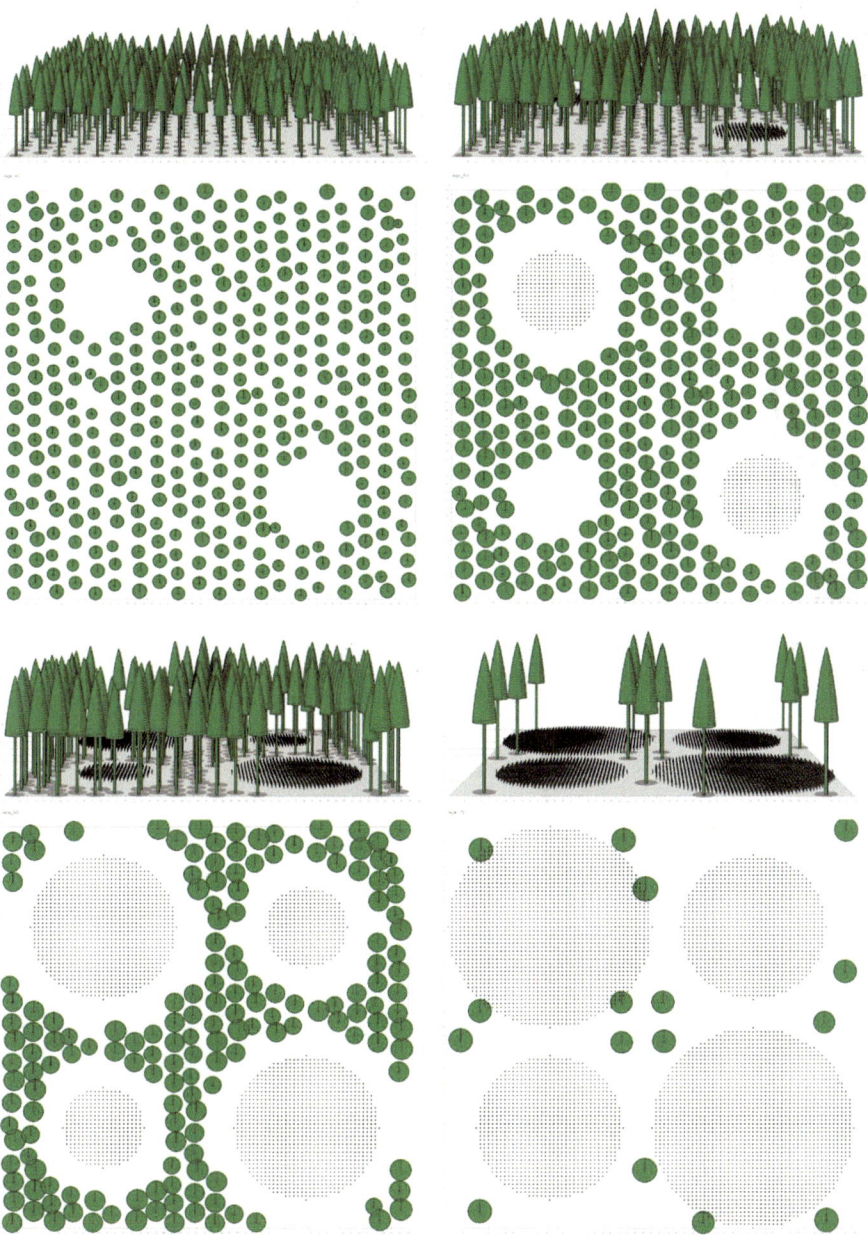

Fig. 2.7 Time series of the gap cutting phase in the group system. The first two gaps with a diameter of 20 m are created at age 40 (subfigure 1) in this even-aged 1-ha stand with little variation in tree heights. These gaps are extended to a diameter of 35 m at age 50 (subfig. 2), when new gaps are also created. All gaps are further extended at age 60 (subfig. 3) and the remaining stand is removed at age 70 (subfig. 4), when the advance regeneration has reached a height of at least 50 cm. The regeneration established after the first gap cuts will now have reached a height of about 2 m

production, depending on treatment. Stand simulators are therefore unlikely to contribute to the design of conversion methods in the near future.

Target diameter harvest has been proposed and tried as a method for converting even-aged Norway spruce stands to the selection system (Sterba 2004; Sterba and Zingg 2001; Reininger 1987). Results so far indicate that the stands have developed towards a multi-layered structure and survived major storm events. In southern Germany, beech has been managed with target diameter harvest; however, this is unlikely to produce multi-layered structures due to the high shade-tolerance of this species (Brunner et al. 2004). Inspired by the practice in Central Europe, experiments with target diameter harvest have been set up in Sweden (Drössler et al. 2015).

Variable-density thinning (VDT) has been developed and practised as an approach for conversion of young even-aged conifer stands in Washington and Oregon during the last 30 years (O'Hara et al. 2012; O'Hara et al. 2010; Brodie and Harrington 2020; Harrington et al. 2005; Carey 2003, 2006; Puettmann et al. 2016; Willis et al. 2018; Comfort et al. 2010; Dodson et al. 2012). VDT results in varying stand density after thinning, from the extremes of openings (*gaps*) to unthinned patches (*skips*), with thinning intensity varying within the thinned matrix between those extremes. Designs are adapted to the objective of the conversion and often planned in detail and marked in the stand before operations. In a simulation study for Finland, Pukkala et al. (2011a) designed a VDT that only harvests four different patches at different times. VDT has been adapted to tree species and thinning practice with harvesters in Norway (Brunner 2024), and a series of trials is currently under way in Norway. Disturbances like wind or snow damage in young stands might also generate stand structures with variable density that can be used to convert to the selection system (Knoke and Plusczyk 2001).

A different concept for conversion of young spruce stands involving heavy but homogeneous thinnings has been tried in Sweden (Drössler et al. 2014), but the results have been contradictory and the treatment plan was modified in the trial (Goude et al. 2022).

In Finland, the approach to conversion is rather pragmatic, given the lack of practical experience and research. The general idea is to initiate regeneration through shelterwood or gap cutting to gradually establish the understorey that may later be used to develop structural complexity (Äijälä et al. 2019; Valkonen 2020). An alternative pathway suggested by (Pukkala et al. 2011b) would be to conduct a shelterwood cutting with very low density in the remaining overstorey. An abundant regeneration of pioneer species could be expected to emerge quite rapidly. The subsequent emergence of an understorey of shade-tolerant species like spruce would pave the way to a structurally complex, mixed-species stand. Another suggested method is *all-aged forestry* (Pukkala 2018), where treatments are tuned to the specific situation, without any strict regimes or methodologies.

In Finland, *thinning from above* is promoted as a conversion method towards the selection system (Pukkala et al. 2014). Thinning from above is applied as the first step towards more complex stand structures, and advance regeneration, undergrowth, and trees of the lower canopy layers are considered valuable assets, and retained and protected in the thinning operation. Originally a thinning method, thinning from above has been tested in experiments in the region and is frequently discussed among researchers (Karlsson and Norell 2005; Nilsson et al. 2010; Hynynen and Kukkola 1989; Mielikäinen and Valkonen 1991). Even though the method resembles target diameter harvesting at an early stage of stand development, it creates homogeneous stand structures in the same way as thinning from below, and is therefore unlikely to convert homogeneous young stands into multi-storeyed stands suitable for the selection system.

Another approach to semi-continuous cover management with conversion connotations is currently practised in Finland with two-storeyed management of pine as an example. The purpose is to initiate the conversion of an even-aged stand towards greater structural complexity. A greater degree of cover is initially maintained, but not at a level required for CCF throughout the rotation. An even-aged mature pine stand is regenerated with the seed tree or shelterwood method. As the new generation emerges, the overstorey is thinned, but a small number of large overstorey pines is retained throughout the next rotation to retain a degree of continuity in large trees. This benefits biodiversity, amenity, and other ecosystem services, and enables the production of very high-quality timber (Valkonen 2020; Äijälä et al. 2019); see Chap. 9, Wood properties and quality. The more or less spatially uneven natural regeneration and the competitive effects of the overstorey trees tend to establish and maintain a degree of structural complexity in the lower storey, which can be enhanced in the precommercial and commercial thinnings during the management cycle, or reduced if that is preferred.

2.1.2.3 Suitability of Stands for Application of and Conversion to CCF

Before initiating application of or conversion to CCF, forest managers need to evaluate whether stands are suitable for these specific methods. Criteria will vary between the silvicultural systems and conversion methods, and no complete list has been developed for any of the methods so far. The criteria addressed here therefore only highlight some of the most important aspects. The criteria have varying importance for the final decision, numerical risk assessments are often difficult for individual requirements, and risk aversion of the forest owner is an important factor when compiling all individual requirements into a final suitability assessment for a given stand.

The most important criterion is a site-adapted species mixture, both now and in future climates. Given that Norway spruce can often be found on sites that are naturally stocked with Scots pine, this considerably limits conversion attempts. Shade-tolerant species are not an absolute prerequisite for CCF, but Scots pine and other light-demanding species, either in pure stands or in mixtures, necessitate modifications to the CCF and conversion methods.

The speed of regeneration is significant for the progress of the conversion work. Frequently, lack of regeneration within short periods is misinterpreted as a site effect, i.e., that only certain vegetation types would permit natural regeneration. Forest managers in Fennoscandia have little experience of natural regeneration in

CCF and therefore often have unrealistic expectations about the speed of regeneration. In subalpine environments, the speed of regeneration is so slow (for example, due to infrequent seed years) that recruitment does not occur within acceptable times for commercial forest management.

During the conversion period it is essential to create individual tree stability by thinning in young stands. Dominant height can be used as an indicator for storm risk, and thinnings should be started at a dominant height of 10–12 m. Crown thinning favours stability of dominant trees more than the commonly practiced thinning from below by removing the dominant competitors of the dominant crop trees. Crown length can be used as an indicator for individual tree stability. The shelterwood system is riskier than the group system and the selection system. Releasing large trees to create a shelter comes with a certain risk, emphasising the need for sufficient thinning history and preparatory cuts. Individual trees damaged by wind or snow in multi-storeyed stands might be acceptable because these types of damage increase variability, compared to when entire stands or a shelter are damaged by the same agents.

The low target diameters in Fennoscandia of only about 40 cm DBH require that conversion efforts begin early. The dominant trees in the stand need to be retained long enough to supply the necessary seed and shelter for the development of a multi-storeyed stand.

Root rot susceptibility is an important limitation, and needs to be assessed before decisions on or conversion to CCF. Browsing can effectively prevent natural regeneration and limit the application of CCF. Saplings are at browsing height for a longer period under a shelter or in small openings in CCF systems. This problem is greatest for moose browsing on Scots pine but might also apply to other species admixed in spruce-dominated stands, including selection and gap systems (Komonen et al. 2020).

2.2 Context for CCF Definitions and Terms

2.2.1 Ecological and Geographical Context

Managed boreal forests in Fennoscandia are dominated by few and native species, mainly Norway spruce and Scots pine. Reluctance to use exotic species (e.g., an almost total ban in Norway, and strict restrictions in Finland and Sweden) means that CCF must use these two species as the main stand components, which is particularly critical, as only one of these species (Norway spruce) is moderately shade tolerant. Further, site conditions, especially shallow soils, permit only Scots pine in many places. Norway spruce often occurs naturally on peat or other waterlogged sites with low stability due to limited rooting space. The low stability of Norway spruce on those sites presents a challenge to CCF.

Especially for the southern parts of the boreal region, other species could be considered as part of the mixture in CCF systems, particularly in a future with a changing climate. Much attention has focused on possibilities and ways to maintain broadleaved species as major stand components. Birch is a commercially valuable species in Finland that could be considered. In the nemoral region in southern Sweden, beech and oak are the most relevant species, but other broadleaved species can also be part of the mixture, e.g., Norway maple.

Simultaneous to the introduction of CCF methods in Fennoscandia, forest management has to adapt forests to the future climate. Brang et al. (2014) show that CCF does not automatically solve all the problems of climate adaptation. For our region, the establishment of mixed-species stands, also involving exotic species, is a major challenge for climate adaptation. CCF methods need to be modified, to prevent them only favouring shade-tolerant species and natural regeneration. Norway spruce is the only naturally occurring shade-tolerant species in the boreal part of the region, which in combination with reluctance to use exotic species is challenging the simultaneous application of CCF and climate adaptation. Norway spruce is already affected by drought periods and bark beetle attacks in the southern part of the region, especially on sites where Scots pine would naturally dominate. The lack of experience regarding climate adaptation and CCF calls for approaches that minimise risks, for example by diversifying species compositions and silvicultural methods. Concepts for addressing these challenges are still lacking, however, and the silvicultural systems presented here in their original form will likely not address the challenges.

Site adaptation of species selection and silvicultural methods is an important prerequisite for CCF. Forest site classification is an essential part of site adaptation, but unfortunately not always practised in Fennoscandia. The only site indicators frequently available are vegetation types and site indices based on dominant height. Due to the low awareness of site adaptation and extensive moose browsing damage on Scots pine, Norway spruce has been planted in large areas on sites more suitable for Scots pine during recent decades. In the context of conversion to CCF and simultaneous climate adaptation, site adaptation needs again to be recognised as an essential principle of forest management. Given the fine-scaled mosaic of site conditions in many forests in Fennoscandia, silvicultural systems presented here will often not be applied on larger areas without modifications.

Knowledge and experience about CCF methods among forestry professionals have never been built up during the last 70 years in the region. Even though the basics of CCF methods have been taught in forestry programmes, the lack of practice has effectively prevented a complete education. Only very few professionals have practised CCF methods and built up the necessary experience. Vitkova et al. (2014) describe a similar situation for Ireland. Vocational education and training about CCF methods has been started during the last decade in response to the growing interest. Mason et al. (2022) compiled knowledge gaps for CCF across Europe, but the group of managers interviewed in this study was heavily biased towards promoters of CCF. Vitkova et al. (2014) describe how political decisions about terminology regarding CCF leave forest managers without clear definitions and knowledge about silvicultural methods, so there is a call for clear definitions and terms.

Applying experiences of CCF from Central Europe to Fennoscandia is challenged by the much less intensive forest management and the much more mechanised forest operations in Fennoscandia. In contrast to Central Europe, trees for removal are not marked by foresters, and selection of trees for removal by harvester operators is hampered by visibility (Kärhä et al. 2021). Individual tree removal by harvesters can cause greater damage to the remaining stand than motor-manual harvesting methods. Diameter premiums are absent in Fennoscandia, and maximum accepted diameters are very low compared to Central Europe, often not exceeding a DBH of 40 cm. The low target diameters in particular are a challenge to the application of traditional CCF, given that regeneration periods and conversion periods need to be shorter than with larger target diameters.

2.2.2   Historical Context

Important background for the current definitions of CCF in Fennoscandia is the historical use of CCF, the use of terminology, and the research efforts related to these silvicultural methods.

Since the first humans migrated to Fennoscandia after the last ice age, people have used the forests for their livelihood. Use has varied with geography, time, population density, and available tools. From early on, people cleared forests for agriculture and picked trees for firewood, house building, or other purposes. Over time, domestic use in addition to industrial development of mining, iron works, tar production, shipbuilding, and sawmilling, and timber export led to overexploitation of forest resources (Leikola 1987; Nygaard and Øyen 2020; Aalde 2000), and with that came the need to find systems of managing the forest in more sustainable ways. Early concerns led to attempts to protect resources through specific forestry decrees already in the seventeenth century (Nygaard and Øyen 2020). However, it was not until the first concepts of systematic silviculture came to Fennoscandia in the middle of the nineteenth century that terminology and content of different silvicultural systems were discussed. However, descriptions of silvicultural systems were often vague, and terminology was often not clearly defined or misunderstood (Lundqvist 2017).

Development in Fennoscandia was clearly inspired by developments in Central Europe. According to Pommerening and Murphy (2004) the popularity of CCF has gone in cycles in Europe, with the first wave of interest starting in the latter part of the nineteenth century. They emphasise that long before an accepted term and corresponding definition were agreed, selection systems were practised in parts of the continent.

In all three countries, there have been bad experiences with earlier versions of CCF in the form of exploitive diameter cuttings, leading to heavily degraded forests in the beginning of the twentieth century. We describe the details in the respective countries below. Today, some forest professionals are worried that a return to CCF methods, especially selective cutting as part of the selection system, may cause us to repeat the mistakes made a hundred years ago. Today's implementation of the selection system must, in a more controlled way, maintain a stand structure and density that leads to sufficient growth and ingrowth to avoid earlier mistakes.

2.2.2.1 Norway

Between around 1500 and 1870, the sawmilling industry in Norway had a high level of activity, and exported large amounts of lumber and timber. In this period, diameter limit cutting (*dimensjonshogst* in Norwegian) was the dominant method. The minimum diameter could vary, but a limit of 20 cm at 1.5 m height was not unusual (Nygaard and Øyen 2020). Although this usually resulted in some sort of continuous cover forest leaving smaller trees, it could also result in clearcuts in areas of more or less even-aged large timber. In addition, smaller timber was used for domestic purposes, not least for firewood. At the same time, providing iron works and mining areas with locally sourced firewood created huge demand for wood, leading to intensive harvests involving clearcuts and sometimes the depletion of larger forest areas, for example around Røros copper mine (Fryjordet 2003).

The lack of regeneration, a low standing volume, and low increment led to the transition to *plukkhogst* (approx. 1900–1940), a selection method where the larger trees were still harvested, but also damaged trees and trees of medium or smaller dimensions (Nygaard and Øyen 2020). In this period, clearcuts were generally considered to be forest degradation, due to negative experiences with excessive harvesting, e.g., around the mining cities. Little focus was put on regeneration and the future growth of the stands. Bøhmer (1922) states that some forests at the time were treated with *bledning* (selection system), described as a more systematic selective cutting that takes future stand development into consideration, as opposed to the exploitive *plukkhogst*. The definitions of methods and terms are, however, rather unclear (Bøhmer 1922), and the transition from diameter limit cutting to different kinds of selective cuttings took place at variable rates over the country.

Even though the new methods were considered to be a better form of silviculture than diameter limit cutting, it left a forest with sparse stocking, though long green crowns made it look healthy; called "the green lie" by Barth (1916). Rising concerns about dwindling forest resources led to the Norwegian National Forest Inventory being initiated in 1919 (Aalde 2000), and to intense discussions about the best way to build up the forest resources. Barth advocated multi-storeyed stands and harvesting by selection systems (*bledning*) in his textbooks (e.g., Barth 1905, 1938), though he also pointed to group cuttings and seed tree stands as suitable treatments in spruce forests. Other researchers, such as Eide, argued for even-aged forestry with clearcutting and planting (Nygaard and Øyen 2020). The brief history is that rotation forest management (*bestandsskogbruket*) gradually took over, totally dominating as a harvesting regime from the 1950s. The method must be seen as successful in terms of restoring forest resources in Norway, tripling the standing volume between 1925 and 2018 (Svensson et al. 2021), though other factors like less intensive use of forests for grazing and firewood also contributed.

Today, the interest in alternatives to clearcutting is again on the rise. Concerns about biodiversity and recreation often lie behind this interest, while others claim that CCF creates better timber quality and production equal to clearcut forestry. However, at present, only small areas are treated with harvesting methods other than clearcutting and planting or seed tree stands (see Sect. 2.2.3).

2.2.2.2   Sweden

In Sweden, selective cutting of individual trees was commonly practised during the late 1800s and early 1900s, with inspiration from forest management in Germany. This selective cutting was called *dimensionshuggning*, *blädning*, or *plockhuggning*. During the first half of the twentieth century, both selective cutting and clearcutting were used and seen as viable options (Lundmark et al. 2013). However, the selective cutting was done in a non-sustainable way from a forest production perspective, since there was heavy cutting without sufficient regeneration. Towards the mid-twentieth century, selective cutting had resulted in low productivity and poorly stocked forests (Holmgren 1959; Nilson 2001). To increase productivity and ensure long-term wood supply, the exploitative selective cutting practice was replaced by clearcutting, which became the primary silvicultural system. This transition was a gradual process where both clearcutting and selective cutting were used and refined during the transition period (Lundmark et al. 2013). This transition to a larger-scale and rational forest management fitted well with the times when increased scale and operational efficiency were promoted in many parts of society. Mechanised forestry also began at this time, also contributing to the shift to the clearcutting system.

After this transition in the mid-twentieth century, CCF methods like the selection system or group system have rarely been applied, and cutting large trees in selective cuttings was more or less outlawed, as it was believed to result in poor productivity. However, as shown by later research, this is not always the case, and selective cutting is now allowed. A CCF method that has been more widely practised is the shelterwood system, particularly natural regeneration through shelterwood for Scots pine and beech.

The interest in various CCF methods is reflected in the long-term forest field trials and the research that has been done (Goude et al. 2022). In Sweden, long-term forestry experiments are heavily focused on clearcut forestry. However, there has always been some interest in alternative management, and experiments in what we today consider CCF were set up in the early twentieth century. The oldest experiment still running was started in the 1920s, a study of selective cutting in Norway spruce. However, many of the old experiments have not been managed properly, or have been damaged in some way, making long-term evaluation difficult. Around 1990, the number of new experiments in CCF increased, due to the growing interest in alternative forest management.

With increased expectations for the forest to provide additional ecosystem services apart from wood production, demand for CCF has increased among stakeholders in recent decades. Assigned by the government to develop and increase the use of clearcut-free forestry (*hyggesfritt skogsbruk*), the Swedish Forest Agency defined the term in 2021 (Appelqvist et al. 2021): "Clearcut-free forestry on forest land intended for wood production implies that the forest is managed in such a way that the land always has a tree cover, without any larger clearcut areas."

2.2.2.3 Finland

Finnish forestry was revolutionised in 2014 when previous legal restrictions prohibiting the use of selection management were lifted as a result of a broad consensus among industrial, governmental, and environmental stakeholders. This was a major change, as CCF had been virtually banned in Finland since the 1950s in a very drastic move to improve forest productivity and the resource base after a 100-year period of outright abuse.

High-grading (harsintahakkuu, määrämittaharsinta in Finnish) in its historically notorious extreme form was widely used in Finland until the 1940s. All merchantable sawlog trees were repeatedly removed, leaving behind a heavily degraded forest with no regard for its future productivity. In the historical silvicultural narrative, the well-meaning sustainable selection regime promoted by state agencies, similar to the contemporary form of selection cutting, degenerated into destructive high-grading, with greater use of wood resources for industrial and domestic purposes when sustainability requirements were still far in the future. Repeated high-grading, along with other unsustainable uses (slash-and-burn agriculture, tar production), had resulted in the serious depletion of forests in the southern half of the country, and degradation in many parts of the north.

After 1945, there was a drastic move away from selective harvesting and natural regeneration due to government-promoted adoption of plantation-type silviculture. This was motivated and justified by aspirations to boost timber production, to secure the resource base for a huge buildup of wood-using industries, especially pulp and paper. In this context CCF and uneven-aged silviculture were seen as enemies to progress that had to be rooted out.

2.2.3 Current Legislation, Policy, and Use of CCF

Recently, legislation and policies in Fennoscandia have changed, both by removing earlier restrictions on CCF and by demanding its increased use. We describe current trends in all three countries in this section.

The percentage of forest area managed by CCF has been compiled across Europe in a recent review (Mason et al. 2022), and this shows a range of <1–6% for the three countries in Fennoscandia. Given that definitions of CCF in that study differ from ours, definitions of CCF vary between the national statistics, and different data sources are used in the three countries to compile these statistics, so it is not possible to derive reliable estimates of the current use of CCF across Fennoscandia. However, in the following sections, we describe and interpret the available national statistics and other sources, to clarify the scale of CCF implementation.

2.2.3.1 Norway

Even though changing forest laws and regulations have focused on protecting forest resources, including regeneration (Girdziusas et al. 2021), and in later years also landscape and biodiversity (Nygaard and Øyen 2020), there have hardly been any regulations regarding which silvicultural system to use in Norway. Today's Forestry Act (https://lovdata.no/lov/2005-05-27-31) simply states that "At harvesting, consideration must be given to the forests' future production and regeneration and at the same time to environmental values." However, the Regulations on Sustainable Forestry (https://lovdata.no/forskrift/2006-06-07-593) now say that "Where biological, economic and technical conditions are suitable, CCF (*lukkede hogster*) should be used, if good stability of remaining trees and satisfactory regeneration can be achieved at the site."

Most forests in Norway are certified through the Norwegian PEFC Standard (PEFC_Norge 2022). The recent version of the standard (revision 2022) emphasises the use of CCF more strongly than before, stating that "In spruce-dominated forests, CCF (*lukkede hogster*) must be used where the conditions are economically and biologically suitable." The standard also stipulates that the group certificate holder (usually the forest owners' associations) must have the necessary expertise about CCF methods, and describes how the proportion of CCF can be increased in the short and long term.

Approximately 10% of the productive forest area in Norway is certified according to FSC (Forest Stewardship Council). This standard is more or less equal to the Norwegian PEFC standard in terms of harvesting methods. It states that "In spruce-dominated forests, selective cutting or small-scale clearcutting should be used where conditions are economically and biologically suitable." In mountainous forests, the use of selective cutting (mountain forest selection cutting) should be used as widely as possible.

Based on data collected from the annual sample of around 1000 regeneration cuts, the use of harvesting methods included in CCF is modest (Landbruksdirektoratet 2022). In 2021, 4% of the sampled area was felled with small clearcuts or strip cuts (including group cuttings), while another 4.3% was felled using shelterwoods, selective cutting, or *mountain forest selection cutting*. In this last group, cutting of windfall is included. The share of these harvesting methods has been rather stable over the last 30 years, seldom exceeding 10% of the total area (Granhus and Eriksen 2017; Granhus et al. 2018). However, following the definition of CCF in this book, a smaller proportion of the harvesting methods qualify, because small clearcuts might be larger than group cuttings (limit of 0.2 ha in the PEFC certification standard) and mountain forest selection cutting often does not maintain a continuous cover. Following the definition in this book, the annual area treated with regeneration cuts in recent decades defined as CCF might be closer to 5%.

2.2.3.2 Sweden

In the current Swedish Forestry Act from 1993, production goals and environmental goals are weighted the same. This marks a major change from previous forestry acts, in which forest productivity was the main focus. One of the intentions of the new forest policy was to favour biodiversity and diversity in management methods. However, diversification in silvicultural systems has been slow. An assessment by the Swedish Forest Agency in 2002 reported that previous management with clearcutting needed to be adjusted, and that other silvicultural systems needed to increase, to attain this goal of giving the same weight to production and the environment (Appelqvist et al. 2021). In the following decades, the Swedish Forest Agency was given many tasks defining alternative forest management, such as CCF, in Swedish conditions and how it could be implemented to complement traditional forest management. The new definition of *hyggesfritt skogsbruk* (clearcut-free forestry), together with the debate in Sweden on how forests can be managed to provide more ecosystem services than just wood production, has increased interest in CCF. This interest is voluntary and not directly pushed by any legislation.

In Sweden, in 2022, a majority (67%) of productive forest land was certified. Of the certified forest area, 75% is certified according to both FSC and PEFC, 19% to only PEFC and 6% to only FSC. In its standard for Swedish forestry, FSC states that 5% of the productive forest should be managed to promote long-term development of natural and/or social values as the primary management goal (FSC_Sweden 2019). CCF methods like selective cutting and gap cutting are given as examples of alternative management to clearcutting, which does not promote natural and social values according to FSC. This request for CCF methods is on top of the 5% that should be set aside to promote natural and biodiversity values. In the PEFC standard, alternative management, like CCF, is recommended when other goals than timber production are promoted, like natural and social values (PEFC_Sverige 2017). However, it is also stated that other management alternatives than clearcut forestry have little research results and experience in Sweden and should be applied in suitable stands only, to avoid jeopardising long-term productivity.

No statistics are available on how much of the forest is being managed through CCF, since selective cuttings are not registered by the Swedish Forest Agency, unlike other harvesting methods. Use of CCF is also difficult to capture in the Swedish National Forest Inventory (NFI), since the inventory does not consider the intention of the forest owner. Estimates from the Swedish NFI have indicated that 2–4% of the forest area is fully layered and with a high enough standing volume for selective cutting to be used appropriately (Hannerz et al. 2017). However, this does not include forests where shelterwoods or gap cutting could be implemented, which would include a much larger proportion of the current forest area.

In 2021, the Swedish Forest Agency conducted a survey, asking forest owners if they applied *hyggesfritt skogsbruk* (clearcut-free forestry) on any part of their forest. Both large industry forest enterprises and smaller family enterprises were included in the survey. The estimated productive forest land managed with *hyggesfritt skogsbruk* (clearcut-free forestry) with a long-term plan to continue was 728,500 ha, i.e.,

3% of the total productive forest land area. Most of this, 487,500 ha, was on land owned by smaller family enterprises (Skogsstyrelsen 2022).

2.2.3.3 Finland

The Finnish Forest Act is rather liberal in allowing the landowner a free selection of silvicultural methods, including those of CCF. The basic operational logic in the application of the Forest Act is that a minimum basal area of viable trees is to be retained in intermediate cuttings, including thinnings, single-tree, and group selection cuttings. The limits are lower for selection cutting. The maximum gap size in CCF is 0.3 ha. Failure to adhere to these limits triggers the obligation to regenerate (which applies to all types of regeneration cuttings). This means that a seedling stand with a sufficient number of stems with an even spatial distribution and a mean height of 0.5 m has to be established within 10–25 years, depending on latitude (faster in the south).

Conversely, there are no legal restrictions on clearcutting or requirements to use CCF instead. Indirectly, clearcutting is virtually banned on a set of ecologically valuable habitats explicitly defined in the Forest Act. Their essential ecological features may not be significantly altered by cuttings, but careful selective cuttings are feasible in some cases. However, the habitats and the requirements are designed to make them economically unimportant, with a very small area limit (legal maximum 2 ha, actual average 0.6 ha), and they constitute less than 1% of the forest area (Siitonen et al. 2021).

In the widely applied Recommendations for Best Practices in Finnish Forestry (Äijälä et al. 2019), CCF methods and instructions are presented alongside the predominant traditional management. Adversities and potential practical problems are invariably listed and emphasised for CCF, but much more seldom for traditional management in the same manner.

Most forests in Finland are certified through the Finnish PEFC Standard (PEFC_Finland_–_Finnish_Forest_Certification_Council 2022). The system generally neither encourages nor discourages the use of CCF, and it seems that the general aspiration among its stakeholders is to uphold the status quo. In some special cases the possibilities of CCF are recognised, for instance in peatland stands with existing structural complexity.

The alternative Forest Stewardship Council (FSC) certification system covers some 10% of the productive forest area in Finland. The use of CCF is directly encouraged, with a requirement that some 5% of productive forest area is to be set aside as special sites and managed without stand-replacing regeneration cuttings. Some other provisions also favour CCF with less specificity.

The proportion of forests managed with CCF is still relatively low compared to the predominant traditional management, which is not surprising as the legal ban on CCF in commercial forests was not lifted until 2014. The proportion of CCF is not well defined or recorded in NFI and statistics. The entry in Mason et al. (2022) for Finland (representing 2019) was based mainly on statistics produced by Suomen metsäkeskus, the state authority responsible for law enforcement, governance, and guidance in forestry. The proportion of CCF is given as the proportion of selection and gap cuttings of all harvesting within each year. As there is no explicit indicator of whether the cutting belongs to CCF or traditional management, let alone which systems are applied as a whole, the entries are very unreliable.

When asked directly in a comprehensive standard survey, 57% of private non-corporate forest owners responded that they apply CCF on all (14%) or part (43%) of their forests (Horne et al. 2020). Metsähallitus (the agency managing state forests in Finland) has long been the frontrunner in applying CCF in its vast multiple-use forests and fragile areas in the north. It has recently (2020) decided to increase the use of CCF from 15% to 25% of its annual area of final harvests, which represents an addition of some 6000 ha annually. Similarly, several major cities have harnessed CCF as the guiding principle in their urban and peri-urban forests, and later extended it to cover all of their forests, with a categorical exclusion of clearcutting. Given the lack of proper statistics, we use the listed indications to estimate the order of magnitude of CCF use according to the definitions in this book as slightly above 5%.

The increased use of CCF on drained peatlands instead of traditional management would very likely bring about major ecological, economic, and social benefits. There are some four million ha of productive forests on peatlands that were drained mainly between 1960 and 1980, and the bulk of them are beginning to approach economic maturity and regeneration cuttings. Recent studies have revealed that the predominant method of clearcutting followed by ditch network maintenance, site preparation, and often planting is associated with much greater detrimental environmental consequences than previously assumed. They arise mainly from runoffs of suspended solids, dissolved organic carbon, and nutrients, resulting in watercourse pollution across the country (Nieminen et al. 2018a, b). In CCF, such drastic measures are mostly absent, mainly because transpiration by the continuous forest cover helps keep the groundwater table level in check without ditch maintenance (Leppä et al. 2020) and because natural regeneration is fully used. In addition to water quality, CCF can reduce the emission of greenhouse gases by stabilising the groundwater table at consistently favourable levels (Korkiakoski et al. 2023; Rissanen et al. 2023). This approach has gained ground among major forestry and environmental stakeholders, and the first steps are being taken. A major advance is under consideration in legislation, i.e. withdrawing government subsidies for ditch network maintenance.

2.3 Conclusions

We define *continuous cover forestry* (CCF) as the group of silvicultural systems that maintain a continuous forest cover. We refrain from giving a detailed definition of a *minimum forest cover* in terms of density, height, homogeneity, and continuity of the cover in our definition of CCF. An overview of definitions for a minimum forest cover in the three countries is given. Silvicultural systems that are, or can be, used in Fennoscandia to maintain a continuous cover are the *selection system,* the *shelterwood system*, and the *group system*. Only the long-term application of one of the silvicultural systems qualifies for CCF, not an individual cut. Conversion to CCF is not included as CCF in our definition. However, conversion will be an important activity in the near future for forest owners that wish to apply CCF.

Bad practices and experiences with selective cutting before 1950 led to CCF being banned, the introduction of rotation forest management, and a loss of experience and willingness to practice CCF in a more controlled way until recently. Approaches and experiences for conversion to CCF are largely lacking in Fennoscandia. Suitability for CCF or conversion to CCF is a complex assessment process with many and varying criteria, for example depending on the silvicultural system to be converted to. Assessment criteria have not yet been developed. Climate adaptation will make it necessary to modify CCF approaches in the region. Recent changes in legislation and policy indicate an increased demand for CCF in Fennoscandia. Despite this rising interest, the current practice is only roughly 5% of the annual area treated with regeneration cuts.

References

Aalde O (2000) 100 år - et omløp i skogen (100 years - a forest rotation). In: Stubsjøen M (ed) Vekst og vern. Det kongelige landbruksdepartement 1900–2000. Det Norske Samlaget, Oslo, pp 85–105

Äijälä O, Koistinen A, Sved J, Vanhatalo K, Väisänen P (2019) Metsänhoidon suositukset (Recommendations for best practices in forest management). Tapion julkaisuja https://tapiofi/wp-content/uploads/2020/09/Metsanhoidon_suositukset_Tapio_2019pdf

Appelqvist C, Sollander E, Norman J, Forsberg O, Lundmark T (2021) Hyggesfritt skogsbruk - Skogsstyrelsens definition (Clearcut-free forestry - defintion by The Swedish Forest Agency). Skogsstyrelsen Rapport 8

Barth A (1905) Skogbrukslære I. Hugstsystemene og skogens naturlige foryngelse (Forestry sciences I. Harvesting systems and the natural regeneration of forests). Grøndahl & Søns boktrykkeri, Oslo

Barth A (1916) Norges skoger med stormskridt mot undergangen (Norway's forests leaping towards their downfall). Tidsskr Skogbr 24(4):123–154

Barth A (1938) Skogskjøtsel på biologisk grunnlag (Forest management on a biological basis). Grøndahl & Søns boktrykkeri, Oslo

Bøhmer JG (1922) Bledningsskog (selection forest). Tidsskr Skogbr 30(5–6):122

Brang P, Spathelf P, Larsen JB, Bauhus J, Boncina A, Chauvin C, Drossler L, Garcia-Guemes C, Heiri C, Kerr G, Lexer MJ, Mason B, Mohren F, Muhlethaler U, Nocentini S, Svoboda M (2014) Suitability of close-to-nature silviculture for adapting temperate European forests to climate change. Forestry 87(4):492–503. https://doi.org/10.1093/forestry/cpu018

Brodie LC, Harrington CA (2020) Guide to variable-density thinning using skips and gaps. Gen. Tech. Rep. PNW-GTR-989. Portland, OR

Brunner A (2024) Variable-density thinning designs for harvester-based operations in Northern Europe. Scand J For Res 39(5):248–256. https://doi.org/10.1080/02827581.2024.2388054

Brunner A, Butler Manning D, Huss J, Rozenbergar D, Diaci J, Schousboe F, Hansen LW (2004) Scenarios of regeneration and stand production of beech under different silvicultural regimes with Regenerator. EU-project Nat-Man "Nature-based management of beech in Europe", Deliverable 18, 34 & 35. Danish Centre for Forest, Landscape and Planning, KVL:90 pp

Brunner A, Hahn K, Biber P, Skovsgaard JP (2006) Conversion of Norway spruce: a case study in Denmark based on silvicultural scenario modelling. In: Hasenauer H (ed) Sustainable Forest management. Growth models for Europe. Springer, Berlin, pp 343–371

Burschel P, Huss J (1997) Grundriß des Waldbaus (basics of silviculture). Parey, Berlin

Carey AB (2003) Biocomplexity and restoration of biodiversity in temperate coniferous forest: inducing spatial heterogeneity with variable-density thinning. Forestry 76(2):127–136. https://doi.org/10.1093/forestry/76.2.127

Carey AB (2006) Active and passive forest management for multiple values. Northwest Nat 87:18–30

Comfort EJ, Roberts SD, Harrington CA (2010) Midcanopy growth following thinning in young-growth conifer forests on the Olympic peninsula western Washington. For Ecol Manag 259(8):1606–1614. https://doi.org/10.1016/j.foreco.2010.01.038

Dodson EK, Ares A, Puettmann KJ (2012) Early responses to thinning treatments designed to accelerate late successional forest structure in young coniferous stands of western Oregon, USA. Can J For Res/Rev Can Rech For 42(2):345–355. https://doi.org/10.1139/x11-188

Drössler L, Eko PM, Balster R (2015) Short-term development of a multilayered forest stand after target diameter harvest in southern Sweden. Can J For Res/Rev Can Rech For 45(9):1198–1205. https://doi.org/10.1139/cjfr-2014-0471

Drössler L, Nilsson U, Lundqvist L (2014) Simulated transformation of even-aged Norway spruce stands to multi-layered forests: an experiment to explore the potential of tree size differentiation. Forestry 87(2):239–248. https://doi.org/10.1093/forestry/cpt037

Egan AF, Waldron K, Raschka J, Bender J (1999) Ecosystem management in the northeast. J For 97(10):24–30

European_Commission (2023) Guidelines on closer-to-nature forest management. Commission Staff Working Document SWD(2023) 284

Fryjordet T (2003) Bergverkenes treforbruk (wood use of the mines). Elverum Trykk AS, Våler

FSC_Sweden (2019) The FSC national forest stewardship standard of Sweden. 03:2019

Girdziusas S, Löf M, Hanssen KH, Lazdina D, Madsen P, Saksa T, Liepins K, Floistad IS, Metslaid M (2021) Forest regeneration management and policy in the Nordic-Baltic region since 1900. Scand J For Res. https://doi.org/10.1080/02827581.2021.1992003

Goude M, Erefur C, Johansson U, Nilsson U (2022) Hyggesfria skogliga fältförsök i Sverige. En sammenställning av tilgängliga långtidsförsök. (Clearcut-free forest experiments in Sweden. A summary of available long-term experiments.). SLU, Enheten for skoglig fältforskning, Rapport 22

Granhus A, Allen M, Bergsaker E (2020) Fjellskoghogst - produksjon, foryngelse og økonomi (Mountain forest selection cutting - production, regeneration, and economy). NIBIO Rapport 6 (72)

Granhus A, Breidenbach J, Eriksen R, Gjertsen AK, Solberg S (2018) Tilstand i foryngelsesfelt. Analyse basert på data fra Resultatkartleggingen, Landsskogtakseringen og Økonomisystem for skogordningene (ØKS) (Status of regeneration areas. Analysis based on data from a regeneration inventoy, the national forest inventory, and the forest subsidy system). NIBIO Rapport, vol 4(159)

Granhus A, Eriksen R (2017) Resultatkontroll skogbruk/miljø. Rapport 2016. (Inventory of results in forestry and environment. Report 2016.). NIBIO Report, vol 159

Gustafsson L, Hannerz M, Koivula M, Shorohova E, Vanha-Majamaa I, Weslien J (2020) Research on retention forestry in northern Europe. Ecol Proc 9(1). https://doi.org/10.1186/s13717-019-0208-2

Hagner M, Lohmander P, Lundgren M (2001) Computer-aided choice of trees for felling. For Ecol Manag 151:151–161

Hånell B, Ottosson-Löfvenius M (1994) Windthrow after shelterwood cutting in Picea abies peatland forests. Scand J For Res 9(3):261–269. https://doi.org/10.1080/02827589409382839

Hanewinkel M, Pretzsch H (2000) Modelling the conversion from even-aged to uneven-aged stands of Norway spruce (*Picea abies* L. Karst.) with a distance-dependent growth simulator. For Ecol Manag 134(1–3):55–70

Hannerz M, Nordin A, Saksa T (2017) Hyggesfritt skogsbruk. En kunskapssammenställning från Sverige och Finland. (Clearcut-free forestry. A compliation of knowledge from Sweden and Finland). SLU, Future Forests Rapportserie 2017:1

Harrington CA, Roberts SD, Brodie LC (2005) Tree and understory responses to variable-density thinning in western Washingston. In: Peterson CE, Maguire DA (eds) Balancing ecosystem values: innovative experiments for sustainable forestry, 2005. Gen. Tech. Rep. PNW-GTR-635. U.S. Department of Agriculture, Forest Service. Pacific Northewest Research Station, pp 97–106

Hasenauer H (2004) Glossary of terms and definitions relevant for conversion. In: Spiecker H, Hansen J, Klimo E, Skovsgaard JP, Sterba H, von Teuffel K (eds) Norway spruce conversion - options and consequences. European Forest Institute research report, 18th edn. Brill, Leiden, pp 5–23

Hertog IM, Brogaard S, Krause T (2022) Barriers to expanding continuous cover forestry in Sweden for delivering multiple ecosystem services. Ecosyst Serv 53:101392. https://doi.org/10.1016/j.ecoser.2021.101392

Hilmers T, Biber P, Knoke T, Pretzsch H (2020) Assessing transformation scenarios from pure Norway spruce to mixed uneven-aged forests in mountain areas. Eur J For Res 139(4):567–584. https://doi.org/10.1007/s10342-020-01270-y

Holmgren A (1959) Skogarna och deras vård i övre Norrland intill år 1930 (Forests and their management in upper Norrland until 1930). Stockholm

Horne P, Karppinen H, Korhonen O, Koskela T (2020) Legitimacy of forest management and practices—Forest Owner 2020. PTT Reports, vol 266

Hynynen J, Kukkola M (1989) Harvennustavan ja lannoituksen vaikutus männikön ja kuusikon kasvuun (influence of thinning method and nitrogen fertilization on the growth and yield of scots pine and Norway spruce stands). Folia Forestalia 731:3–17

Kärhä K, Ovaskainen H, Palander T (2021) Decision-making among harvester operators in tree selection and need for advanced harvester operator assistant systems (AHOASs) on thinning sites. Paper presented at the COFE-FORMEC 2021 - Forest Engineering Family: Growing Forward Together, Corvallis, OR, U.S.A., September 276–30, 2021

Karlsson K, Norell L (2005) Predicting the future diameter of stems in Norway spruce stands subjected to different thinning regimes. Can J For Res/Rev Can Rech For 35(6):1331–1341

Knoke T, Plusczyk N (2001) On economic consequences of transformation of a spruce (*Picea abies* (L.) karst.) dominated stand from regular into irregular age structure. For Ecol Manag 151:163–179

Koivula M, Kuuluvainen T, Hallman E, Kouki J, Siitonen J, Valkonen S (2014) Forest management inspired by natural disturbance dynamics (DISTDYN) - a long-term research and development project in Finland. Scand J For Res 29(6):579–592. https://doi.org/10.1080/0282758 1.2014.938110

Komonen A, Paananen E, Elo M, Valkonen S (2020) Browsing hinders the regeneration of broadleaved trees in uneven-aged forest management in southern Finland. Scand J For Res 35(3–4):134–138. https://doi.org/10.1080/02827581.2020.1761443

Korkiakoski M, Ojanen P, Tuovinen JP, Minkkinen K, Nevalainen O, Penttilä T, Aurela M, Laurila T, Lohila A (2023) Partial cutting of a boreal nutrient-rich peatland forest causes radically less short-term on-site CO_2 emissions than clear-cutting. Agric For Meteorol 332. https://doi.org/10.1016/j.agrformet.2023.109361

Kuuluvainen T, Angelstam P, Frelich L, Jogiste K, Koivula M, Kubota Y, Lafleur B, Macdonald E (2021) Natural disturbance-based forest management: moving beyond retention and continuous-cover forestry. Front For Glob Change 4. https://doi.org/10.3389/ffgc.2021.629020

Landbruksdirektoratet (2022) Kartlegging av foryngelse og miljøhensyn ved hogst (Inventory of regeneration and environmental considerations at final harvest). Rapport 2021

Larsen JB, Angelstam P, Bauhus J, Carvalho JF, Diaci J, Dobrowolska D, Gazda A, Gustafsson L, Krumm F, Knoke T, Konczal A, Kulluvainen T, Mason B, Motta R, Plötzelsberger E, Rigling A, Schuck A (2022) Closer-to-nature rorest management. From science to policy, vol 12. European Forest Institute. https://doi.org/10.36333/fs12

Leikola M (1987) Leading ideas in Finnish silviculture. Silva Fenn 21:332–341

Leppä K, Hökkä H, Laiho R, Launiainen S, Lehtonen A, Mäkipää R, Peltoniemi M, Saarinen M, Sarkkola S, Nieminen M (2020) Selection cuttings as a tool to control water table level in boreal drained peatland forests. Front Earth Sci 8. https://doi.org/10.3389/feart.2020.576510

Lundmark H, Josefsson T, Östlund L (2013) The history of clear-cutting in northern Sweden - driving forces and myths in boreal silviculture. For Ecol Manag 307:112–122. https://doi.org/10.1016/j.foreco.2013.07.003

Lundqvist L (2017) Tamm review: selection system reduces long-term volume growth in Fennoscandic uneven-aged Norway spruce forests. For Ecol Manag 391:362–375. https://doi.org/10.1016/j.foreco.2017.02.011

Lundqvist L, Cedergren J, Eliasson L (2014) Blädningsbruk (Selection system). Skogsskötselserien 11:1–57

Mason WL, Diaci J, Carvalho J, Valkonen S (2022) Continuous cover forestry in Europe: usage and the knowledge gaps and challenges to wider adoption. Forestry 95(1):1–12. https://doi.org/10.1093/forestry/cpab038

Matthews JD (1989) Silvicultural systems. Clarendon Press, Oxford

Mielikäinen K, Valkonen S (1991) Harvennustavan vaikutus varttuneen metsikön tuotokseen ja tuottoihin Etelä-Suomessa (Effect of thinning method on the yield of middle-aged stands in southern Finland). Folia Forestalia 776:22

Nieminen M, Hökkä H, Laiho R, Juutinen A, Ahtikoski A, Pearson M, Kojola S, Sarkkola S, Launiainen S, Valkonen S, Penttilä T, Lohila A, Saarinen M, Haahti K, Mäkipää R, Miettinen J, Ollikainen M (2018a) Could continuous cover forestry be an economically and environmentally feasible management option on drained boreal peatlands? For Ecol Manag 424:78–84. https://doi.org/10.1016/j.foreco.2018.04.046

Nieminen M, Palviainen M, Sarkkola S, Laurén A, Marttila H, Finér L (2018b) A synthesis of the impacts of ditch network maintenance on the quantity and quality of runoff from drained boreal peatland forests. Ambio 47(5):523–534. https://doi.org/10.1007/s13280-017-0966-y

Nilson K (2001) Regeneration dynamics in uneven-aged spruce forests with special emphasis on single-tree selection. Acta Universitatis Agriculturae Sueciae, vol 209. Sveriges Lantbruksuniversitet, Umeå

Nilsson U, Ekö P-M, Elfving B, Fahlvik N, Johansson U, Karlsson K, Lundmark T, Wallentin C (2010) Thinning of scots pine and Norway spruce monocultures in Sweden: effects of different thinning programmes on stand level gross- and net stem volume production. Studia Forestalia Suecica 219:46

Nygaard PH, Øyen BH (2020) Skoghistorisk tilbakeblikk med vekt på utviklingen av bestandssskogbruket i Norge (Forest historical review with emphasis on the development of even-aged forestry in Norway). NIBIO rapport, vol 45

O'Hara KL, Leonard LP, Keyes CR (2012) Variable-density thinning and a marking paradox: comparing prescription protocols to attain stand variability in coast redwood. West J Appl For 27(3):143–149. https://doi.org/10.5849/wjaf.11-042

O'Hara KL, Nesmith JCB, Leonard L, Porter DJ (2010) Restoration of old forest features in coast redwood forests using early-stage variable-density thinning. Restor Ecol 18:125–135. https://doi.org/10.1111/j.1526-100X.2010.00655.x

PEFC_Finland_-_Finnish_Forest_Certification_Council (2022) Sustainable Forest Management — Requirements 2022. PEFC FI 1002:63

PEFC_Norge (2022) Norsk PEFC skogstandard. PEFC N 02:2022 (Norwegian PEFC forestry standard). Oslo

PEFC_Sverige (2017) Svenska PEFC:s Skogsstandard. PEFC SWE 002:4 (Swedish PEFC forestry standard)

Pommerening A, Murphy ST (2004) A review of the history, definitions and methods of continuous cover forestry with special attention to afforestation and restocking. Forestry 77(1):27–44

Puettmann KJ, Ares A, Burton JI, Dodson EK (2016) Forest restoration using variable density thinning: lessons from Douglas-fir stands in western Oregon. Forests 7(12). https://doi.org/10.3390/f7120310

Pukkala T (2018) Instructions for optimal any-aged forestry. Forestry 91(5):563–574. https://doi.org/10.1093/forestry/cpy015

Pukkala T, Lahde E, Laiho O (2011a) Variable-density thinning in uneven-aged forest management - a case for Norway spruce in Finland. Forestry 84(5):557–565. https://doi.org/10.1093/forestry/cpr020

Pukkala T, Lähde E, Laiho O (2011b) Metsän jatkuva kasvatus (Continuous cover forestry). Joensuu

Pukkala T, Lähde E, Laiho O (2014) Stand management optimization — the role of simplifications. For Ecosyst 1(1):3. https://doi.org/10.1186/2197-5620-1-3

Reininger H (1987) Zielstärkenutzung oder die Plenterung des Altersklassenwaldes (Target diameter harvesting, or selection cutting of even-aged stands). Österreichischer Agrarverlag, Wien, 163 pp.

Reventlow DOJ, Nord-Larsen T, Biber P, Hilmers T, Pretzsch H (2021) Simulating conversion of even-aged Norway spruce into uneven-aged mixed forest: effects of different scenarios on production, economy and heterogeneity. Eur J For Res 140(4):1005–1027. https://doi.org/10.1007/s10342-021-01381-0

Rissanen AJ, Ojanen P, Stenberg L, Larmola T, Anttila J, Tuominen S, Minkkinen K, Koskinen M, Mäkipää R (2023) Vegetation impacts ditch methane emissions from boreal forestry-drained peatlands - Moss-free ditches have an order-of-magnitude higher emissions than moss-covered ditches. Front Environm Sci 11. https://doi.org/10.3389/fenvs.2023.1121969

Rouvinen S, Kuuluvainen T (2005) Tree diameter distributions in natural and managed old Pinus sylvestris-dominated forests. For Ecol Manag 208(1–3):45–61

Schütz J-P (2001) Der Plenterwald und weitere Formen strukturierter und gemischter Wälder (The selection forest and other forms of structured mixed forests). Parey Buchverlag, Berlin

Siitonen J, Määttä K, Punttila P, Syrjänen K (2021) Metsälain arvioinnin jatkoselvitys 10 §:n muutosten vaikutuksista monimuotoisuuden turvaamiseen (Review on the effects of changes to the 10 § of the Forest Act on safeguarding biodiversity). Luonnonvara- ja biotaloudentutkimus, vol 6/2021. Helsinki

Skogsstyrelsen (2022) Åtgärder i skogbruket 2021 (Measures in forestry in 2021). Statistikprodukt, vol JO0312

Sterba H (2004) Equilibrium curves and growth models to deal with forests in transition to uneven-aged structure - application in two sample stands. Silva Fenn 38(4):413–423

Sterba H, Zingg A (2001) Target diameter harvesting - a strategy to convert even-aged forests. For Ecol Manag 151(1–3):95–105

Svensson A, Eriksen R, Hylen G, Granhus A (2021) Skogen i Norge. Statistikk over skogforhold og skogressurser i Norge for perioden 2015-2019. (the forest in Norway. Statistics about forest conditions and forest resources in Norway for the period 2015-2019.). NIBIO Rapport 142(7):1–53

Valkonen S (2020) Metsän jatkuvasta kasvatuksesta (On continuous cover forestry). Metsäkustannus and Luonnonvarakeskus, Helsinki:127

Vitkova L, Dhubhain AN, Upton V (2014) Forestry professionals' attitudes and beliefs in relatino to and understanding of continuous cover forestry. Scott For 68(3)

Willis JL, Roberts SD, Harrington CA (2018) Variable density thinning promotes variable structural responses 14 years after treatment in the Pacific northwest. For Ecol Manag 410:114–125. https://doi.org/10.1016/j.foreco.2018.01.006

Chapter 3
Regeneration

Mikolaj Lula, Kjersti Holt Hanssen, Martin Goude, Hannu Hökkä, Sauli Valkonen, Andreas Brunner, Pasi Rautio, Charlotta Erefur, and Aksel Granhus

Abstract

- In the context of continuous cover forestry (CCF), natural regeneration is the preferred form of regeneration, but it is a long-lasting and complex process. Shelter density has a large effect on the regeneration process and results.
- The selection system, particularly suited for shade-tolerant species like Norway spruce, relies on continuous regeneration and ingrowth into larger size classes.

M. Lula (✉)
Swedish University of Agricultural Sciences (SLU), Alnarp, Sweden
e-mail: mikolaj.lula@slu.se

K. H. Hanssen · A. Granhus
Norwegian Institute of Bioeconomy Research (NIBIO), Ås, Norway
e-mail: kjersti.hanssen@nibio.no; aksel.granhus@nibio.no

M. Goude
Swedish University of Agricultural Sciences (SLU), Simlångsdalen, Sweden
e-mail: martin.goude@slu.se

H. Hökkä
Natural Resources Institute Finland (Luke), Oulu, Finland
e-mail: hannu.hokka@luke.fi

S. Valkonen
Natural Resources Institute Finland (Luke), Helsinki, Finland
e-mail: sauli.valkonen@luke.fi

A. Brunner
Norwegian University of Life Sciences, Ås, Norway
e-mail: andreas.brunner@nmbu.no

P. Rautio
Natural Resources Institute Finland (Luke), Rovaniemi, Finland
e-mail: pasi.rautio@luke.fi

C. Erefur
Swedish University of Agricultural Sciences (SLU), Umeå, Sweden
e-mail: charlotta.erefur@slu.se

© The Author(s) 2025
P. Rautio et al. (eds.), *Continuous Cover Forestry in Boreal Nordic Countries*,
Managing Forest Ecosystems 45, https://doi.org/10.1007/978-3-031-70484-0_3

Regeneration and ingrowth rates vary significantly among stands, influenced by site and historical factors, with no clear relationship to current stand conditions.

- In the group system, edge trees influence regeneration by providing seeds, checking weed growth, and exerting competition. Regeneration in gaps is generally satisfactory for both Norway spruce and Scots pine. However, seedlings usually grow slower, especially close to the gap edges.
- The shelterwood system promotes regeneration through a successive, uniform opening of the canopy. Shelter trees provide seeds, and reduce seedling damage and competition from ground vegetation. On the other hand, the remaining overstorey shelter trees reduce seedling growth.
- Conversion to the selection system initiates regeneration in young stands, aiming for slow and steady regeneration. Given the rapid growth and crown closure in young stands, frequent manipulation of shelter density is essential during conversion, for example by opening small gaps.

Keywords Natural regeneration · Seedling · Shelter density · Edge effect · Ingrowth

3.1   Introduction

Natural regeneration is generally considered a prerequisite for the profitability of continuous cover forestry (CCF), as artificial regeneration by direct seeding or planting is difficult and expensive where trees are present. In this chapter we will first briefly summarise the regeneration process in boreal forests, with a main focus on natural regeneration, which is most often used in the context of CCF. We will further review research results about regeneration in three silvicultural systems under CCF and during conversion to CCF. In Fennoscandia, spruce and pine are often found in more or less pure stands on the sites they are best adapted to, so we will separately review the existing knowledge for these tree species under each silvicultural system. Large differences in the regeneration processes among sites, differing in mineral soil vs. drained peatland, or lower vs. higher elevations and sites close to arctic treelines, make it necessary to sometimes look at these sites separately.

Terms and definitions
- *Regeneration* is a term reflecting (1) a process of renewal or reestablishment of a forest after disturbance or timber harvest, or (2) the result of this process in the form of small, young trees.
- In this chapter, *"established seedlings"* refers to the end of the seedling stage, after which growth of small saplings or trees is no longer considered a regeneration process.
- *Seedling recruitment* is the process by which new individuals enter an existing population.
- *Ingrowth* is the growth of small trees past a certain size threshold into the tree stratum.
- *Advance regeneration* refers to seedlings or saplings already present in a stand before an active regeneration phase begins.

3.1.1   The Regeneration Process in Boreal Forests

Norway spruce and Scots pine have different successional strategies. Norway spruce is a late-successional species. In natural forests, it establishes in gaps created by death or windthrow of one or more trees. It is a shade-tolerant species, which can exist as advance regeneration even under a dense canopy of mature trees of shade-tolerant or light-demanding species (Engelmark and Hytteborn 1999). The dense canopy (high leaf area and low crowns) in mature Norway spruce stands makes for a dark understorey (Goude et al. 2022). As understorey light is an important factor affecting seedling establishment, survival, and growth, the dense canopy of Norway spruce stands is considered a competitive advantage, restricting other tree species' establishment underneath.

Scots pine is more of a light-demanding pioneer tree species, which regenerates well after medium- and large-scale disturbances such as forest fires or windthrows create openings in natural forests. Scots pine rapidly grows taller in high-light conditions (Engelmark and Hytteborn 1999). Norway spruce and Scots pine reach their distribution limits in Fennoscandia both at high latitudes and altitudes. Regeneration processes often limit their ranges. Seed production is infrequent, and maturation of seeds often fails in subalpine or subarctic conditions (Kullman 1996; Mork 1968). Short growth periods, low summer temperatures, and frequent summer frosts further limit seedling establishment (Kielland-Lund 1981; Mikola 1971).

The reproductive cycle of Norway spruce takes 2 years, from cone initiation to seed dispersal. Pollen-cone buds are initiated in the first summer of the cycle. Pollen release and fertilisation usually take place the following May, followed by seed ripening during summer and autumn. Seeds are subsequently dispersed between autumn of the second year and winter of the third year (Karlsson 2000).

The reproductive cycle of Scots pine lasts 3 years. This is because fertilisation, and consequently seed ripening, is delayed until the third year. The seeds are then dispersed from April until June of the fourth year (Sarvas 1962; Koski 1991).

Under optimal conditions, seed production of Norway spruce is irregular, with good seed years at intervals of several years. Scots pine produces seeds almost every year, although the amount of seeds varies greatly among years. However, seed-year intervals vary largely with site conditions, both edaphic and climatic. As summer temperatures are critical for seed production, there is a latitudinal and elevation gradient resulting in infrequent seed years at higher latitudes and elevations in Fennoscandia for both species, sometimes separated by up to 30 years (Hagner 1965; Mork 1968). For both species, the variation in seed production among years is influenced by weather conditions over the entire reproductive cycle.

Nutrients and light availability are other important factors triggering both species' reproduction. Seed production is therefore much higher in trees with large crowns and little competition from neighbours (Hagner 1965). Seed production increases with increasing site index (Sarvas 1962) and responds positively to fertilisation with nitrogen, or fertilisers combining nitrogen, phosphorus, and potassium (NPK, Mikola 1971). Differences among trees in seed production are also determined by local variation in site fertility (Sarvas 1962) and genetic factors (Koski 1978). Seed production increases several years after release cutting (Karlsson and Örlander 2002) in response to increased light, heat, water, and mineral nutrients.

Temperature and moisture control the germination of viable seeds. After establishment, light and nutrient availability determine seedling survival and growth. Predation and pathogens also kill seedlings. Mechanical site preparation is one of the most common silvicultural tools to improve soil properties, which influences seed storage and germination, seedling growth, and survival. However, the treatment's effects vary widely among the specific methods and sites (Löf et al. 2012). Mechanical site preparation creates different seedbed types; bare mineral soil is usually the most favourable for emergence (e.g. Kyrö et al. 2022; Oleskog and Sahlén 2000).

Mature shelter trees serve as a seed source (Beland et al. 2000), while reducing pine-weevil damage to seedlings (Von Sydow and Örlander 1994), frost risk (Langvall and Örlander 2001; Lofvenius 1995), and competition from ground vegetation (Hagner 1962). However, shelter trees also have negative effects, such as reduced growth due to competition (Erefur et al. 2011). Manipulating the shelter density influences understorey microclimate and light environment, and belowground competition. Composition, abundance, and succession rate in understorey plant communities are also heavily affected by shelter density (Beland et al. 2000; Hagner 1962).

3.2   The Selection System

3.2.1   The Selection System in Norway Spruce-Dominated Forests

3.2.1.1   The Selection System in Norway Spruce-Dominated Forests on Mineral Soil

A stand managed by selection cutting must contain a reserve of undergrowth to supply ingrowth into the smallest tree size class which will eventually replace trees that have been harvested, damaged by harvesting, or died. For a sustained reserve, regeneration must occur regularly or at least in bursts at shorter or longer intervals. Site conditions (moisture, fertility, vegetation) tend to control regeneration and may vary with stand structure and treatments, sometimes alternating between favourable and unfavourable periods. Moister sites have shown the greatest potential for regeneration, particularly drained and undrained peatlands (Lukkala 1946; Hökkä et al. 2011, 2012).

Seedling turnover is a major driver of regeneration in spruce selection stands. Few seedlings survive their first year, with reported mortality as high as 80–86% (Valkonen and Maguire 2005) or even 90–99% (Arnborg 1947; Leemans 1991). Later, mortality declines sharply. Average annual mortalities of 2–8% have been recorded in different studies, with lower mortality for spruce and higher for pine and birch (Lundqvist 1991; Nilson and Lundqvist 2001; Lundqvist and Nilson 2007; Eerikäinen et al. 2014).

The seedling density varies greatly among and within selection stands. Around 2000 seedlings/ha germinated each year in the Finnish ERIKA single-tree selection experiment harvested 2–4 times since the 1980s, but as mentioned above, most did not survive the first year. When looking at more-established seedlings, the average density for 11–130 cm-tall spruce was 5000 to 25,000 stems/ha (Saksa and Valkonen 2011; Saksa 2004).

The spatial distribution of regenerated seedlings is also usually uneven (Bøhme 1957; Granhus et al. 2021). In the ERIKA plots, almost half of the 4 m² sample plots completely lacked regeneration (Saksa 2004; Saksa and Valkonen 2011) despite the rather high average density level. The variation in seedling density may be credited to local variation in soil moisture, vegetation and other seedbed properties, as well as stand density.

Seedlings grow very slowly in spruce-selection stands as the canopy cover is always dense. Average seedlings grow around 2–4 cm/year (Lundqvist 1991; Eerikäinen et al. 2014). At this growth rate, it takes on average 40–60 years for a spruce seedling to reach 130 cm and constitute ingrowth into the smallest diameter class. Seedling density is highly variable, both between and within stands. It seems obvious that stand history, especially past density, structure, and treatments, is reflected in growth of regeneration and cohorts of small trees. Within a stand, the plants with the best condition, vitality, and height growth have better chances to survive and reach lower-canopy layers than those with lower vitality and slower current growth. Understorey trees with a long-pointed crown with lots of healthy needles or leaves show good vitality. Current height growth is also a good indicator. A 5–10 cm leader shoot over a 3-year average in undergrowth spruce seems to signal a good response capacity to postharvest resource availability increases, and a marked height-growth spurt usually appears within a few years (Koistinen and Valkonen 1993; Mielikäinen and Valkonen 1995).

Due to the dynamic components of emergence, survival, and growth of seedlings and saplings, in Fennoscandia there generally seems to be sufficient ingrowth to replace losses of trees due to logging and natural mortality in spruce-dominated selection stands (Bøhme 1957; Lundqvist 1993; Eerikäinen et al. 2014; Andersson 2015; Moan Mn 2021). When tree diameter distribution is used as a key characteristic in management, it becomes essential to know how many new trees join each diameter class each period. With a minimum diameter at breast height of 0.1 cm, average spruce ingrowth has been 10–70 stems/ha/yr in comparable conditions in Finland, Sweden, and Norway (Lundqvist 1991, 1993; Saksa 2004; Eerikäinen et al. 2014; Andersson 2015; Moan Mn 2021) or as high as 170 stems/ha/yr in some studies (Lähde et al. 2002). Ingrowth of shade-intolerant trees has been minimal compared with spruce. In stands with on average 77% of standing volume consisting of spruce, ingrowth was 3 stems/ha/yr for birch and 0.1 stems/ha/yr for pine with a threshold diameter of 0.1 cm (Eerikäinen et al. 2014).

Ingrowth can be highly variable, and is often poorly correlated with current stand conditions (Lundqvist 2004, 2007; Moan Mn 2021). Indeed, current ingrowth in a selection-managed stand is deeply rooted in the seedling emergence, survival, and growth processes, usually dating back several decades. Therefore, ingrowth rates tend to vary between stands and studies (Lundqvist 2017). Studies looking into the relationship between stand density (e.g., standing volume) and ingrowth have found no significant relationship or a small negative relationship with standing volumes below 300 m^3/ha (Lundqvist 2004, Lundqvist and Nilson 2007; Ahlström and Lundqvist 2015; Moan Mn 2021). This might be explained by increased seedling mortality associated with harvesting damage and large reductions in stand density (Lundqvist 2017). This increased seedling mortality counterbalances the otherwise-expected post-harvest ingrowth and height growth increases in the understorey. In contrast, Eerikäinen et al. (2014) showed a very clear relationship with stand density and seedling growth, implying that ingrowth will be much slower under high standing volumes. Minimisation of harvesting damage to undergrowth and small trees is paramount (Valkonen et al. 2020). Furthermore, current knowledge and experience do not permit us to assess what seedling densities are required to maintain sufficient ingrowth levels and sustain stand structure under given conditions. One principle is obvious though: the density of trees in the regeneration and undergrowth size classes must be much larger than the density of small trees, which must in turn exceed that of mid-size trees, and so on, because of slow growth and high mortality of seedlings, saplings, and small trees due to logging damage or natural causes (Valkonen et al. 2020). Only one tree in a small group may eventually survive and grow to become a large, mature individual. Promoting regeneration and development of an undergrowth reserve in selection management is essential. Some larger trees (diameter > 25 cm) must also be retained to produce seeds (Saksa 2004; Nygren et al. 2017).

3.2.1.2 The Selection System in Norway Spruce-Dominated Forests on Drained Peatland

High and constant soil moisture and abundant cover of *Sphagnum* moss (Place 1955; Sarasto and Seppälä 1964; Wood and Jeglum 1984) enhance spruce natural regeneration in peatland sites. Regeneration of uneven-aged spruce stands on peatlands has been investigated in ongoing experiments in Finland for only 5 years, so no published results are yet available. Preliminary results suggest that cutting to a low basal area of 9–13 m^2/ha increases seedling establishment in post-cutting years. However, such heavy cutting also increases the risk of wind and snow damage and prolongs the cutting cycle. About one-third of the seedlings had been established after the selection cutting. In untreated control plots, one-fifth of the seedlings had been established after cutting of the treated stands. One-third of the seedlings were found on *Sphagnum* surfaces, which indicates how *Sphagnum* enhances seedling emergence.

At the above-mentioned experimental sites, the post-cutting growth of spruce seedlings has not yet been measured, but observations from the experimental sites suggest that there is large variation, which in some sites is explained by a severe lack of potassium. In sites with more balanced nutrition, strong growth responses can be seen for all spruce seedlings after 3 years of stunted post-cutting growth.

Lehtonen et al. (2023) found immediate recovery of photosynthetic capacity of previously-suppressed spruce trees on a fertile peatland in southern Finland after selection cutting removing 70% of the initial 278 m^3/ha (with 71 m^3/ha retained). The photosynthetic capacity of the trees was studied by measuring the carbon isotope ($\delta^{13}C$) composition of increment cores (Lehtonen et al. 2023). A larger response of diameter growth was found after 5 years had passed since cutting.

At some drained peatland sites, the ground vegetation may indicate high production potential with continuous advance spruce recruitment, although Scots pine and downy birch form the dominant canopy layer. In peatlands, nitrogen availability increases over time after drainage due to peat decomposition, which, in turn, improves spruce survival and growth. In such stands, density of dominant pine and birch can be reduced by selection cutting to achieve uneven-aged spruce-dominated stands (Saarinen et al. 2020). Later on, nutrient imbalances likely need to be controlled by fertilising with phosphorus and potassium or wood ash (Paavialainen and Päiväven 1995; Saarinen et al. 2020).

3.2.1.3 The Selection System in Norway Spruce-Dominated Forests in Mountain Areas

The main difference for selection management in mountain areas is the less-favourable climate due to high altitude, mostly affecting seed production (Mork 1968). Most results about regeneration in mountain forests are from a method called mountain forest selection cutting, which reduces stand density to very low levels and combines selection and gap cuttings. Regeneration in selection stands in mountain areas shows similar patterns to selection stands elsewhere. The regeneration density varies greatly, with a positive effect of site quality; more seedlings grow at sites with richer vegetation (ferns or herbs) compared to sites with bilberry and lingonberry-heather (Nilsen 1988; Granhus et al. 2020). The cutting rate and residual stand basal area showed no significant effect on either recruitment or seedling development, indicating again that these processes depend on more than current stand conditions.

The harsher conditions in these subalpine areas reduce spruce recruit density. Studies in southeastern Norway have reported 500–800 stems/ha for spruce <3 m tall (Øyen and Nilsen 2002, 2004;). The annual ingrowth into the 5 cm diameter at breast height category in stands in southeastern Norway was reported to be on average 46 trees/ha/yr (Granhus et al. 2020). Moan Mn (2021) reported ingrowth rates comparable to stands at lower altitudes in a selection-system plot in a mountain forest at 800 m.

3.2.2   The Selection System in Scots Pine-Dominated Forests

The challenge with using the selection system to regenerate pine-dominated stands is that this light-demanding species requires large gaps or low density stands to regenerate and grow. This would require sparsely-stocked stands, resulting in lowered production.

A study of four multi-layered Scots-pine stands (77–99% of basal area) in northern and central Sweden showed that a multi-layered stand structure could be created but was not sustainable in the long run (Lundqvist et al. 2019). When basal area exceeded 12–13 m^2/ha, ingrowth past 1.3 m stopped along with regeneration of new seedlings. In comparison, Moan Mn (2021) reported ingrowth rates between 15 and 20 pine stems/ha/year in pure Scots pine selection-system plots at a rather high stand basal area, approaching 20 m^2/ha at the end of the observation period.

Rautio et al. (2023) studied the effect of stem density and site preparation on natural regeneration and seedling growth in mature pine forests in Lapland. Even though new seedlings emerged in their unthinned control stands, this was much less than in stands thinned to 50 stems/ha. Regeneration density in stands thinned to 150 or 250 stems/ha did not differ from that in unthinned control stands. In unthinned stands and in stands with 150 or 250 stems/ha, over 70% of the seedling-monitoring plots were empty. This suggests that although there is some regeneration under closed canopies, it is patchy and most seedlings in these patches will not survive. Patchy pine regeneration has also been observed by Karlsson and Nilsson (2005). All in all, these results suggest that creating and maintaining an uneven-aged stand using only pine can be difficult, and the long-term sustainability of multi-layered pine stands is questionable.

3.3   The Group System

Depending on the size of the gaps, this is the CCF system most similar to clearcutting when it comes to regeneration. Although there is less ground-level light in a group felling than in a clearcut, it is on average brighter than under the canopy of a selectively-cut stand (Hanssen 2007). Belowground competition is also lower, and regeneration measures like soil scarification and planting are easier. Still, it is common to use natural regeneration after group felling.

The species, height and density of trees surrounding a gap influences regeneration within it, depending on site conditions, size and shape of the gap, and the regenerating species. The edge trees may both benefit (providing seeds, reducing competition from brush and weed) and harm (exerting above- and below-ground competition) regeneration. Seed dispersal of Norway spruce and Scots pine diminishes with distance from the edge (Hesselman 1938), but seed supply will usually suffice throughout CCF-sized gaps (Lehto 1956; Hanssen 2003; Valkonen and Siitonen 2016; Hallikainen et al. 2019). Furthermore, there are large within-gap gradients in the competition for light, water, and nutrients by the dominant trees surrounding a gap (Kuuluvainen et al. 1993; de Chantal et al. 2003). At high latitudes, sunlight mostly falls in the northern part of the gap (de Chantal et al. 2003), while competition for belowground resources is highest at gap edges and decreases toward the centre (Smith et al. 1997). The location of seedlings within a gap affects their emergence, survival, and growth, and spatial patterns within the gap differ among those demographic phases.

The group system initiates regeneration in individual gaps, series of gaps, or systematic grids of gaps. In the initial regeneration phases, gaps are isolated and surrounded by closed mature stands. In later phases, new gaps are created next to earlier gaps. Shading by adjacent tall trees means gaps created later have edge effects and environmental conditions different from neighbouring earlier patches. Existing studies have only investigated the establishment and development of regeneration in early gaps, which were surrounded by mature stands. The development of later regeneration phases in the group system might deviate substantially from those patterns. Later phases of the group system frequently remove the mature stand in larger patches or use periods of shelter.

There are few studies of regeneration in gap cuttings, and their coverage is uneven among regions, site types, treatments, and stand properties. Many region-site combinations are still not covered. In Finland, large experimental studies have recently begun in key areas, but it will still take several years to see conclusive results, especially in the north.

3.3.1 The Group System in Norway Spruce-Dominated Forests

3.3.1.1 The Group System in Norway Spruce-Dominated Forests on Mineral Soil

3.3.1.1.1 Seedling Density

According to experimental results, Norway spruce has regenerated rather well in stands across Finland (Table 3.1). Site preparation has enhanced seedling densities markedly on mineral soils in northern areas, but not much on fertile sites in the south, where the proliferation of vegetation has tended to counteract the benefits

Table 3.1 Average spruce-seedling densities and heights in gap-cutting studies conducted in Finland

Region, site, and site preparation	Reference no.	Gap shape and size	Seedling density of main species (stems/ha)	Density of other conifer species + birch (stems/ha)	Main crop seedling density[a], (stems/ha)	Mean height of crop seedlings, time since treatment
South, mineral, submesic, no site prep	1	Square, 40 × 40 m	7600	4000	1300	60 cm, 10–11 year
South, mineral, submesic, disc trenching	1	Square, 40 × 40 m	6700	12,300	1300	80 cm, 10–11 year
Kainuu (north), mineral, submesic, no site prep	2	Shape variable, diameter ~ 10–60 m	2400	5900	1700	50 cm, 13–15 year
Kainuu (north), mineral, submesic, disc trenching	2	Shape variable, diameter ~ 10–60 m	9300	13,900	2000[b]	50 cm, 13–15 year
South, mineral, Mesic and submesic, various site prep	3	Shape variable, diameter 10–60 (80) m	11,400	5800	1770	105 cm, 7–10 year
North, drained peatland, fertile, no site prep	4	Circular, diameter 10–25 m + 0.2–0.3 ha patch cuts	10,400	3200	2300	73–84 cm, 10 year

[a] Main crop seedlings are defined as seedlings that would be retained in a pre-commercial thinning, i.e. healthy seedlings with adequate spacing. See Valkonen et al. (2011)
[b] The number of main crop seedlings was 2300 stems/ha when planted pines were included and prioritised in the survey. There were such huge numbers of natural seedlings that the average number of crop seedlings would have exceeded the target of 2000 stems/ha without the planted seedlings
References: 1 = Valkonen et al. (2011), 2 = Valkonen and Siitonen (2016), 3 = Valkonen (2019), 4 = Hökkä and Repola (2018)

Table 3.2 Average pine-seedling densities and heights in gap cutting studies conducted in Finland and Sweden

Region, site, and site preparation	Reference no.	Gap shape and size	Seedling density of main species (stems/ha)	Density of other conifer species + birch, (stems/ha)	Main crop seedling density[a] (stems/ha)	Mean height of crop seedlings, time since treatment
N Karelia, Finland, mineral, xeric and subxeric, various site prep	1	Circular, diam. 10–60 (80) m	13,000	3400	1530	39 cm, 6–9 year
Northern Finland, mineral, xeric and subxeric, patch scarification	2	Circular, diam. 20, 40 and 80 m	21,700	7300	~2000[b]	9 cm, 5 year
Northern Sweden, mineral, Mesic, no scarification	3	Square, 40 × 40 m	~ 7800[a]	Not applicable[c]	3200[c]	60 cm, 14 year

[a] See definition in Table 3.1. The Swedish study operates with closer spacing for crop trees than the Finnish studies
[b] Calculated using the study's model with the proportion of site preparation at 10–20%
[c] The Swedish study was established in a mixed pine-spruce stand, and no distinction is made between seedling species. Pine constituted 45% of the main crop seedlings, birch 36%, and spruce 19%
References: 1 = Valkonen (2019), 2 = Hallikainen et al. (2019), 3 = Goude et al. (2022)

much sooner. In southern Finland on the most mesic sites, the establishment of large gaps >40 m wide has resulted in the intensive proliferation of brush and weed, hindering regeneration (Downey et al. 2018; Valkonen 2019).

In Sweden, studies of chequered-gap systems have also shown good regeneration results, for instance in a trial with 40 × 40 m gaps in mixed pine-spruce forest in Gällivare, northern Sweden (Ackemo 2018; Goude et al. 2022, see also Table 3.2).

Seedling density is highest in the 5–10 m closest to the gap edge (Hanssen 2003; Valkonen et al. 2011; Goude et al. 2022). Brush and weed may be less prolific near the edge, so seedlings thrive better there. Goude et al. (2022) also found that there were more seedlings in the north of the gap compared to the south, although the density was generally sufficient everywhere.

Unlike studies in southern Finland, in spruce stands in Kainuu, northern Finland (Valkonen and Siitonen 2016) with a lower fertility and a cooler climate, ground vegetation barely changed and spruce-seedling density was similar throughout gaps. As discussed above, studies show a relatively small influence of the diminishing seed rain density toward gap centres in such small gaps (10–60 m diameter). It was concluded that the spruce-seed rain from edge stands may be sufficient to restock gaps up to 0.5–1.0 ha if the other main factors are favourable.

3.3.1.1.2   Seedling Growth

Negative edge effects on spruce-seedling growth have been found in some (Borgstrand 2014; Valkonen et al. 2011), but not all studies (Goude et al. 2022). In southern Finland, Valkonen et al. (2011) found seedling height increased substantially from stand edges into the gap centre up to a distance of at least 20 m for all major species.

Even though seedling growth in gaps is more vigorous than under a canopy (Granhus et al. 2003; Hanssen 2003, gap diameter 25–50 m), growth is usually less than in open clearcuts (Valkonen et al. 2011, gap diameter 40 m). However, there is a balance between competition from and facilitation by edge trees. Their positive effects include checking weed growth and frost protection. Thus, some studies show similar growth of spruce seedlings in gaps and on clearcuts (Borgstrand 2014, gap size 30 × 45 m).

3.3.1.2   The Group System in Norway Spruce-Dominated Forests on Drained Peatland

In drained-peatland spruce stands, regeneration after group felling has been studied for 10 years at two sites in northern Finland (Hökkä et al. 2011, 2012). The diameter of the cut gaps was 10–25 m, the latter being about equal to dominant stand height. Regeneration in four 0.2–0.3 ha (40 × 50 m to 50 × 60 m) patch clearcuts was also investigated.

Advance regeneration formed a significant part of the seedling stock that developed in the gaps (Fig. 3.1). Five years after cutting, surviving advance regeneration (>10 cm tall) accounted for almost half of all spruce seedlings. Within 3–5 years, many small spruce seedlings (<10 cm) were found in all gap sizes. During the monitoring period, two good seed years occurred.

Downy birch accounted for 25–57% of all seedlings 5 years after cutting (Hökkä et al. 2011). The proportion of birch increased with gap size, suggesting that its establishment requires increased light availability. Very few Scots pine seedlings were found in the gaps. Ten years after gap cutting, spruce-seedling density varied between 5500 and 12,500 stems/ha (Hökkä and Repola 2018). In the 0.2–0.3 ha gaps, the 10-year regeneration result was poorer than on smaller canopy gaps, i.e., there were 850 spruce and 560 birch crop seedlings/ha. This suggests that in bigger gaps, seedling establishment takes longer, the resulting stand is more irregular, and the proportion of downy birch is higher than in smaller gaps (Hökkä and Repola 2018).

Establishment was poor and slow in the most productive shallow-peated and herb-rich sites due to aggressive growth of tall herbs, ferns, and grass. In sites with thicker peat and *Sphagnum*, more rapid establishment of spruce seedlings took place. Patch scarification appeared to hamper seedling establishment (Hökkä et al. 2012). Site preparation destroyed part of the advance regeneration and fast-growing herbs and grasses occupied the patches, effectively out-competing spruce seedlings.

Fig. 3.1 Regeneration in a small gap in Oulu (Finland), 12 years after cutting. Photo: Hannu Hökkä

Ten years after cutting, spruce crop seedling density in gaps averaged 2200/ha and their height varied between 73 and 84 cm (Hökkä and Repola 2018). The tallest seedlings were in the largest gaps. Taller advance-regeneration seedlings had faster height growth. High seedling densities, including competing birch seedlings, indicate that pre-commercial thinning is needed to promote good development of the established seedling stand.

An inventory study (Pulliainen 2019) revealed that the quality of advance regeneration in gaps was variable. Many seedlings had leader shoots that had diebacks. Almost one-third of the seedlings were infected with some sort of rot, making them unsuitable to produce saw timber in the future.

3.3.2 The Group System in Scots Pine-Dominated Forests

3.3.2.1 The Group System in Scots Pine-Dominated Forests on Mineral Soil

3.3.2.1.1 Seedling Density

For Scots pine, studies have also shown good seedling densities after group felling (Hallikainen et al. 2019; Ackemo 2018; Goude et al. 2022, Table 3.2). Site preparation may enhance the emergence substantially. Exposing 10–20% of the mineral soil guaranteed abundant and evenly-distributed regeneration in a study by Hallikainen et al. (2019). In an experiment in North Karelia, seedling densities were only moderate, probably due to their measurement coming shortly after the treatment (Valkonen 2019). A later substantial increase in seedling densities is expected, and it was concluded that pure-pine stands on infertile sites are very favourable for natural regeneration of pine, including in patches established by site preparation.

3.3.2.1.2   Seedling Growth

Pine seedling growth is noticeably influenced by surrounding stands (Axelsson et al. 2014; Borgstrand 2014; Hallikainen et al. 2019). However, Goude et al. (2022) found no edge effects on seedling growth (pine, birch, and spruce) in 40 × 40 m gaps in northern Sweden after 14 years. Borgstrand (2014) found seedlings in northern parts of gaps to grow 30% faster compared to other areas in gaps. However, compared to a reference clearcut, the pines in the gaps grew about 30% less.

An average stand dominated by pine or spruce clearly reduces seedling growth for at least 20 m beyond its edge (Ruuska et al. 2008; Valkonen et al. 2011). The edge influence compounds as the seedlings grow taller. A study by Ruuska et al. (2008) found that in 5–10 m tall pine sapling stands, stem density, height, and volume all decreased strongly toward the surrounding established pine stands in southern Finland. Adjacent to the edge stand, there was a zone a few metres wide where few or no pine saplings survived or grew.

3.3.2.2   The Group System in Scots Pine-Dominated Forests on Drained Peatland

Pine regeneration after group felling on peatland in Finland has only been studied in a few field experiments, with 20–40 m-wide strip-formed gaps. The regeneration success has so far only been followed for a few years.

Tentative results show that, especially on dwarf-shrub-dominated sites, the biggest regeneration problem is uneven seedling establishment among gaps. It is caused by dense ground coverage of raw humus, forest moss, and dwarf shrubs. Over half of the seedlings were located on *Sphagnum* surfaces or machine tracks. Light site preparation disrupting raw-humus and forest-moss surfaces could result in more even seedling establishment.

3.3.3   The Group System in Broadleaf-Dominated Forests

Downy and silver birch are shade-intolerant species. Conifers, especially spruce, strongly suppress shade-intolerant seedlings, and large-diameter gaps, at least 20 m (Valkonen et al. 2011) and maybe over 50 m, are required to allow birch development near their centres. More experimental results are needed for more general conclusions and recommendations.

Shade-tolerant broadleaved species, for instance European beech (*Fagus sylvatica*), could possibly expand into today's boreal forest areas in southern Fennoscandia given climate change (Kramer et al. 2010). For beech, both the group and shelterwood systems are suitable silvicultural methods.

3.4   The Shelterwood System

Compared to the seed-tree cutting, the shelterwood cutting(s)provides several benefits in addition to seed sources (see Sect. 3.1). The shelterwood system also has higher stocking levels and longer-duration retention.

A common shelterwood density in Fennoscandia is about 10–12 m^2/ha after the first regeneration cut. It should be noted however, that optimal shelter density varies among species, sites and management objectives. Generally, higher retention levels are desirable on more-fertile sites with high competition from ground vegetation, and on sites prone to frost or waterlogging. On less-fertile sites, a lower density of shelter trees may be a sufficient seed source and enhance establishment of light-demanding tree species.

Regeneration in shelterwood systems can either be planted or naturally recruited. The next generation often consists of several years of seedling cohorts, with the entire regeneration process spanning around 10–30 years; the faster end of this range is typical of Scots pine and the slower end for Norway spruce.

3.4.1   The Shelterwood System in Norway Spruce-Dominated Forests

3.4.1.1   The Shelterwood System in Norway Spruce-Dominated Forests on Mineral Soil

Norway spruce is generally well suited to regeneration under shelter. However, irregular seed years, high susceptibility of the shelter trees to wind damage, and shallow root systems restricting mechanical site preparation complicate shelterwood regeneration. In Fennoscandia, there have been relatively few controlled studies that assess the suitability of shelterwood regeneration of Norway spruce. Furthermore, little is known about the recruitment patterns and regeneration dynamics in such stands.

In northern Sweden, Hagner (1962) surveyed 58 Norway spruce and Scots pine shelterwood cuttings (totalling 10,000 ha), which they defined broadly as shelters of varying densities, from widely-spaced seed trees to closed forests. Based on the collected data and an estimate of the seed production after shelterwood cuttings across

Sweden, Hagner (1962) concluded that Norway spruce, with a combination of a shelter and mechanical site preparation, may naturally regenerate at satisfactory levels across the entire country. However, seed years play a decisive role in the success of the regeneration process.

Sikström (1997) examined 52 shelterwood cuttings in southern and central Sweden, representing a wide range in site fertility (site indices 18–30 m). Higher regeneration success in the southern locations (65% of sites had a density $\geq$ 4000 Norway spruce seedlings/ha, compared to 38% in the north) was thought to be due to more favourable climatic conditions, higher seed production, and more common fresh-moist and moist soils.

Leinonen et al. (1989) studied the regeneration success of spruce shelterwood cutting on mineral soil sites in south-central Finland. The amount of regeneration stocking was inventoried before cutting, during the summer after cutting, and 1 year after cutting. The mean retained volume after cutting was 120 m³/ha (observed range 39–220 m³/ha) and stem density 186/ha. Prior to cutting, the density of acceptable seedlings was 1440/ha and 1 year after cutting 1546/ha. The result was found unsatisfactory, and the method considered risky, although the monitoring period (two growing seasons) was very short.

Several authors recommend combining a shelter basal area of around 10 m²/ha with mechanical site preparation as a tradeoff between satisfactory seedling survival and growth (Leinonen et al. 1989; Nilsson et al. 2002; Örlander and Karlsson 2000). Higher seedling survival under shelter trees compared to open clearcut areas was mainly due to: (1) less-extreme temperatures that cause frost and frost heaving (Langvall and Örlander 2001; Lofvenius 1995), (2) reduced competition from ground vegetation (Hagner 1962), and (3) reduced pine-weevil damage to seedlings (Petersson 2004).

According to Nilsson et al. (2002) denser shelters (basal area 13.2–28.2 m²/ha) and mechanical site preparation promote emergence of Norway spruce seedlings in southern Sweden. However, Örlander and Karlsson (2000) concluded that denser shelters depress seedlings' height growth. In Norway, Skoklefald (1989) compared planting and natural regeneration in shelterwoods (250 trees/ha) with clearcutting (50 × 90 m) with or without site preparation in a bilberry spruce stand in SE Norway. After 11 years, natural regeneration was clearly denser under shelterwoods with mechanical site preparation. However, seedlings germinating after the initial cut were only 17 cm tall after 11 years, compared to 42 cm in the clearcut.

Overstorey depression of seedling growth is most likely due to the combination of above- and below-ground competition from shelter trees. Although light availability does not limit germination and initial seedling growth, it may limit growth in the longer run. Örlander and Karlsson (2000) reported relatively slow growth of all sized advance-growth seedlings during the first 3–4 years following the release cutting, but it was most pronounced for the shortest (< 100 cm) seedlings (see also Skoklefald 1967). Delayed growth of taller (>100 cm) seedlings was explained by needles needing time to adapt to brighter light, whereas small seedlings probably struggled with both needle adaptation and dry humus. Shade-grown seedlings suddenly exposed to bright light may suffer photosynthetic damage, something known as the release effect. This effect is generally greater in shade-tolerant species (Grossnickle 2000) like Norway spruce. To avoid post-release seedling damage, the shelter should be removed gradually. Örlander and Karlsson (2000) tested shelter densities from 0.9–32.7 m²/ha, finding that 24% of seedlings under 20 cm died if shelter basal area was ≤7 m/ha. The corresponding mortality under denser (basal areas ≥12 m²/ha) shelters was only 1%. Mortality was attributed to release effects (25%), pine weevils (28%), and unknown factors (47%). In Norway, Skoklefald (1989) reported high mortality of Norway-spruce seedlings after overstorey removal, reducing the seedling density by about 80%. Skoklefald (1967) reported 16 and 38% mortality after overstorey removal in two shelterwoods in southeastern Norway, with small seedlings (≤ 10 cm) clearly being the most vulnerable.

Pre-cutting height and leading-shoot length are good predictors of post-release seedling survival (Örlander and Karlsson 2000; Skoklefald 1967). Örlander and Karlsson (2000) showed that, for shelters of basal area 12–33 m²/ha, seedling survival probability decreases dramatically, when pre-cutting seedling heights are below 50 cm. This is likely because small seedlings are more susceptible to the release effect and pine weevils compared to larger seedlings. Therefore, the authors recommended that the seedlings should not be released until a sufficient number are at least 50 cm tall.

On the other hand, taller seedlings suffer more logging-related damage (Skoklefald 1967; Sikström and Glöde 2000). Sikström and Glöde (2000) reported that around one-third of all saplings 1–1.5 m tall (at densities of 6400–25,400/ha) suffered serious logging damage. The proportion of damaged seedlings increased with increasing shelterwood stem volume (132–234 m³/ha). The most common causes of seedling damage were burial under slash (32–58%), machinery (12–31%), felling (3–20%) and tree dragging (up to 4%). It should be noted, however, that damage sources were often difficult to identify.

3.4.1.2 The Shelterwood System in Norway Spruce-Dominated Forests on Drained Peatland

Drained peatlands (regardless of the peat thickness) promote coniferous-seedling establishment relatively well. Dense cover of advance regeneration before shelterwood cuttings was reported in several studies (Moilanen et al. 2011; Hånell 1993; Örlander and Karlsson 2000). The shelterwood system is a possible strategy to mitigate the negative effects of clearcutting. For instance, shelterwood retention may reduce post-harvest water-level rises, competition from ground vegetation, frost and frost heaving.

Hånell (1993) and Holgen and Hånell (2000) investigated effects of two shelterwood densities (140 and 200 stems/ha) for natural and artificial regeneration on highly-productive peatlands (peat depth 1.3 m) in nine mature Norway spruce forests across Sweden. Stocking levels of large (≥ 10 cm) naturally-regenerated Norway spruce seedlings (five growing seasons after the cut) were on average 4500 and 9000 seedlings/ha in stands of 140 and 200 stems/ha, respectively (Hånell

1993). The results from underplanting showed that mechanical site preparation with mounding promoted both seedling survival and height growth (Holgen and Hånell 2000). At the end of the 6-year study period, 43% and 38% of shelter trees were blown down in 140 and 200 stems/ha stands, respectively (Hånell and Ottosson-Löfvenius 1994). These results do not mean shelterwood cutting in peatland spruce forests should be avoided due to windthrow risks. However, to minimise the risks, higher shelterwood densities were recommended.

Moilanen et al. (2011) compared regeneration results after different regeneration methods in a thick-peated drained spruce mire in eastern-central Finland. Fifteen years after shelterwood cutting, spruce-, birch- and pine-seedling densities were 7000, 16,000 and 200 stems/ha, respectively. Of those, the crop-seedling densities for spruce, birch and pine were 2215, 355, and 25 stems/ha, respectively. The crop-seedling density results were comparable to treatments using different soil-preparation methods (mounding, patch scarification) and planting with spruce or pine.

Only a limited number of Fennoscandian shelterwood experiments have been carried out in fertile Norway spruce-dominated drained-peatland forests. The few published studies suggest that the shelterwood system provides quick and abundant natural regeneration of spruce and downy birch on drained-peatland sites, even without site preparation.

3.4.1.3 The Shelterwood System in Norway Spruce-Dominated Forests in Mountain Areas

The shelterwood system for Norway spruce in mountain areas shows many of the same benefits and disadvantages seen elsewhere for establishment, survival and growth. Elfving (1990) studied a Norway-spruce-dominated shelterwood (240 stems/ha) and an adjacent clearcut area near the treeline in northern Sweden (latitude 63.28° N, elevation 550 m), both planted with 4-year-old seedlings at 2×2 m spacing. At the end of the 27-year observation period, the average sapling height under the shelter was 2.24 m, approximately 0.75 m shorter than average saplings on the clearcut. The observed slowing of height growth corresponds to 5 years of development, and a loss of 18 m^3/ha over the rotation. However, the volume loss in the new generation was more than offset by the increment of the shelter trees (44 m^3/ha during 27 years). At the time of the inventory, natural regeneration consisted of 1300 spruce and 700 birch seedlings/ha. The average height of naturally-regenerated Norway-spruce seedlings was 1.77 m. The reported survival rates for planted seedlings were 90% and 59% in the shelterwood and the clearcut, respectively. Higher seedling survival under the shelterwood was likely due to a reduced risk of frost damage.

3.4.2   The Shelterwood System in Scots
             Pine-Dominated Forests

Successful implementation of the shelterwood system in Scots pine stands may yield very dense regeneration, sometimes reaching tens of thousands of seedlings/ha. In a study in southern Sweden, Beland et al. (2000) reported that 4 years after regeneration cutting and mechanical site preparation, natural regeneration of Scots pine yielded 53,000 and 90,000 seedlings/ha in shelters with basal areas of 12 and 15 m^2/ha, respectively. Kyrö et al. (2022) and Lula (2022) found comparable results. Although several reports show a clear positive effect of shelter density on seedling densities (e.g. Beland et al. 2000), a study conducted by Rautio et al. (2023) in northern boreal Finland found the opposite trend. This is likely due to shading by shelter trees limiting seed germination and seedling survival.

Thick humus layers and competing ground vegetation are among the most important constraints on natural regeneration, especially in southern Fennoscandia. Competition from ground vegetation generally increases in more-productive sites. Therefore, natural regeneration is usually applied on low- to medium-fertility sites, where mechanical site preparation is also often recommended. Mechanical site preparation promotes seedling establishment (Kyrö et al. 2022), height growth (Hagner 1962), and increases regeneration homogeneity (Fries 1979).

Kyrö et al. (2022) observed 1000–4400 and 32,000–92,000 seedlings/ha on intact and prepared ground, respectively. However, seedling emergence and survival on intact ground depended on the ground cover's species composition. The benefit of mechanical site preparation diminishes over time, primarily due to the gradual invasion of ground vegetation. However, as suggested by Beland et al. (2000) and Hagner (1962), this process is slower under a shelter, giving several years of seedling cohorts a chance to grow into future stands. Several studies have reported steadily-increasing seedling density up to 10 years after the regeneration cut (Beland et al. 2000; Kyrö et al. 2022; Lula 2022; Rautio et al. 2023). Hence, gradual and constant seedling emergence seems possible on favourable seedbeds, as some viable seeds are produced in most years.

Kyrö et al. (2022) reported that seedling mortality declined from a maximum in the first year to a fraction of the initial level 8 or more years after emergence. Therefore, these authors concluded that regeneration success can only be assessed after seedling mortality has stabilised sufficiently after emergence of the main seedling cohort. Beland et al. (2000) and Lula (2022) reported that the seedling cohort established in the first year after the regeneration cut constitutes >50% of the entire 5–6-years post-cutting seedling population.

In naturally-regenerated stands, dispersed seeds land on all types of seedbeds present. Mechanical site preparation creates different seedbed types; bare mineral soil is the most favourable for emergence, but is not necessary for seedling survival. High mortality on mineral soil, however, is outweighed by abundant emergence, resulting in a constant seedling-density increase following mechanical site preparation (Kyrö et al. 2022). A mixture of mineral soil and humus is often perceived as the most favourable seedbed for natural regeneration, as it provides an optimal compromise among satisfactory seedling emergence, survival, and growth (Beland et al. 2000; Kyrö et al. 2022). Low seed production and insufficient seedling recruitment are two of the major limitations of natural Scots-pine regeneration in northern Fennoscandia. The frequency and viability of seed years generally decrease at higher latitudes and elevations (Valkonen 1992; Henttonen et al. 1986). Seed years occur irregularly at intervals of 2–5 years south of 60° N, and up to 100 years at the northern treeline (~ 70° N, Heikinheimo 1937; Numminen 1982; Valkonen 2000). Therefore, timing site preparation for a good seed year is recommended (Karlsson and Örlander 2000). Release cuttings lead to several years of increased cone production (Karlsson and Örlander 2002), consequently promoting pine regeneration. Furthermore, compared to fully-stocked stands, year-to-year variation in seed crops is generally lower in the released stands (Heikinheimo 1937).

Shelterwoods can help reduce the risk of pine weevil (*Hylobius abietis* L.) damage to both planted and naturally regenerated conifer seedlings. The riskiest time for pine weevil damage is when removing the shelter, as pine weevils are attracted by the odour of newly-cut stumps (Sundkvist 1994). To reduce this risk, Wallertz et al. (2005) recommend that shelter trees not be removed until most seedlings have reached the safe sizes of 9 mm basal diameter for Norway spruce and 12 mm for Scots pine.

Shelter trees inhibit growth of the next generation, especially for light-demanding species like Scots pine. The high stocking levels and long retention periods of the shelterwood system result in slower regeneration compared to seed trees. Scots pine thus has shorter shelter periods than Norway spruce. Beland et al. (2000) suggested that removal of the dense shelter with a basal area of approximately 15 m²/ha should begin when seedlings reach about 0.5 m tall, and the overstorey should be maintained until regeneration reaches about 6 m. In a contemporary Finnish shelterwood variant for pine, some overstorey trees are retained throughout the rotation to promote ecosystem services and production of very-high-quality timber (Äijälä et al. 2019; Valkonen 2020). The system generates within-stand tree-size stratification through differences in growth rates, leading to more heterogeneous stands (Lundqvist et al. 2019).

In Norway, the shelterwood system has been practised in Scots pine in some places (see picture in Chap. 2). The practice was catalysed by the impossibility of successfully regenerating pine due to excessive moose-browsing damage. Harvesting old pine trees successively created an open shelter, although the name "shelterwood" was not used. High-density natural regeneration of pine established under the shelter and has now been released and developed into young stands. These examples illustrate that pine can be successfully managed using the shelterwood system, and not just with the seed-tree method.

3.5 Conversion to Continuous Cover Forestry

Regeneration is an important process in the conversion to CCF. Applying the shelterwood and group systems in stands not previously managed this way needs no specific adaptation of the regeneration management, as conversion starts with the regeneration cuts. Conversion to the selection system, on the other hand, is a long process toward a multi-storied stand, and regeneration is essential in this conversion. Conversion to the selection system requires initiating regeneration at an earlier stage than usual of stand development.

A proposed method for conversion to the selection system is variable-density thinning (VDT, Brodie and Harrington 2020), which creates gaps in young stands to initiate regeneration. Another method is more homogeneous, open shelters, for example created after snow damage (Knoke and Plusczyk 2001). Openings enable natural regeneration and give good results if they stay open long enough. Rapid growth and development of young stands closes gaps and canopies, and further interventions are essential to continue regenerating seedlings and saplings. VDT creates horizontal small-scale variation in tree size rather than vertical stratification, so seems more appropriate for young stands at the beginning of the conversion period.

Low seed production in young stands is usually caused by high stand density and small crowns (Hagner 1965; Nygren et al. 2017). Reducing stand density promotes seed production, but in some methods only for edge trees next to openings.

Artificial regeneration is often used alone or in combination with natural regeneration in conversion methods, particularly to introduce new species (e.g., Knoke and Plusczyk 2001), especially in the context of climate adaptation.

A conversion concept tried in Sweden (Drössler et al. 2014), based mostly on heavy thinning, suffered low regeneration during the first 10 years. The trials have been modified to maintain more open stands (Goude et al. 2022).

Target-diameter harvest can also be used as a conversion treatment. Experiments in Sweden have used it to study early-regeneration establishment (Drössler et al. 2015, 2017). In response to the harvests, seedling and sapling density and growth have increased, not only of Norway spruce, but also of more light-demanding species.

Conversion to the selection system is a topic only addressed during recent decades, and so far little covered by regeneration studies in Fennoscandia. The few results available from other regions are not especially relevant for the Nordic region due to species and site differences, but VDT has been tried in Washington, Oregon, and Chile (Dodson et al. 2012; Donos et al. 2020; Puettmann et al. 2016). VDT creates openings that allow regeneration to establish and develop while also promoting further development of spatially-variable advance regeneration, in a way similar to the group system.

3.6   Conclusions

In the context of CCF, the preferred form of regeneration is natural regeneration, but this is a slow and complex process. Managing natural regeneration requires insight into all stages and processes, and above all patience. Common to all CCF silvicultural systems are the varying effects of shelter density (in the group system this refers to the adjacent tall trees) on the regeneration outcomes. Managing shelter density is therefore an essential method to optimise regeneration.

In the selection system, continuous germination of new seedlings and ingrowth into larger size classes is required to replace harvested trees for long-term sustainable management. Regeneration and ingrowth rates vary significantly between stands, influenced by historical factors, with no clear relationship to current stand conditions, particularly stand density. The system is more suitable for shade-tolerant species like Norway spruce, while pine-dominated stands struggle to achieve and sustain regeneration due to their shade intolerance and high stand densities.

In the shelterwood system, a successive, uniform opening of the canopy promotes regeneration. Shelter trees provide seeds, while reducing damage to seedlings and competition from ground vegetation. On the other hand, they can also have an adverse effect on seedlings by directly competing with them. Evidence from the reviewed literature from Fennoscandia shows that both Scots pine and Norway spruce can be successfully regenerated using the shelterwood system. Although the system is generally more suitable for Scots pine, high densities of shelter trees, especially if retained over a long period, may hinder artificial or natural regeneration of light-demanding Scots pine seedlings.

In the group system, regeneration is initiated in individual gaps, series of gaps, or systematic grids of gaps. The edge trees influence regeneration through providing seeds, checking weed growth, and exerting competition. Fennoscandian studies on regeneration in individual gaps generally show satisfactory seedling density for both Norway spruce and Scots pine. However, seedling growth is usually slowed, especially close to the gap edges. The edge effect increases as seedlings grow taller, making long-term survival difficult for light-demanding pine saplings nearest the edge.

Conversion to the selection system initiates regeneration in young stands, and tries to maintain a continual regeneration process with sustained slow recruitment. Given the rapid growth and crown closure of young stands, frequent manipulation of shelter density is essential during conversion. Initiating regeneration in small gaps, for example using variable-density thinning, might therefore be easier than a homogeneous shelter approach.

References

Ackemo J (2018) Naturlig trädföryngring och epifytiska hänglavar 10 år efter en avverkning i schackruteform (Natural tree regeneration and epiphytic lichens 10 years after a Chequered-Gap-cutting). Master thesis, Swedish University of Agricultural Sciences

Ahlström MA, Lundqvist L (2015) Stand development during 16–57 years in partially harvested sub-alpine uneven-aged Norway spruce stands reconstructed from increment cores. For Ecol Manag 350:81–86

Äijälä O, Koistinen A, Sved J, Vanhatalo K, Väisänen P (eds) (2019) Metsänhoidon suositukset (best practices for forest management). Tapion julkaisuja, Helsinki, p 252

Andersson S (2015) Tilvekst på bestands- og enkelttrenivå ti år etter selektiv hogst etter KONTUS-prinsippet (Stand and single-tree growth ten years after selective cuttings by the Kontus-principle). Master Thesis, Norwegian University of Life Sciences, pp 70

Arnborg T (1947) Föryngringsundersökningar i mellersta Norrland (Regeneration studies in central Norrland). Norrlands Skogsvårdsförbunds Tidskrift, pp 247–293

Axelsson EP, Lundmark T, Högberg P, Nordin A (2014) Belowground competition directs spatial patterns of seedling growth in boreal pine forests in Fennoscandia. Forests 5(9):2106–2121

Beland M, Agestam E, Ekö P, Gemmel P, Nilsson U (2000) Scarification and seedfall affects natural regeneration of scots pine under two shelterwood densities and a clear-cut in southern Sweden. Scand J For Res 15(2):247–255

Bøhme JG (1957) Bledningsskog II. Tidsskr Skogbr 65:203–247

Borgstrand E (2014) Plantors och träds tillväxt efter schackrutehuggning och i konventionellt trakthyggesbruk (Seedling and tree growth after Chequered-Gap-Shelterwood-Cutting and in conventional clear-cutting system). Master thesis, Swedish University of Agricultural Sciences 2014:23

Brodie LC, Harrington CA (2020) Guide to variable-density thinning using skips and gaps, vol 37. US Department of Agriculture, Forest Service, Pacific Northwest Research Station, Portland, OR

de Chantal M, Leinonen K, Kuuluvainen T et al (2003) Early response of *Pinus sylvestris* and *Picea abies* seedlings to an experimental canopy gap in a boreal spruce forest. For Ecol Manag 176:321–336

Dodson EK, Ares A, Puettmann K (2012) Early responses to thinning treatments designed to accelerate late successional forest structure in young coniferous stands of western Oregon, USA. Can J For Res Revue Canadienne De Recherche Forestiere 42(2):345–355

Donos PJ et al (2020) Short-term effects of variable-density thinning on regeneration in hardwood-dominated temperate rainforests. For Ecol Manag 464

Downey M, Heikkinen J, Valkonen S (2018) Natural tree regeneration and vegetation dynamics across harvest gaps in Norway spruce dominated forests in southern Finland. Can J For Res 48:524–534

Drössler L, Eko PM, Balster R (2015) Short-term development of a multilayered forest stand after target diameter harvest in southern Sweden. Can J For Res 45(9):1198–1205

Drössler L, Fahlvik N, Wysocka NK, Hjelm K, Kuehne C (2017) Natural regeneration in a multi-layered Pinus sylvestris-Picea abies Forest after target diameter harvest and soil scarification. Forests 8(2)

Drössler L, Nilsson U, Lundqvist L (2014) Simulated transformation of even-aged Norway spruce stands to multi-layered forests: an experiment to explore the potential of tree size differentiation. Forestry 87(2):239–248

Eerikäinen K, Valkonen S, Saksa T (2014) Ingrowth, survival and height growth of small trees in uneven-aged *Picea abies* stands in southern Finland. For Ecosyst 1(5):10

Elfving B (1990) Granplantering under gles högskärm i fjällskog (Spruce planting under sparse high canopy in mountain forest). Sveriges Skogsvardsforbunds Tidskrift 5:1–8

Engelmark O, Hytteborn H (1999) Coniferous forests. Acta phytogeographica suecica 84:55–74

Erefur C, Bergsten U, Lundmark T, De Chantal M (2011) Establishment of planted Norway spruce and Scots pine seedlings: effects of light environment, fertilisation, and orientation and distance with respect to shelter trees. New Forests 41(2):263–276

Fries J (1979) Natural regeneration within the Siljansfors experimental park. Sveriges Skogsvaardsfoerbunds Tidskrift

Goude M, Erefur C, Johansson U, Nilsson U (2022) Hyggesfria skogliga fältforsök i Sverige. En sammenställning av tillgängliga långtidsförsök (Clear-cut free forest field trials in Sweden. A compilation of available long-term trials). SLU Report 22

Granhus A, Allen M, Bergsaker E (2020) Fjellskoghogst—produksjon, foryngelse og økonomi (Mountain selective cutting - production, regeneration and economy). NIBIO Raport 72

Granhus A, Brække FH, Hanssen KH, Haveraaen O (2003) Effects of partial cutting and scarification on planted *Picea abies* at mid-elevation sites in south-East Norway. Scand J For Res 18:237–246

Granhus A, Ødegård E, Bergseng E, Bergsaker E (2021) Lukkede hogster - produksjon, foryngelse og økonomi (CCF - production, regeneration and economy). NIBIO Rapport 7(148):42

Hagner S (1962) Naturlig föryngring under skärm. En analys av föryngringsmetoden, dess möjligheter och begränsningar i mellannorrländskt skogsbruk (Natural regeneration under canopy. An analysis of the regeneration method, its possibilities and limitations in mid-northern forestry). Meddelande från Statens skogsforskningsinstitut 52(4):1–263

Hagner S (1965) Om fröproduktion, fröträdsval och plantuppslag i försök med naturlig föryngring (About seed production, seed tree selection and seedling selection in experiments with natural regeneration). Studia Forestalia Suecica 27:113

Hallikainen V, Hyppönen M, Hökkä H et al (2019) Natural regeneration after gap cutting in scots pine stands in northern Finland. Scand J For Res 34:115–125

Hånell B (1993) Regeneration of *Picea abies* forests on highly productive peatlands—clearcutting or selective cutting? Scand J For Res 8(1–4):518–527

Hånell B, Ottosson-Löfvenius M (1994) Windthrow after shelterwood cutting in *Picea abies* peatland forests. Scand J For Res 9(1–4):261–269

Hanssen KH (2003) Natural regeneration of *Picea abies* on small clear-cuts in SE Norway. For Ecol Manag 180:199–213

Hanssen KH (2007) Endringer i mikroklima ved lukkede hogster (Changes in microclimate with CCF). In: Nygaard PH and Fløistad IS (eds). Foryngelse for et bærekraftig skogbruk (Regeneration for sustainable forestry). Forskning fra Skog og landskap 2(7):17–21

Heikinheimo O (1937) Metsäpuiden siementämiskyvystä II (of the seeding capacity of forest trees II). Commun Inst For Fenn 24(4):67

Henttonen H, Kanninen M, Nygren M, Ojansuu R (1986) The maturation of *Pinus sylvestris* seeds in relation to temperature climate in northern Finland. Scand J For Res 1(1–4):243–249

Hesselman H (1938) Fortsatta studier över tallens och granens fröspridning samt kalhyggets besåning (Continued studies on pine and spruce seed dispersal on clear-cuts). Medd Stat Skogförsöksanstalt 31:1–64

Hökkä H, Repola J (2018) Pienaukkohakkuun uudistumistulos Pohjois-Suomen korpikuusikossa 10 vuoden kuluttua hakkuusta (Regeneration results in small gaps on fertile peatlands in Northern Finland 10 years after cutting). Metsätieteen aikakauskirja 2018–7808, p 17

Hökkä H, Repola J, Moilanen M et al (2011) Seedling survival and establishment in small canopy openings in drained spruce mires in northern Finland. Silva Fenn:45633–45645

Hökkä H, Repola J, Moilanen M et al (2012) Seedling establishment on small cutting areas with or without site preparation in a drained spruce mire—a case study in northern Finland. Silva Fenn 46:695–705

Holgen P, Hånell B (2000) Performance of planted and naturally regenerated seedlings in *Picea abies*-dominated shelterwood stands and clearcuts in Sweden. For Ecol Manag 127(1–3):129–138. https://doi.org/10.1016/S0378-1127(99)00125-5

Karlsson C (2000) Effects of release cutting and soil scarification on natural regeneration in Pinus sylvestris shelterwoods. Swedish University of Agricultural Sciences

Karlsson C, Örlander G (2000) Soil scarification shortly before a rich seed fall improves seedling establishment in seed tree stands of Pinus sylvestris. Scan J For Res 15(2):256–266

Karlsson C, Örlander G (2002) Mineral nutrients in needles of *Pinus sylvestris* seed trees after release cutting and their correlations with cone production and seed weight. For Ecol Manag 166(1–3, 183):–191

Karlsson M, Nilsson U (2005) The effects of scarification and shelterwood treatments on naturally regenerated seedlings in southern Sweden. For Ecol Manag 205:183–197

Kielland-Lund J (1981) Hva er fjellskog? (What is a mountain forest?). Tidsskr Skogbr 89:46–61

Knoke T, Plusczyk N (2001) On economic consequences of transformation of a spruce (*Picea abies* (L) Karst) dominated stand from regular into irregular age structure. For Ecol Manag 151:163–179

Koistinen E, Valkonen S (1993) Models for height development of Norway spruce and scots pine advance growth after release in southern Finland. Silva Fennica 27(3):179–194

Koski V (1978) Results of long-time measurements of the quantity of flowering and seed crop of forest trees

Koski V (1991) Generative reproduction and genetic processes in nature. Genetics of Scots pine:59–72

Kramer K, Degen B, Buschbom J, Hickler T, Thuiller W, Sykes MT, De Winter W (2010) Modelling exploration of the future of European beech (Fagus sylvatica L.) under climate change—range, abundance, genetic diversity and adaptive response. For Ecol Manag 259(11):2213–2222

Kullman L (1996) Norway spruce present in the Scandes Mountains, Sweden at 8000 BP: new light on Holocene tree spread. Glob Ecol Biogeogr Lett:94–101

Kuuluvainen T, Hokkanen TJ, Järvinen E et al (1993) Factors related to seedling growth in a boreal scots pine stand: a spatial analysis of a vegetation-soil system. Can J For Res 23:2101–2109

Kyrö M, Hallikainen V, Valkonen S, Hyppönen M, Puttonen P, Bergsten U, Winsa H, Rautio P (2022) Effects of overstory tree density, site preparation, and ground vegetation on natural scots pine seedling emergence and survival in northern boreal pine forests. Can J For Res 52(5):860–869

Lähde E, Laiho O, Norokorpi Y, Saksa T (2002) Development of Norway spruce dominated stands after single-tree selection and low thinning. Can J For Res 32:1577–1584

Langvall O, Örlander G (2001) Effects of pine shelterwoods on microclimate and frost damage to Norway spruce seedlings. Can J For Res 31(1):155–164

Leemans R (1991) Canopy gaps and establishment patterns of spruce (*Picea abies* (L) karst) in two old-growth coniferous forests in Central Sweden. Vegetatio 93:157–165

Lehtonen A, Leppä K, Rinne-Garmston KT, Sahlstedt E, Schiestl-Aalto P, Heikkinen J, Young GH, Korkiakoski M, Peltoniemi M, Sarkkola S (2023) Fast recovery of suppressed Norway spruce trees after selection harvesting on a drained peatland forest site. For Ecol Manag 530:120759

Lehto J (1956) Tutkimuksia männyn luontaisesta uudistumisesta Etelä-Suomen kangasmailla. Summary (studies on the natural reproduction of scots pine on the upland soils of southern Finland). Acta For Fenn 66:1–106

Leinonen K, Leikola M, Peltonen A, Räsänen PK (1989) Kuusen luontainen uudistaminen Pirkka-Hämeen metsälautakunnassa (natural regeneration of Norway spruce in the Pirkka-Häme forestry agency). Acta Forestalia Fennica 209

Löf M, Dey DC, Navarro RM, Jacobs DF (2012) Mechanical site preparation for forest restoration. New For 43:825–848

Lofvenius MO (1995) Temperature and radiation regimes in pine shelterwood and clear-cut area

Lukkala OJ (1946) Korpimetsien luontainen uudistaminen (the natural regeneration of the riparian forests). Communicationes Instituti Forestalis Fenniae 34(3):150

Lula M (2022) Regeneration methods and long-term production for Scots pine on medium fertile and fertile sites. Thesis. Southern Swedish Forest Research Centre, Swedish University of Agricultural Sciences, Alnarp

Lundqvist L (1991) Some notes on the regeneration on six permanent plots managed with single-tree selection. For Ecol Manag 46:49–57

Lundqvist L (1993) Changes in the stand structure on permanent *Picea abies* plots managed with single-tree selection. Scand J For Res 8:510–517

Lundqvist L (2004) Stand development in uneven-aged sub-alpine *Picea abies* stands after partial harvest estimated from repeated surveys. Forestry 77:119–129

Lundqvist L, Ahlström MA, Axelsson EP, Mörling T, Valinger E (2019) Multi-layered scots pine forests in boreal Sweden result from mass regeneration and size stratification. For Ecol Manag 441:176–181

Lundqvist L, Nilson K (2007) Regeneration dynamics in an uneven-aged virgin Norway spruce forest in northern Sweden. Scand J For Res 22:304–309

Lundqvist L (2017) Tamm review: selection system reduces long-term volume growth in Fennoscandic uneven-aged Norway spruce forests. For Ecol Manag 391:362–375

Mielikäinen K, Valkonen S (1995) Kaksijaksoisen kuusi–koivu-sekametsikön kasvu (growth of a biennial spruce-birch mixed stand). Folia Forestalia 2:81–97

Mikola P (1971) Reflexion of climatic fluctuation in the forestry practices of northern Finland. Reports from the Kevo Subarctic Research Station 8:115–121

Moan Mn Å (2021) Effects of stand structure and stand density on volume growth and ingrowth in selectively cut stands in Norway. Master Thesis, Norwegian University of Life Sciences, pp 61

Moilanen M, Issakainen J, Vesala H (2011) Metsän uudistaminen mustikkaturvekankaalla—luontaisesti vai viljellen? (Forest regeneration in blueberry bog - natural or cultivated?). Metlan työraportteja 192. http://urn.fi/URN:ISBN:978-951-40-2287-6

Mork E (1968) Ecological investigations in the mountain forest at Hirkjølen experimental area. Nor For Res Inst Report 93:463–614

Nilsen P (1988) Fjellskoghogst i granskog - gjenvekst og produksjon etter tidligere hogster (Selective cutting in mountain spruce forests - regeneration and production after earlier cuttings). Raport Nor Inst Skogforsk 2(88):1–26

Nilson K, Lundqvist L (2001) Effects of stand structure and density on development of natural regeneration in two *Picea abies* stands in Sweden. Scand J For Res 16:253–259

Nilsson U, Gemmel P, Johansson U, Karlsson M, Welander T (2002) Natural regeneration of Norway spruce, scots pine and birch under Norway spruce shelterwoods of varying densities on a Mesic-dry site in southern Sweden. For Ecol Manag 161(1–3):133–145

Numminen E (1982) Pohjois-Lapin metsäpuiden siementuotanto (Forest tree seed production in Northern Lapland)

Nygren M, Rissanen K, Eerikäinen K, Saksa T, Valkonen S (2017) Norway spruce cone crops in uneven-aged stands in southern Finland. For Ecol Manag 390:68–72

Oleskog G, Sahlén K (2000) Effects of seedbed substrate on moisture conditions and germination of Pinus sylvestris seeds in a clearcut. Scand J For Res 15(2):225–236

Örlander G, Karlsson C (2000) Influence of shelterwood density on survival and height increment of *Picea abies* advance growth. Scand J For Res 15(1):20–29. https://doi.org/10.1080/02827580050160439

Øyen BH, Nilsen P (2002) Growth effects after mountain forest selective cutting in Southeast Norway. Forestry 75:401–410

Øyen BH, Nilsen P (2004) Growth and recruitment after mountain forest selective cutting in irregular spruce forest. A case study in northern Norway. Silva Fenn 38:383–392

Paavialainen E, Päiväven J (1995) Peatland forestry—ecological principles. Springer Verlag, p 248

Petersson M (2004) Regeneration methods to reduce pine weevil damage to conifer seedlings

Place ICM (1955) The influence of seedbed conditions on the regeneration of spruce and balsam fir. Canada Department of Northern Affairs and Natural Resources. Forestry Branch, Bull 117:87

Puettmann KJ, Ares A, Burton JI, Dodson EK (2016) Forest restoration using variable density thinning: lessons from Douglas-fir stands in Western Oregon. Forests 7(12)

Pulliainen L (2019) Taimikon laatu ja jatkokehityksen edellytykset Lounais-Lapin turvemaakuusikoiden pienaukkohakkuissa (The quality of spruce seedlings and the condition for further development of spruce mire stands in southwest Lapland canopy gaps). Master's Thesis. Helsingin yliopisto. Maatalous-metsätieteellinen tiedekunta, metsätieteiden osasto, p 67

Rautio P, Hallikainen V, Valkonen S, Karjalainen J, Puttonen P, Bergsten U, Winsa H, Hyppönen M (2023) Manipulating overstory density and mineral soil exposure for optimal natural regeneration of scots pine. For Ecol Manag 539:120996

Ruuska J, Siipilehto J, Valkonen S (2008) Effect of edge stands on the development of young *Pinus sylvestris* stands in southern Finland. Scand J For Res 23:214–226

Saarinen M, Valkonen S, Sarkkola S, Nieminen M, Penttilä T, Laiho R (2020) Jatkuvapeitteisen metsänkasvatuksen mahdollisuudet ojitetuilla turvemailla (Possibilities for continuous cover forestry in drained peatland forests). Metsätieteen aikakauskirja 2020–10372 p. 21. https://doi.org/10.14214/ma.10372

Saksa T (2004) Regeneration process from seed crop to saplings - a case study in uneven-aged Norway spruce-dominated stands in southern Finland. Silva Fenn 38(4):371–381

Saksa T, Valkonen S (2011) Dynamics of seedling establishment and survival in uneven-aged boreal forests. For Ecol Manag 261(8):1409–1414

Sarasto J, Seppälä K (1964) Männyn kylvöistä ojitettujen soiden sammal- ja jäkäläkasvustoihin (On sowing of pine in moss and lichen vegetation on drained swamps). Suo 15(3):54–58

Sarvas R (1962) Investigations on the flowering and seed crop of Pinus silvestris. Metsatieteellisen tutkimuslaitoksen julkaisuja 53(4)

Sikström U (1997) Avgång i skärmen och plantetablering vid föryngring av gran under högskärm (Decreasing in shelterwood and establishment of plants spruce under shelterwood). Skogforsk Arbetsrapport nr 369, p 147

Sikström U, Glöde D (2000) Damage to Picea abies regeneration after final cutting of shelterwood with single-and double-grip harvester systems. Scand J For Res 15(2):274–283

Skoklefald S (1967) Fristilling av naturlig gjenvekst av gran (Release of natural Norway spruce regeneration). Medd Norske Skogforsves 23:381–409

Skoklefald S (1989) Planting og naturlig föryngelse av gran under skjerm og på snauflate (Planting and natural regeneration of Norway spruce under shelterwood and on clear-cut area). Norsk Inst Skogforsk, Rapp 6, pp 39

Smith DM, Larson BC, Kelty MJ, Ashton PMS (1997) The practice of Silviculture - applied Forest ecology, no Ed 9. Wiley & Sons, New York

Sundkvist H (1994) Extent and causes of mortality in *Pinus sylvestris* advance growth in northern Sweden following overstorey removal. Scand J For Res 9(1–4):158–164

Sydow F, Örlander G (1994) The influence of shelterwood density on *Hylobius abietis* (L.) occurrence and feeding on planted conifers. Scand J For Res 9(1–4):367–375

Valkonen S (1992) Metsien uudistaminen korkeilla alueilla Pohjois-Suomessa (Regeneration of forests in high areas in Northern Finland). Metsäntutkimuslaitos

Valkonen S (2000) Effect of retained scots pine trees on regeneration, growth, form, and yield of forest stands. For Syst 9:121–145

Valkonen S (2019) Pienaukkojen ja osittaishakkuuaukkojen taimettuminen Häiriödynamiikka -hankkeen tutkimusalueilla (Tree regeneration in cut gaps of variable sizes in the experimental sites of the DistDyn project). Luonnonvara- ja biotalouden tutkimus, 69/2019

Valkonen S (ed) (2020) Metsän jatkuvasta kasvatuksesta (On Continuous-Cover Forestry). Metsäkustannus and Luonnonvarakeskus, Helsinki, p 127

Valkonen S, Aulus Giacosa L, Heikkinen J (2020) Tree mortality in the dynamics and management of uneven-aged Norway spruce stands in southern Finland. Eur J For Res 139(6):989–998. https://link.springer.com/article/10.1007/s10342-020-01301-8

Valkonen S, Koskinen K, Mäkinen J et al (2011) Natural regeneration in patch clear-cutting in *Picea abies* stands in southern Finland. Scand J For Res 26:530–542

Valkonen S, Maguire DA (2005) Relationship between seedbed properties and the emergence of spruce germinants in recently cut Norway spruce selection stands in southern Finland. For Ecol Manag 210:255–266

Valkonen S, Siitonen J (2016) Tree regeneration in patch cutting in Norway spruce stands in northern Finland. Scand J For Res 31:271–278

Wallertz K, Örlander G, Luoranen J (2005) Damage by pine weevil *Hylobius abietis* to conifer seedlings after shelterwood removal. Scand J For Res 20(5):412–420

Wood JE, Jeglum JK (1984) Black spruce regeneration trials near Nipigon, Ontario: Planting versus seeding, lowlands versus upland, clearcut versus stripcut. Canadian Forestry Service, Sault Ste. Marie, Information Report O-X-361 p 19

Chapter 4
Growth and Yield

Simone Bianchi, Andreas Brunner, Kjersti Holt Hanssen, Hannu Hökkä, Urban Nilsson, Nils Fahlvik, and Jari Hynynen

Abstract

- There is still a lack of knowledge on growth and yield (G&Y) in continuous cover forestry (CCF). Most published studies are on the selection system with Norway spruce.
- Published comparisons of the selection system with rotation forestry (RF) show contrasting results. Generally, there seems to be a trend toward faster stand growth in RF.
- However, there are many uncertainties due to several confounding factors, such as stand-density effects, site-quality classification, and/or growth models used.

S. Bianchi (✉)
Natural Resources Institute Finland (Luke), Helsinki, Finland
e-mail: simone.bianchi@luke.fi

A. Brunner
Norwegian University of Life Sciences (NMBU), Ås, Norway
e-mail: andreas.brunner@nmbu.no

K. H. Hanssen
Norwegian Institute of Bioeconomy Research (NIBIO), Ås, Norway
e-mail: kjersti.hanssen@nibio.no

H. Hökkä
Natural Resources Institute Finland (Luke), Oulu, Finland
e-mail: hannu.hokka@luke.fi

U. Nilsson
Swedish University of Agricultural Sciences (SLU), Alnarp, Sweden
e-mail: urban.nilsson@slu.se

N. Fahlvik
Forestry Research Institute of Sweden (Skogforsk), Svalöv, Sweden
e-mail: nils.fahlvik@skogforsk.se

J. Hynynen
Natural Resources Institute Finland (Luke), Savonlinna, Finland
e-mail: jari.hynynen@luke.fi

© The Author(s) 2025
P. Rautio et al. (eds.), *Continuous Cover Forestry in Boreal Nordic Countries*,
Managing Forest Ecosystems 45, https://doi.org/10.1007/978-3-031-70484-0_4

Most studies do not properly account for all these factors, making it difficult to generalise their results.
- The optimal stand density trade off for the selection system between stand growth and recruitment should be better investigated. Preliminary results show this could strongly affect stand growth.
- There is even less knowledge related to G&Y during conversion, a potential bottleneck for full implementation of CCF in the region.

Keywords Growth modelling · Site quality · Stand density · Stand structure · Stand dynamics

4.1 Introduction

4.1.1 Main Drivers of Forest Growth and Yield

Growth and yield (G&Y) studies investigate the dynamics of forest increment, usually to formulate growth models able to predict the future development of forest stands, and thus support decision making for their management. G&Y has been widely studied in even-aged stands of the clearcutting system. However, it has been less studied in continuous cover forestry (CCF) where there are additional intrinsic challenges, briefly mentioned in this introduction and expanded on in later sections.

In even-aged stands, trees have similar ages, the structure is relatively homogeneous, and average stand variables are highly correlated to stand age. Age is often used directly or indirectly as a predictor in modelling. An important example of this is the site index, a system widely used to evaluate stand productivity based on the height of the dominant trees at a given age (Skovsgaard and Vanclay 2008). However, this index is not directly applicable to CCF, as we will see in Sect. 4.2.3.

Stand density is also an important driver of growth and yield. In CCF, stand density varies greatly in space, much more than in even-aged stands. It is often not even possible to accurately assess stand density with simple metrics in structurally-complex multi-layered stands.

Ingrowth is a major contribution to total production in CCF. This complicates G&Y studies. Regeneration and ingrowth are complex processes (see Chap. 3) that vary greatly in space and time, and are too poorly understood to be represented in process-based models. Empirical models of regeneration or ingrowth need extensive data which often does not exist, and need to incorporate many random effects. G&Y models without such tools are incomplete for CCF studies (cf. Ekholm et al. 2023).

In Chap. 12 an analysis of various disturbances will be presented. G&Y studies often do not include losses due to episodic events, like windthrow, snow breakage, or bark-beetle attacks, which are likely to be less significant in CCF than in even-aged stands. On the other hand, during conversion of even-aged stands to CCF windthrow risk may be higher (see Sect. 4.3). The limited knowledge about damage further hinders the comparison of G&Y between CCF and even-aged stands.

4.1.2 Forest Growth Models for CCF

All factors mentioned in Sect. 4.1.1 influence what types of forest growth models can be suitable for CCF stands. Kuuluvainen et al. (2012) suggest developing "growth models general enough to describe both [even-aged and CCF] management alternatives relatively well," whereas Lundqvist (2017) argues that it "is difficult to know whether the model used can handle different stand structures and silvicultural systems equally well." Forest planning needs simulators adequate for stands managed under any system, without over-predicting yields from one system relative to the other (see Chap. 5). At a minimum, forest growth models need to comprise tools for regeneration or ingrowth, growth, and mortality.

Stand-level models, where only average stand characteristics are simulated, cannot be used for CCF due to the high variation of tree age, size, and spatial distribution. The best alternative is individual-tree models, which track each tree in the simulation unit. Another important characteristic of individual-tree models is their approach to spatial distributions. In distance-independent models, resource competition is based only on average stand conditions. If asymmetric competition is considered, i.e., where larger individuals obtain more resources and suppress the growth of smaller individuals (Weiner 1990), it is only based on competitors' size and not on their location. In distance-dependent or spatially explicit models, resource competition is also a function of the subject tree's location relative to its competitors. Some studies find that distance-dependent models only slightly improve on distance-independent models, even in spatially- and structurally-complex stands, questioning if their higher complexity is worth the small improvements (Kuehne et al. 2019; Bianchi et al. 2020). However, those studies only simulated single growing periods. Assuming that competition pressure is equal for all trees of the same size within an irregular stand may have big consequences in long-term simulations. For example, simulated tree growth would not be influenced by local density, and trees released from competition due to nearby harvesting would not increase their growth more than trees further from newly opened gaps. This issue should be investigated more.

In Finland, Pukkala et al. (2013, 2021) prepared distance-independent tree-level models claimed to be suitable for both CCF and even-aged forestry, comprising tools for simulating ingrowth, growth, and mortality. The most recent models (Pukkala et al. 2021) have been fitted to Finnish national forest inventory (NFI) plots, and a version based on Swedish NFI plots is in preparation. Both models have been used widely for simulation studies comparing CCF and even-aged forestry (e.g, Parkatti et al. 2019; Österberg et al. 2023). Bianchi et al. (2023) independently validated their basal area increment component together with a new alternative fitted to most recent Finnish NFI plots, confirming that all those models could be used both in even-aged forestry and CCF, although with slight differences in accuracy.

Other existing forest growth models in Fennoscandia are fitted and targeted to even-aged stands. In Finland, MELA and MOTTI are simulators using the same growth models. They use stand- and tree-level tools to simulate full-rotation development. MOTTI implements regeneration models by Eerikäinen et al. (2007) for uneven-aged stands and CCF, based on the permanent experimental ERIKA plots. They cover establishment and height development of the established seedlings. For trees past the regeneration phase, there are calibration functions for diameter and height growth to adapt the original growth models to CCF (Lee et al. 2024). In Sweden, the models of Elfving and Nyström (2010) within HEUREKA, originally suited only for even-aged stands, were independently validated with CCF data by Fagerberg et al. (2022), showing some biases. CCF results from HEUREKA are less precise than for the clearcutting system (Lämås et al. 2023). In Norway, the models of Andreassen and Øyen (2002), Bollandsås and Næsset (2009), and Øyen et al. (2011) could be used on CCF data although they need a site index, whose limitations for CCF were previously noted. Also in Norway, a transition-matrix model from Bollandsås et al. (2008) was used to compare CCF and even-aged stands in Parkatti et al. (2019). All these simulators use distance-independent models.

4.2   The Selection System

Productivity of forest stands managed with the selection system may vary over time, among sites, with stand density, and with stand structure. All of these factors affect all productivity-related processes (growth, mortality, and ingrowth), and must be considered in growth and yield studies. Comparing productivity under the selection system and even-aged forestry is a topic of great interest that needs to consider these effects (Sect. 4.2.6).

The selection system in Fennoscandia has mostly been studied for stands dominated by Norway spruce, as it is the most shade-tolerant of the important timber species. In this chapter, we therefore focus mostly on spruce and only briefly discuss this system in Scots pine or mixed-species stands. In addition, mountain forest selection cutting is practised in Norway and Sweden (Sect. 4.2.7).

4.2.1   Variation in Time

In contrast to clearcutting and most other CCF methods, the selection system maintains a long-term target stand structure with frequent cuts. Therefore, stand density and productivity vary less over time compared to other systems. Current annual increment (CAI) is defined as the annual volume increment for a specific and short period, while mean annual increment (MAI) is the total volume increment averaged over the full rotation. For the clearcutting system, MAI is calculated over a rotation from planting to final felling to provide the long-term productivity estimate, while

CAI varies greatly within the cycle (Fig. 4.1). For the selection system, CAI increases slightly as density increases within a period between two cuts, but stays relatively constant among cut periods in optimal management and remains very similar to MAI (Fig. 4.1). Given optimal management, including adequate ingrowth, observation periods of only a few decades might suffice for estimating long-term productivity in the selection system. This possibility is frequently overlooked in studies comparing the selection system with the clearcutting system (Ekholm et al. 2023).

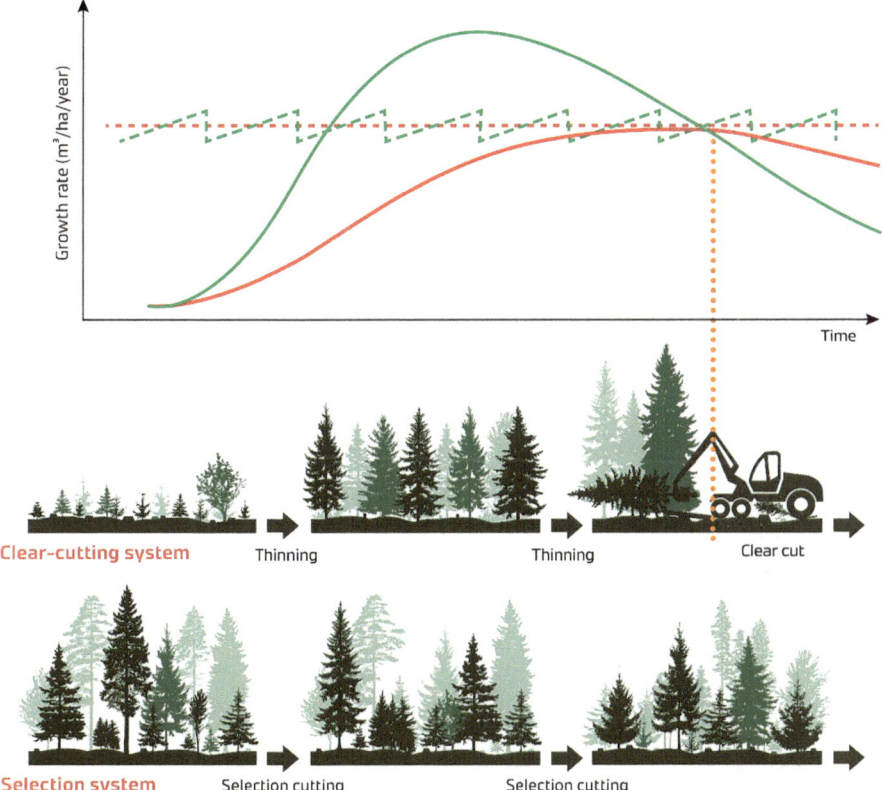

Fig. 4.1 Comparison of forest growth between the clear-cutting and selection systems during their corresponding cycles. Top: how current annual increment (CAI, green) and mean annual increment (MAI, red) vary between clearcut forestry (continuous line) and the selection system (dashed line) over time. In this example, MAI was set to be the same for both systems over the full cycle. Bottom: simplified phases of stand development and removals. Redrawn from Ekholm et al. (2023) and Routa and Huuskonen (2022)

4.2.2   Site Quality Effects

Site quality, defined as "the combination of physical and biological factors characterizing a particular geographic location or site," has large effects on forest productivity (Skovsgaard and Vanclay 2008). While site quality is a descriptive assessment, site productivity is "a quantitative estimate of the potential of a site to produce plant biomass" (Skovsgaard and Vanclay 2008). For even-aged stands, site productivity has been often quantified with a dendrocentric classification such as site index (SI), the dominant stand height at a given age. SI was developed for stands managed with thinning from below, mainly removing non-dominant trees without affecting the dominant height. SI cannot be easily applied to uneven-aged CCF stands; tree age is unknown or does not relate in the same way with average stand characteristics, and selection harvesting cutting affects the stand dominant height. There could be further confounding effects due to increased species diversity in CCF (del Río et al. 2016). Variation of G&Y with site quality is so large that other effects, like stand density, cannot be studied without correcting for site quality effects.

In the Norwegian selection system experiments, SI has been estimated with a number of alternative methods developed for the clearcutting system in that country (Andreassen 1994). In Finland and Sweden, there are models for calculating SI based on environmental data, including climate, geographical variables, and ground vegetation, known as the geocentric approach (Hägglund and Lundmark 1977; Hynynen et al. 2002). Such an age-independent system could in theory be applied to CCF stands, although its application should be properly validated.

Some studies outside Fennoscandia have explored height-diameter relationships for site quality estimation, calculating indices where dominant diameter replaces age in determining site index. The dominant stand height is calculated at a reference dominant diameter (Huang and Titus 1993; Wang 1998; Beltran et al. 2016; Duan et al. 2018; Castaño-Santamaría et al. 2023). In two studies (Beltran et al. 2016; Castaño-Santamaría et al. 2023), the diameter-based index was well correlated with the age-based index, but not in a third study (Wang 1998). However, although these studies used uneven-aged stand data, all were unmanaged. After selection cutting, the relationship between dominant height and dominant diameter may be different. Stand density may have further confounding effects on dominant diameter. In contrast, the tree-level height-diameter relationship is more stable over time in the selection system, with a sigmoid shape (Pretzsch 2009). The asymptote of this relationship indicates maximum dominant height and might therefore indicate site quality. These concepts need to be developed further for Fennoscandian conditions. An alternative approach stemming from age-based site indices uses dynamic dominant height models derived from differential equations (Salas-Eljatib 2020). Riofrío et al. (2023) transformed an age-based site index formula into a differential time-based formula, independent of the actual age, to create a site index based on dominant-height growth rate. The geocentric approach was applied by Hennigar et al. (2017), modelling total biomass increment of naturally-regenerated stands with a non-linear function whose asymptote was estimated mainly from environmental variables. Such an asymptote could be computed for any site providing an estimate of its maximum productivity.

4.2.3   Stand Density Effects

Managing the selection system requires information about stand density effects on growth and ingrowth. Optimal management not only has to maintain the desired stand structure, but also a stand density that leads to maximum stand growth and sufficient ingrowth.

Lundqvist (2017) reviewed 14 selection system field studies starting from the 1980s, finding a positive relationship between standing volume and volume growth in all cases. The standing volume in those stands ranged up to about 300 m³/ha. He suggested that the repeatedly-harvested study plots had yet not reached a maximum growth rate; the density-growth relationship had not peaked. He also noted that the lack of site quality corrections for these datasets might make the density-growth relationship appear linear rather than asymptotic. Ekholm et al. (2023) showed a similar positive density-growth relationship in their Fig. 5, partly based on the same data as in Lundqvist (2017). A similar trend is shown in the Swedish report from Goude et al. (2022) up to 200–300 m³/ha, after which the increase in growth with increased standing volume tapered off. Most likely the optimal stand density for volume growth varies with site quality. However, too little data is currently available to quantify this.

Variation in growth between individual saplings in selection system plots can only be partly explained by neighbourhood stand density variables (Lundqvist 2017). However, if mean stand density changes, regeneration and therefore ingrowth might change in reaction. Moan (2021) found no correlation of mean stand density with ingrowth in the Norwegian series across a rather wide range of densities below 300 m³/ha. Lundqvist (2017) summarises a number of ingrowth studies and reports inconsistent stand density effects, and calculates ingrowth rates above 8–10 cm diameter at breast height to be around 10 trees/ha/yr. Using the results from Moan (2021) more broadly could justify a simple non-spatial ingrowth modelling approach. The Swedish report from Goude et al. (2022) also seems to show no stand density effect on ingrowth in their Fig. 6. The Finnish models of Eerikäinen et al. (2007) found distance-dependent competition to significantly affect ingrowth, but stand-level basal area was not significantly correlated with regeneration below 28 m²/ha.

These studies should jointly underlie investigating the optimal stand density for both stand growth and ingrowth. They suggest maximum growth at a standing volume of around 300 m³/ha, up to which the effect of stand density seems to not be significantly correlated with ingrowth. Further research is needed to validate these results.

Recently, the selection system has also been applied to Scots pine forests, for example in recreation areas. In the context of climate adaptation, it will be necessary to include mixtures of less shade-tolerant species in the selection system. To allow sufficient ingrowth of the light-demanding species, stand density needs to be reduced with the consequence of reduced growth. Thinning appears to reduce Scots pine growth more strongly than Norway spruce growth based on the density-growth relationship observed in thinning experiments in the region (Mäkinen and Isomäki 2004).

4.2.4 Stand Structure Effects

The stand density effect on growth is further modified by the tree size distribution and the tree-size-growth relationship (Forrester 2019). In other words, given the same stand density and tree growth rate, the size distribution has a profound effect on stand growth. Lundqvist (2017) suggests that the stand structure of older CCF studies, more two-storied than full-storied, may have caused slower growth.

Stands with irregular structure have been sometimes shown to have significantly higher tree-level growth, perhaps due to better resource use efficiency (Lei et al. 2009; Bianchi et al. 2020). However, as discussed by Forrester (2019), those results must still be analysed in the complementary framework of stand density, structure, and size growth to avoid biased conclusions. For example, Moan (2021) found no significant relationships between structural diversity and stand volume growth.

4.2.5 Comparison with Clearcutting Forestry

Selection system and clearcutting system productivity can only be directly compared on the same site, under optimal management, and over a full rotation. This type of experiment has never been started, let alone completed, in Fennoscandia. All studies comparing selection and clearcutting system productivity have either used short-term experimental data or growth model predictions. For these comparisons, all factors that have been addressed in the previous sections need to be controlled or corrected for simultaneously. However, most studies of selection system productivity in Fennoscandia lack these corrections, and sometimes crucial variables like site quality are not even reported in the publications (see Sect. 4.2.3). This leads to highly contradictory published results, from underyielding to overyielding in the selection system compared to the clearcutting system (Andreassen 1994; Lundqvist 2017), and limits possible conclusions and generalisations. For example, a review by Kuuluvainen et al. (2012) found that almost half of studies show faster long-term growth in uneven-aged stands (using a definition broader for CCF than in this book) than in RF. They examined seven simulation studies and four field studies, but without considering or discussing any of the above-mentioned confounding factors.

The review and earlier studies by Lundqvist (2017) have addressed and tried to consider the density effect when analysing results from previously published studies. Problems quantifying site quality correctly for both systems appear to be a major obstacle to drawing valid conclusions. Lundqvist (2017) noted that selection system field studies often report slower growth than for the clearcutting system. However, despite an extensive description of stand density effects in an earlier section of the review, the author did not correct for it in the growth comparison section. Eventually, he suggested that selection systems should be managed with moderate harvests at relatively short intervals to maintain a larger growing stock. This could achieve sustainable volume growth approaching the site productivity estimated by environmental factors. However, Lundqvist also pointed out that the current SI system in Sweden underestimates stand productivity in newer even-aged stands (Tegnhammar 1992; Elfving and Nyström 1996; Yue et al. 2014). If true, this would suggest a bigger difference in long-term growth rates between the two systems. Lundqvist (2017) also describes in detail how systematic, but small, selection system experiments were established in the early twentieth century. Time series published for the experiments in Norway (Andreassen 1994) clearly illustrate how insufficient their diameter distributions and stand densities were during their first 60 years. Results and data from these experiments are still used for other studies (e.g. Lundqvist 2017; Moan 2021) but need to be interpreted carefully or original results often need to be corrected.

Hynynen et al. (2019) compared selection system productivity from the ERIKA experiments (see Chap. 3) with: (1) mid-rotation empirical CAI from thinning experiments, and (2) MAI from model simulations of even-aged stands at the same sites. In both cases, even-aged stands showed higher productivity. In comparison (1), selected thinning experiments were geographically close to the CCF stands, and with similar site fertility according to the vegetation type. Since a difference in the growing conditions was still possible, a modelling framework was applied to remove the effects of site quality and stand density. However, the empirical comparison was confounded by using CAI from thinning experiments instead of MAI (see Sect. 4.2.2). The model comparison (2) showed a 15% higher MAI in the clearcutting system. Using the same data, the study also suggested a faster stand-level reaction after thinning than after selection cutting.

Moan (2021) simultaneously corrected for stand density and site quality effects in data from two experimental series in Norway. She found that selection system plots managed at an optimal density had much smaller MAI differences from even-aged forests than reported by Andreassen (1994) for earlier data from one of the experimental series without correction for stand density effects.

Ekholm et al. (2023) discussed several limitations in comparisons carried out both using empirical studies and model simulations that are likely responsible for the large variation in the published results and make conclusions about productivity differences impossible. Use of CAI instead of MAI (see Sect. 4.2.1), unclear site quality effects (see Sect. 4.2.2), missing density corrections (see Sect. 4.2.3), poor experimental design (see Sect. 4.1), poor ingrowth models in simulators (see Sect. 4.1), and missing validation of growth models are discussed. However, many of the studies reviewed there do not investigate selection systems but thinning methods in transformation treatments to the selection system (see Sect. 4.5).

We must also highlight that there are continuous advancements in soil preparation methods (see Hjelm et al. 2019 and references within), fertilisation, and improved regeneration material that increase growth in clearcutting systems. A review of forest tree breeding programmes in Scandinavia found that they can currently increase volume growth by 10–25% (Jansson et al. 2017) but it might be much higher in the future (Rosvall and Wennström 2008). However, such advancements cannot be applied as easily to the selection system.

To conclude, interpretation of data on selection system productivity is difficult, mostly due to inadequate descriptions of density and site quality effects in published studies. Although the trend leans toward faster growth in the clearcutting system, the first attempts to correct for density effects alone (Lundqvist 2017; Ekholm et al. 2023) illustrate that productivity of the selection system might be closer to the productivity of clearcutting system. Simulation models have not been validated against empirical data from selection system stands, have poor representations of ingrowth, and show biased predictions compared with empirical data for the selection system (Ekholm et al. 2023).

4.2.6   Mountain Forest Selection Cutting

In general, harsh climate and low site indices in mountainous areas of Fennoscandia hinder regeneration and restrict investments in planting. This has led to the practice of mountain forest selection cutting ("fjellskoghogst"), where selection cutting is often combined with other types of CCF, especially group cuttings. The harvesting level has traditionally been high, removing damaged trees as well as most merchantable timber. Despite long harvesting cycles, low stand density reduces growth and yield, and regeneration is usually slow due to the cold climate (Chap. 3).

Reducing stand density to very low levels at harvesting inevitably leads to low forest growth. Lundqvist (2004) found that heavy partial harvests in sub-alpine stands in northern Sweden, leaving standing volumes below 50 m^3/ha, resulted in volume increments less than half the estimated site productivity. Øyen and Nilsen (2004) studied mountain-forest growth after selection cutting in four irregular spruce forests in Nordland, Norway. The volume increment amounted to approximately two-thirds of the yield table figures for dense even-aged stands. They interpreted the reductions as resulting from a low initial density after heavy cuttings (removing 50–65% of the standing volume) two to three decades before the inventory.

Granhus et al. (2020) studied growth 13–44 years after selection cutting in 16 mountain forest stands in Norway. Average harvested basal area was about 50%, ranging from 20% to 80%. The volume increment in the post-felling period varied considerably among sites but on average met the estimated production capacity for even-aged forests at the same site index. However, this was only the case many decades after the cuttings, with considerably lower growth for decades. Both basal area and volume growth between felling and measurement were positively correlated with post-felling basal area, and negatively correlated with basal-area-weighted mean diameter.

4.2.7 Selection Systems on Peatlands

There are no published studies from the Nordic countries based on empirical growth data from peatland forests treated with the selection system. In a recent simulation study, profitability of the selection system has been compared to the clearcutting system in peatland spruce stands in Finland (Juutinen et al. 2021). Stand development was predicted with the EFIMOD process model (Shanin et al. 2016). EFIMOD is a spatially-explicit individual-tree growth model, developed to simulate carbon- and nitrogen-pool dynamics in a tree-soil system, along with population dynamics of forest stands. The model was originally developed for mineral soil stands. In Juutinen et al. (2021) the EFIMOD model was calibrated to data from a peatland stand not managed with the selection system. When the retained basal area was above 10 m²/ha, the MAI of selection system simulations was higher than for the clearcutting system on the same site. Because the calibration data originated from a naturally established drained peatland spruce stand, the simulated MAI most likely underestimated what the clearcutting system can achieve when sites are prepared and planted using present-day methods and material.

Peatland forests have two specific features needing consideration in all kinds of management: soil water level and nutrient availability. Changes in stand volume due to growth and cutting impact the water table level, which in turn may affect stand growth. In general, deeper water tables enhance growth (Hökkä et al. 2008), while high water tables (within 35 cm of the soil surface) may hinder growth. Especially if the retained volume is very low, as in heavy selection cutting, the technical drainage may require maintenance. These feedback effects and need for ditch network maintenance (DNM) where retained basal area is very low were accounted for in the Juutinen et al. (2021) study. In such situations, DNM significantly and profitably increased stand growth, but because of additional costs, the profitability was less than in scenarios where DNM was not needed. Peat-soil nitrogen content is disproportionately high compared to phosphorus and potassium. Specifically, potassium deficiency significantly reduces needle mass and growth. Hence, fertilisation with wood ash may be needed every 40 years no matter the stand-management method.

Six field experiments have been established in Finland from 2014–2016 in Norway-spruce-dominated stands to investigate the response of peatland forests to the selection system. The retained basal area after cutting varies from 6–17 m²/ha. Results from those experiments cover only the first five-year growth period and are not yet published.

4.3   The Shelterwood System

In Sweden, Lula et al. (2021) have used growth models to study production and profit in planted and naturally regenerated stands under dense Scots pine shelterwoods, using starting values from experiments on relatively fertile sites in southern Sweden. They compared different initial tree densities, including a shelterwood maintained at around 15 m^2/ha, but production after natural regeneration was on average about 10–20% lower compared to planting. However, wood quality was found to be higher after natural regeneration than after planting (Agestam et al. 1998), which could economically offset the loss of production. Lula et al. (2021) estimated the cost and income of shelterwoods and the future stands and concluded that shelterwoods and natural regeneration could be an economically feasible alternative to clearcutting and planting due to low regeneration costs.

Since the study cited above was based on relatively fertile sites, little is known about long-term production and profit on less fertile sites in northern Scandinavia where abundant Scots pine stands are available for the shelterwood system. Elfving and Jakobsson (2006) concluded that the negative effect of retained Scots pine trees on understorey tree growth was relatively higher on low fertility sites than on more fertile sites. However, the retention time was between 30–90 years and overstory density low, so it is not totally comparable to a shelterwood that is removed when the new regeneration is established but is much denser than green-tree retention. Therefore, establishing experiments for comparing long-term growth of planted and naturally-regenerated stands along a fertility gradient should be prioritised.

The risk for windthrow increases after shelterwood cutting and may significantly affect total volume production. Örlander (1995) investigated shelterwood stability in southern Sweden. He found shelter tree windthrow rates of 18% for Scots pine vs. 35% for Norway spruce. Nilsson et al. (2006) investigated 22 Scots pine shelterwoods from all regions of Sweden. On average, windthrow was 9% in the south and north and 18% in central Sweden. Sikström (1997) studied damage to shelter trees in 52 Norway spruce-dominated shelterwoods. The shelterwoods were measured 2–6 growing seasons after cutting, showing on average 15% windthrow of shelter trees. Different weather conditions and variable length between cutting and measurement could have caused these differences among studies. In Finland, Pukkala et al. (2016) showed that shelterwood stands have higher windthrow risk than clearcutting system stands, which in turn are at higher risk than selection system stands. Such risks must be considered in G&Y forecasting.

4.4   The Group System

The only available results related to G&Y in gap cutting are on their regeneration dynamics, which are discussed in Chap. 3. For G&Y studies, the edge effects on the growth of both adult trees and the regeneration are important and need to be considered in experimental designs and modelling approaches (e.g., Shell et al. 2022). Edge trees next to gaps also have higher mortality risks.

Strip cutting in pine peatland relies on seed production from the edge stand to naturally regenerate the clearcut strip, for example 16–25 m wide in Stenberg et al. (2022). As a pioneer species, Scots pine needs light for seedling establishment and growth, so as a variation of the group system, strip clearcutting may provide such conditions. Strip cutting is chosen over larger clearcuts in pine on drained peatlands to mitigate the water table rise (Stenberg et al. 2022) commonly observed after peatland clearcuts (e.g., Heikurainen 1970). Rising water tables after stand removal cause problems by releasing nutrients (specifically phosphorus) and metals (iron, aluminium) from the surface peat into watercourses (Kaila et al. 2015). By cutting only half of the peatland area and keeping the other part forested, a sufficiently deep water table (at least 35 cm below ground) can be maintained without additional ditch cleaning.

Ahtikoski et al. (2022) used a simulation to compare the financial performance of rotation forestry (clearcutting with site preparation, planting, and regular ditch cleaning) and cutting in 20, 35, and 50 m-wide strips in drained peatland Scots pine stands in southern, central, and northern Finland. Growth of sapling and edge stands was modelled using a stand-level model that accounted for edge stand height and the variation in shading when different strip widths were used. Changes in ditch depth and mean water table depth due to cuttings were incorporated to guarantee that water tables remained sufficiently far from the soil surface. The second strip cutting (of the edge stands) was done when the sapling stand reached a volume of 80 m³/ha. The simulation showed that the geographical location and strip width had a significant impact on the cutting removals and the length of the conversion phase before the steady state in strip cutting management. The wider the strip and the more southern the location the faster was the development of the sapling stand. However, with 50 m strip width the clearcut area is so wide that water level may rise and cause harmful loads to water courses before the sapling stand has established. Strip cutting is thus balancing between good growth of the sapling stand and sufficiently deep water level (Ahtikoski et al. 2022).

4.5   Conversion to CCF

Finding efficient conversion methods is crucial for adoption of CCF in northern Europe, where even-aged forest management has long been dominant. At present, a lack of experience in transforming from the clearcutting system to CCF is a major obstacle for wider scale introduction of CCF (Mason et al. 2022). Concepts for conversion of even- to uneven-aged forests have been discussed and general strategies to achieve this goal formulated (Schütz 2001; Nyland 2003). These concepts normally include repeated thinning, target diameter cutting, and patch cutting. The overall goal of the cuttings is to increase dimensional variation by manipulating the existing stand's size distribution and favouring regeneration. Recruitment is essential for the long-term transition to CCF (Hanewinkel and Pretzsch 2000). Conversion concepts typically include greater initial removals than in conventional thinnings to initiate regeneration (Drössler et al. 2014; Juutinen et al. 2018). However, heavy cuttings might decrease volume production (Hanewinkel and Pretzsch 2000) and increase the risk of storm damage losses (Wallentin 2007). Schütz (2001) stressed the importance of timely onset of conversion measures to allow stabilising measures and to assure the present stand lives long enough to complete the transition. In the Nordic context this consideration is even more important because target diameters of only around 40 cm limit the life time of shelter trees much more.

There is a great need for empirical data to evaluate and formulate conversion concepts adapted to regional conditions in northern Europe. There are ongoing efforts to establish field trials but currently few experiments are available, and they have been monitored for a limited time. One experimental series exists in Sweden to study the conversion of an even-aged Norway spruce stand to a multi-layered forest (Drössler et al. 2014). A heavy first thinning (60% of basal area) was carried out focusing on removing medium-sized trees and leaving the smallest and the largest ones. Simulations of continuous conversion measures during a 50-year period indicated a transition to a multi-layered stand in central but not southern Sweden. The estimated volume production of the conversion forestry was one-third lower than a conventional clearcutting system (Drössler et al. 2014). Recent measurements of the series showed adequate regeneration in central Sweden but no regeneration in the south (Goude et al. 2022). This calls for modifications of the treatment to adapt the conversion strategies to local conditions.

Another experiment was established in a structured mature stand dominated by Scots pine in southern Sweden (Drössler et al. 2012, 2015, 2017). The treatments included target diameter cutting with or without silvicultural actions to favour regeneration. Continuous target diameter cutting was simulated for 50 years and compared to an alternative scenario where the present stand was clearcut and replaced by a Norway spruce plantation. The estimated basal area growth was one third lower for scenarios with target diameter cutting compared to replanting with spruce (Drössler et al. 2012). Later revisions of the experiment indicated that simulated growth in the target diameter cutting scenarios might have been underestimated (Drössler et al. 2012; Goude et al. 2022). The revisions reveal an increasing share of spruce in the regeneration after initial establishment of birch in the new gaps.

In northern Sweden, larger scale conversion to CCF was tested in field experiments by cutting chequered patches (0.135–0.16 ha) in 35–120 year-old coniferous stands from 2005 to 2012 (Erefur 2010; Goude et al. 2022). Production and the abundance and growth of the regeneration have been followed. Preliminary results indicate adequate regeneration in the patches with a more hindering edge effect on tree growth in smaller gaps (Goude et al. 2022).

Due to the rarity of conversion experiments in northern Europe and their variety of initial stand and site conditions, it is difficult to draw firm conclusions. Conversion might also span several decades (Schütz 2001; Drössler et al. 2014), requiring long-term studies. Growth modelling could help elaborate different conversion alternatives (Sterba 2004).

Several papers have been recently published summarising 30 years of experiments in Washington and Oregon (USA) dealing with G&Y in variable-density thinning conversions (e.g. Roberts and Harrington 2008; Dodson et al. 2012; Willis et al. 2018). They are difficult to transfer to Fennoscandian conditions due to differences in the conversion objective, tree-marking methods, and tree species. New experiments with this method adapted to the Nordic conditions are being established in Norway.

4.6   Conclusions and Future Research Needs

Knowledge is still lacking on the many drivers of G&Y in CCF. Most published studies are on the selection system with Norway spruce. Furthermore, the few field experiments are sometimes managed with different criteria (some are even misclassified in terms of system), resulting in different or unknown tree size distributions.

Most growth modelling tools, from site quality classification to forest growth simulators, are targeted at even-aged stands. Site quality assessment is crucial for choosing the best species or treatment for a given site. Alternative, age-independent site classifications are being published and should be further developed for Fennoscandia. Existing forest growth simulators (such as MOTTI and HEUREKA) are being patched with additional tools for CCF, but work is in progress, and they have not been independently validated.

Many empirical studies show stand growth in the selection system increasing with stand density until values around 300 m^3/ha. Stand density also affects regeneration, and high density may not provide sustainable levels of ingrowth. However, empirical studies have found no significant relationship between stand density and ingrowth up to 300 m^3/ha. Thus, more research is needed into investigate what stand density optimises the growth and ingrowth simultaneously.

Lack of knowledge on aspects of G&Y such as site quality, stand density, and ingrowth makes it difficult to compare CCF and the clearcutting system productivity in field experiments. Similarly, a lack of proper growth models limits simulation studies. There are contrasting results in the published literature that are difficult to reconcile. However, the trend leans toward higher growth in the clearcutting system. New CCF field experiments should comprise adjacent thinned and unthinned clearcutting system plots to minimise site quality effects. Growth models should better analyse ingrowth with larger datasets and investigate the feasibility of spatial or non-spatial approaches for long-term simulations. Mortality also affects G&Y and needs to be represented with empirical models for all silvicultural systems.

Some important topics, such as peatland CCF and the conversion to CCF have been even less studied. There are ongoing efforts to establish more experiments (peatland in Finland, conversion in Norway and Sweden), but existing data are insufficient.

References

Agestam E, Ekö PM, Johansson U (1998) Timber quality and volume growth in naturally regenerated and planted Scots pine stands in SW Sweden

Ahtikoski A, Hökkä H, Siipilehto J (2022) Strip cutting management in scots pine stands on peatlands—a financial comparison to rotation forestry. Scand J For Res 37:119–129. https://doi.org/10.1080/02827581.2022.2055135

Andreassen K (1994) Development and yield in selection forest. Meddelelser fra Skogsforsk 47:1–37

Andreassen K, Øyen B-H (2002) Economic consequences of three silvicultural methods in uneven-aged mature coastal spruce forests of Central Norway. Forestry 75:483–488. https://doi.org/10.1093/forestry/75.4.483

Beltran HA, Chauchard L, Velásquez A et al (2016) Diametric site index: an alternative method to estimate site quality in Nothofagus obliqua and N. alpina forests. Cerne 22:345–354. https://doi.org/10.1590/01047760201622032207

Bianchi S, Myllymaki M, Siipilehto J et al (2020) Comparison of spatially and nonspatially explicit nonlinear mixed effects models for Norway spruce individual tree growth under single-tree selection. Forests 11:1338. https://doi.org/10.3390/f11121338

Bianchi S, Siipilehto J, Repola J et al (2023) Individual tree basal area increment models suitable for different stand structures in Finland. For Ecol Manag 549:121467. https://doi.org/10.1016/j.foreco.2023.121467

Bollandsås OM, Buongiorno J, Gobakken T (2008) Predicting the growth of stands of trees of mixed species and size: a matrix model for Norway. Scand J For Res 23:167–178. https://doi.org/10.1080/02827580801995315

Bollandsås OM, Næsset E (2009) Weibull models for single-tree increment of Norway spruce, scots pine, birch and other broadleaves in Norway. Scand J For Res 24:54–66. https://doi.org/10.1080/02827580802477875

Castaño-Santamaría J, López-Sánchez CA, Ramón Obeso J, Barrio-Anta M (2023) Development of a site form equation for predicting and mapping site quality. A case study of unmanaged beech forests in the Cantabrian range (NW Spain). For Ecol Manag 529:120711. https://doi.org/10.1016/j.foreco.2022.120711

del Río M, Pretzsch H, Alberdi I et al (2016) Characterization of the structure, dynamics, and productivity of mixed-species stands: review and perspectives. Eur J Forest Res 135:23–49. https://doi.org/10.1007/s10342-015-0927-6

Dodson EK, Ares A, Puettmann KJ (2012) Early responses to thinning treatments designed to accelerate late successional forest structure in young coniferous stands of western Oregon, USA. Can J For Res 42:345–355. https://doi.org/10.1139/x11-188

Drössler L, Ekö PM, Balster R (2015) Short-term development of a multilayered forest stand after target diameter harvest in southern Sweden. Can J For Res 45:1198–1205. https://doi.org/10.1139/cjfr-2014-0471

Drössler L, Fahlvik N, Ekö P-M (2012) Stand structure and future development of a managed multi-layered forest in southern Sweden: Eriksköp-A case study. Institutionen för sydsvensk skogsvetenskap, Sveriges lantbruksuniversitet

Drössler L, Fahlvik N, Wysocka NK et al (2017) Natural regeneration in a multi-layered Pinus sylvestris-Picea abies Forest after target diameter harvest and soil scarification. Forests 8:35. https://doi.org/10.3390/f8020035

Drössler L, Nilsson U, Lundqvist L (2014) Simulated transformation of even-aged Norway spruce stands to multi-layered forests: an experiment to explore the potential of tree size differentiation. Forestry 87:239–248

Duan G, Gao Z, Wang Q, Fu L (2018) Comparison of different height–diameter modelling techniques for prediction of site productivity in natural uneven-aged pure stands. Forests 9:63. https://doi.org/10.3390/f9020063

Eerikäinen K, Miina J, Valkonen S (2007) Models for the regeneration establishment and the development of established seedlings in uneven-aged, Norway spruce dominated forest stands of southern Finland. For Ecol Manag 242:444–461. https://doi.org/10.1016/j.foreco.2007.01.078

Ekholm A, Lundqvist L, Petter Axelsson E et al (2023) Long-term yield and biodiversity in stands managed with the selection system and the rotation forestry system: a qualitative review. For Ecol Manag 537:120920. https://doi.org/10.1016/j.foreco.2023.120920

Ekö P, Agestam E (1994) A comparison of naturally regenerated and planted scots pine (Pinus sylvestris L.) on fertile sites in southern Sweden. For Landscape Res 1:111–126

Elfving B, Jakobsson R (2006) Effects of retained trees on tree growth and field vegetation in Pinus sylvestris stands in Sweden. Scand J For Res 21:29–36. https://doi.org/10.1080/14004080500487250

Elfving B, Nyström K (1996) Stability of site index in scots pine (Pinus sylvestris, L.) Plantations over year of planting in the period 1900–1977 in Sweden. Growth trends in European forests: studies from 12 countries 71–77

Elfving B, Nyström K (2010) Growth modelling in the Heureka system. Department of Forest Ecology and Management. Swedish University of Agricultural Sciences, Umeå, p 97

Erefur C (2010) Regeneration in continuous cover forestry systems. Doctoral Thesis, Department of Forest Ecology and Management, Swedish University of Agricultural Sciences

Fagerberg N, Olsson J-O, Lohmander P et al (2022) Individual-tree distance-dependent growth models for uneven-sized Norway spruce. Forestry 95:634–646. https://doi.org/10.1093/forestry/cpac017

Forrester DI (2019) Linking forest growth with stand structure: tree size inequality, tree growth or resource partitioning and the asymmetry of competition. For Ecol Manag 447:139–157. https://doi.org/10.1016/j.foreco.2019.05.053

Goude M, Erefur C, Johansson U, Nilsson U (2022) Hyggesfria skogliga fältförsök i Sverige : en sammanställning av tillgängliga långtidsförsök (Clearcut-free forest experiments in Sweden. A summary of available long-term experiments.). Rapport (Sveriges lantbruksuniversitet, Enheten för skoglig fältforskning)

Granhus A, Allen M, Bergsaker E (2020) Fjellskoghogst–produksjon, foryngelse og økonomi (Mountain forest selection cutting - production, regeneration, and economy). NIBIO Rapport

Hägglund B, Lundmark J-E (1977) Site index estimation by means of site properties. Swedish College of Forestry, Stockholm

Hanewinkel M, Pretzsch H (2000) Modelling the conversion from even-aged to uneven-aged stands of Norway spruce (Picea abies L. Karst.) with a distance-dependent growth simulator. For Ecol Manag 134:55–70. https://doi.org/10.1016/S0378-1127(99)00245-5

Heikurainen L (1970) The effect of thinning, clear cutting, and fertilization on the hydrology of peatland drained for forestry. Society of Forestry in Finland, Helsinki

Hennigar C, Weiskittel A, Allen HL, MacLean DA (2017) Development and evaluation of a biomass increment based index for site productivity. Can J For Res 47:400–410. https://doi.org/10.1139/cjfr-2016-0330

Hjelm K, Nilsson U, Johansson U, Nordin P (2019) Effects of mechanical site preparation and slash removal on long-term productivity of conifer plantations in Sweden. Can J For Res 49:1311–1319. https://doi.org/10.1139/cjfr-2019-0081

Hökkä H, Repola J, Laine J (2008) Quantifying the interrelationship between tree stand growth rate and water table level in drained peatland sites within Central Finland. Can J For Res 38:1775–1783. https://doi.org/10.1139/X08-028

Huang S, Titus SJ (1993) An index of site productivity for uneven-aged or mixed-species stands. Can J For Res 23:558–562. https://doi.org/10.1139/x93-074

Hynynen J, Eerikäinen K, Mäkinen H, Valkonen S (2019) Growth response to cuttings in Norway spruce stands under even-aged and uneven-aged management. For Ecol Manag 437:314–323. https://doi.org/10.1016/j.foreco.2018.12.032

Hynynen J, Ojansuu R, Hökkä H et al (2002) Models for predicting stand development in MELA system. Metsäntutkimuslaitos, Vantaa

Jansson G, Hansen JK, Haapanen M, Kvaalen H, Steffenrem A (2017) The genetic and economic gains from forest tree breeding programmes in Scandinavia and Finland. Scand J For Res 32:273–286

Juutinen A, Ahtikoski A, Mäkipää R, Shanin V (2018) Effect of harvest interval and intensity on the profitability of uneven-aged management of Norway spruce stands. Forestry 91:589–602. https://doi.org/10.1093/forestry/cpy018

Juutinen A, Shanin V, Ahtikoski A et al (2021) Profitability of continuous-cover forestry in Norway spruce dominated peatland forest and the role of water table. Can J For Res 51:859–870. https://doi.org/10.1139/cjfr-2020-0305

Kaila A, Laurén A, Sarkkola S et al (2015) Effect of clear-felling and harvest residue removal on nitrogen and phosphorus export from drained Norway spruce mires in southern Finland. Boreal Envinron Res 20:693–706

Kuehne C, Weiskittel AR, Waskiewicz J (2019) Comparing performance of contrasting distance-independent and distance-dependent competition metrics in predicting individual tree diameter increment and survival within structurally-heterogeneous, mixed-species forests of northeastern United States. For Ecol Manag 433:205–216. https://doi.org/10.1016/j.foreco.2018.11.002

Kuuluvainen T, Tahvonen O, Aakala T (2012) Even-aged and uneven-aged Forest Management in Boreal Fennoscandia: a review. Ambio 41:720–737. https://doi.org/10.1007/s13280-012-0289-y

Lämås T, Sängstuvall L, Öhman K et al (2023) The multi-faceted Swedish Heureka forest decision support system: context, functionality, design, and 10 years experiences of its use. Front For Glob Change 6:1163105

Lee D, Repola J, Bianchi S et al (2024) Calibration models for diameter and height growth of Norway spruce growing in uneven-aged stands in Finland. For Ecol Manag 558:121783. https://doi.org/10.1016/j.foreco.2024.121783

Lei X, Wang W, Peng C (2009) Relationships between stand growth and structural diversity in spruce-dominated forests in New Brunswick, Canada. Can J For Res 39:1835–1847

Lula M, Trubins R, Ekö PM et al (2021) Modelling effects of regeneration method on the growth and profitability of scots pine stands. Scand J For Res 36:263–274. https://doi.org/10.1080/02827581.2021.1908591

Lundqvist L (2004) Stand development in uneven-aged sub-alpine Picea abies stands after partial harvest estimated from repeated surveys. Forestry 77:119–129

Lundqvist L (2017) Tamm review: selection system reduces long-term volume growth in Fennoscandic uneven-aged Norway spruce forests. For Ecol Manag 391:362–375. https://doi.org/10.1016/j.foreco.2017.02.011

Mäkinen H, Isomäki A (2004) Thinning intensity and growth of Norway spruce stands in Finland. Forestry 77:349–364. https://doi.org/10.1093/forestry/77.4.349

Mason WL, Diaci J, Carvalho J, Valkonen S (2022) Continuous cover forestry in Europe: usage and the knowledge gaps and challenges to wider adoption. Forestry 95:1–12. https://doi.org/10.1093/forestry/cpab038

Moan MÅ (2021) Effects of stand structure and stand density on volume growth and ingrowth in selectively cut stands in Norway. Master's Thesis, Norwegian University of Life Sciences, Ås

Nilsson U, Örlander G, Karlsson M (2006) Establishing mixed forests in Sweden by combining planting and natural regeneration—effects of shelterwoods and scarification. For Ecol Manag 237:301–311. https://doi.org/10.1016/j.foreco.2006.09.053

Nyland RD (2003) Even-to uneven-aged: the challenges of conversion. For Ecol Manag 172:291–300

Örlander G (1995) Stormskador i sydsvenska tallskärmar (Storm damage in southern Swedish pine shelterwoods). Skog och forskning 3:52–56

Österberg N, Parkatti V-P, Tahvonen O (2023) Comparing stand growth models in optimizing mixed-species forest management. Scand J For Res 38:353–366. https://doi.org/10.1080/02827581.2023.2227095

Øyen B-H, Nilsen P (2004) Growth and recruitment after mountain forest selective cutting in irregular spruce forest. A case study in northern Norway. Silva Fenn 38:383–392

Øyen B-H, Nilsen P, Bøhler F, Andreassen K (2011) Predicting individual tree and stand diameter increment responses of Norway spruce (Picea abies (L.) karst.) after mountain forest selective cutting. Forestry Studies 55:33–45. https://doi.org/10.2478/v10132-011-0100-z

Parkatti V-P, Assmuth A, Rämö J, Tahvonen O (2019) Economics of boreal conifer species in continuous cover and rotation forestry. Forest Policy Econ 100:55–67. https://doi.org/10.1016/j.forpol.2018.11.003

Pretzsch H (2009) Forest dynamics, growth, and yield. Springer, Berlin

Pukkala T, Lähde E, Laiho O (2013) Species interactions in the dynamics of even- and uneven-aged boreal forests. J Sustain Forest 32:371–403. https://doi.org/10.1080/10549811.2013.770766

Pukkala T, Laiho O, Lähde E (2016) Continuous cover management reduces wind damage. For Ecol Manag 372:120–127. https://doi.org/10.1016/j.foreco.2016.04.014

Pukkala T, Vauhkonen J, Korhonen KT, Packalen T (2021) Self-learning growth simulator for modelling forest stand dynamics in changing conditions. Forestry 94:333–346. https://doi.org/10.1093/forestry/cpab008

Riofrío J, White JC, Tompalski P et al (2023) Modelling height growth of temperate mixedwood forests using an age-independent approach and multi-temporal airborne laser scanning data. For Ecol Manag 543:121137. https://doi.org/10.1016/j.foreco.2023.121137

Roberts SD, Harrington CA (2008) Individual tree growth response to variable-density thinning in coastal Pacific northwest forests. For Ecol Manag 255:2771–2781. https://doi.org/10.1016/j.foreco.2008.01.043

Rosvall O, Wennström U (2008) Förädlingseffekter för simulering med Hugin i SKA 08 (Effect of correction for simulation with Hugin in SKA 08). Skogforsk

Routa J, Huuskonen S (2022) Jatkuvapeitteinen metsänkasvatus: Synteesiraportti (Continuous cover forestry: synthesis report). Luonnonvara- ja biotalouden tutkimus 40/2022. Luonnonvarakeskus. Helsinki

Salas-Eljatib C (2020) Height growth–rate at a given height: a mathematical perspective for forest productivity. Ecol Model 431:109198. https://doi.org/10.1016/j.ecolmodel.2020.109198

Schütz J-P (2001) Opportunities and strategies of transforming regular forests to irregular forests. For Ecol Manag 151:87–94

Shanin V, Valkonen S, Grabarnik P, Mäkipää R (2016) Using forest ecosystem simulation model EFIMOD in planning uneven-aged forest management. For Ecol Manag 378:193–205. https://doi.org/10.1016/j.foreco.2016.07.041

Shell AB, Sharma A, Vogel JG, Willis JL (2022) Converting plantations to uneven-aged stands: effects of harvest type, group opening size, orientation, and canopy position on residual tree development. Trees For People 10:100342. https://doi.org/10.1016/j.tfp.2022.100342

Sikström U (1997) Avgång i skärmen och plantetablering vid föryngring av gran under högskärm (Losses in the shelter and regeneration establishment when rejuvenating spruce under shelterwood). Skogforsk, Arbetsrapport

Skovsgaard J, Vanclay JK (2008) Forest site productivity: a review of the evolution of dendrometric concepts for even-aged stands. Forestry 81:13–31

Stenberg L, Leppä K, Launiainen S et al (2022) Measuring and modeling the effect of strip cutting on the water table in boreal drained Peatland pine forests. Forests 13:1134

Sterba H (2004) Equilibrium curves and growth models to deal with forests in transition to uneven-aged structure—application in two sample stands. Silva Fenn 38:10.14214/sf.409

Tegnhammar L (1992) On the estimation of site index for Norway spruce. Rapport-Sveriges Lantbruksuniversitet, Institutionen foer Skogstaxering (Sweden)

Wallentin C (2007) Thinning of Norway spruce. Southern Swedish Forest Research Centre, SLU

Wang GG (1998) Is height of dominant trees at a reference diameter an adequate measure of site quality? For Ecol Manag 112:49–54. https://doi.org/10.1016/S0378-1127(98)00315-6

Weiner J (1990) Asymmetric competition in plant populations. Trends Ecol Evol 5:360–364. https://doi.org/10.1016/0169-5347(90)90095-U

Willis JL, Roberts SD, Harrington CA (2018) Variable density thinning promotes variable struc-
    tural responses 14 years after treatment in the Pacific northwest. For Ecol Manag 410:114–125.
    https://doi.org/10.1016/j.foreco.2018.01.006
Yue C, Mäkinen H, Klädtke J, Kohnle U (2014) An approach to assessing site index changes
    of Norway spruce based on spatially and temporally disjunct measurement series. For Ecol
    Manag 323:10–19

Chapter 5
Forest Planning and Continuous Cover Forestry

Lauri Mehtätalo, Annika Kangas, Jeannette Eggers, Tron Eid,
Kyle Eyvindson, Johanna Lundström, Jouni Siipilehto, and Karin Öhman

Abstract

- Forest planning requires unbiased and sufficient information on current forest resources, their anticipated dynamics under different management scenarios, and the objectives of the decision maker.
- Forest planning systems need to be adapted to improve their potential to deal with continuous cover forestry (CCF).
- The current forest planning systems and associated models can be adapted to group systems by treating each group as a separate calculation unit.
- In the selection system, currently available growth models may not realistically describe the growth reaction of trees, which causes additional uncertainty in forest-planning calculations. Furthermore, field-data collection based on airborne laser scanning alone is not sufficient for planning of CCF, and additional field measurements are needed.

L. Mehtätalo (✉) · A. Kangas
Natural Resources Institute Finland (Luke), Joensuu, Finland
e-mail: lauri.mehtatalo@luke.fi; annika.kangas@luke.fi

J. Eggers · J. Lundström · K. Öhman
Swedish University of Agricultural Sciences (SLU), Umeå, Sweden
e-mail: jeannette.eggers@slu.se; johanna.lundstrom@slu.se; karin.ohman@slu.se

T. Eid
Faculty of Environmental Sciences and Natural Resource Management, Norwegian University of Life Sciences, Ås, Norway
e-mail: tron.eid@nmbu.no

K. Eyvindson
Faculty of Environmental Sciences and Natural Resource Management, Norwegian University of Life Sciences, Ås, Norway

Natural Resources Institute Finland (Luke), Helsinki, Finland
e-mail: kyle.eyvindson@nmbu.no

J. Siipilehto
Natural Resources Institute Finland (Luke), Helsinki, Finland
e-mail: jouni.siipilehto@luke.fi

© The Author(s) 2025
P. Rautio et al. (eds.), *Continuous Cover Forestry in Boreal Nordic Countries*,
Managing Forest Ecosystems 45, https://doi.org/10.1007/978-3-031-70484-0_5

- Tree-level measurements by drones open interesting opportunities for forest planning, which might be especially useful under CCF.

Keywords Decision support · Software · Forest inventory · Simulation · Forest dynamics · Optimization · Forecasting

5.1 Introduction

5.1.1 What Is Forest Planning?

The use of a certain forest area is affected by the available forest resources, their future production possibilities, the objectives of the decision maker (e.g. forest owner, company, stakeholders or society), and restrictions set by laws, agreements and guidelines. Besides wood production, forests provide a multitude of other goods, services and values, including non-wood products such as berries and mushrooms, biodiversity, carbon storage and sequestration, recreation opportunities, and a basis for reindeer husbandry. The objectives include those related to the production of all goods, services, and values. Forest planning can provide valuable information for the decision maker on how to manage the forest to reach the objectives.

From an economic point of view, harvesting decisions are particularly influenced by forecasts of the value of the felled and remaining trees in each stand, as well as the interest rate used by the decision maker. Economic analyses are often based only on the revenues from timber sales and silvicultural costs, but can include markets for other products, such as biodiversity, carbon sinks and recreation. Other goals, such as those related to the timing of harvest revenue, the harvested volume, the structure of the landscape, biodiversity, carbon sinks, and other goods, services and values discussed above have been considered by using different methodological approaches like multi-attribute utility theory (Kangas et al. 2015).

Forest planning can encompass a wide range of time spans and spatial scales. In the narrowest sense, forest planning is the search for a treatment schedule (a sequence of activities for a stand over the entire planning horizon) for an individual stand in the near future. In the broadest sense, forest planning designs long-term scenario analyses about the development of forest resources under different harvest levels at the national level. In tactical forest planning, forest management is planned for a landowner's forest property for the next 10–20 years. From the forest planning situations described above, it is important to distinguish between genuine planning, where the landowner sets the management objectives and accepts and implements the plan, and calculations for public operators which can only be implemented indirectly through various forest policy incentives.

The starting point of forest planning is the clarification of the decision maker's goals. After that, the planning often proceeds in the following steps: (1) Based on the inventory data, a detailed description of the current state of the forest is produced, allowing (2) simulation of alternative treatment schedules for each stand to sufficiently cover the range of acceptable management alternatives, and (3) the optimal combination of treatment schedules for all stands is identified by solving an optimisation problem with mathematical programming or some heuristic method. Steps (2) and (3) could involve interactive processes where new alternative treatment schedules are designed, and updated preference information on the forest owner's goals becomes more precise and integrated into the optimisation approach (Eyvindson et al. 2018). The simulation of alternative treatment schedules in step (2) requires models that can predict the dynamics of all decision variables over time under different management alternatives, as well as realistic algorithms to describe the silvicultural operations.

However, the method consisting of simulation and optimisation, is not the only planning approach. If the management objectives do not interlink the individual stands' optima, the management plan can be produced by optimising the management of each stand separately, using approaches where simulation and optimisation are integrated into a single step. Stand-level optimisation can also be used when there are management goals that make stand-level decisions interdependent (for instance the requirement of a steady flow of timber or revenue). This requires the stand-level objective functions to be augmented with a penalty function that measures deviations from the forest-level constraints (see Hoganson and Rosc 1984; Pukkala et al. 2009). The penalty function is gradually adjusted during the process, and the stand-level optimisations are repeated after every adjustment, until the forest-level constraints or targets are met.

In addition, for large-scale scenario analyses, only one scenario may be simulated for each management unit using a rule-based approach, and optimisation is not used.

5.1.2 Current Forest Planning Approaches

The forest resource data used for planning may be based on sample plots with measured tree diameters and heights. Such data is often used in country-level scenario analyses, where the national forest inventory (NFI) sample plots are commonly used as input data, and not all forest stands need to be covered. When all forest stands of a certain area do need to be covered, tree-level diameter and height measurements and/or tree maps are often considered too costly for the increased accuracy they would provide. In such a case, only stand-level characteristics, such as the basal area or number of stems, mean diameter, mean height and mean tree age by tree species and canopy layer are recorded (either using remote sensing and or field cruising as described below) and used as input for forest planning. However, increased availability of drone- or smartphone-based inventory methods may alter this situation soon by allowing low-cost tree-level inventories.

The stand characteristics of interest are used to generate a tree list using models that predict the parameters of a diameter-distribution model and height-diameter curve (Mehtätalo and Kangas 2005) separately for all species and canopy layers of the stand. Other approaches can be used that estimate all or some of these parameters mathematically (Mehtätalo and Lappi 2020, Chap. 11). Thereafter, a systematic sample of trees can be generated from those models to provide an artificial tree list for the stand or plot.

In the systems used in Sweden and Finland, forest development is described using tree-level growth-model systems (see Chap. 4), which are sometimes calibrated using the most recent NFI plot data. Tree-level models are favoured because they allow flexible simulation of different harvest operations. Based on the tree stocks the models produce, a diverse set of structural indicators can be computed for biodiversity, such as area of mature broadleaf-rich forest, volume of deadwood, etc. (e.g. Kangas and Pukkala 1996). Furthermore, static predictive models have been fitted to various ecosystem services, such as berry (Kilpeläinen et al. 2016) and mushroom yield (Tahvanainen et al. 2018), amenity values, and habitat suitability (e.g. Tikkanen et al. 2007). The inputs for these models are usually ordinary stand characteristics whose values are predictions from the growth simulator. Based on these models, the effects of forest management on various values can be estimated and used in planning calculations. After final felling, a new stand is generated based on the commonly applied guidelines for artificial regeneration and empirical knowledge about the amount of natural regeneration in planted or seeded areas (e.g. Wikberg 2004). If a seed tree or shelterwood method is used, natural regeneration is the only source of the new tree generation.

The collection of stand-level forest resource information for strategic and tactical forest planning in Nordic countries is often based on remote sensing and field data using the area-based approach (ABA, Næsset et al. 2004). In the ABA, the forest stand characteristics of the training plots (usually some 500 plots per campaign) are explained with the features of the remote sensing data (laser scanning and aerial photographs). The models are then used to predict the characteristics for all areas covered by the remote sensing data. Such data is collected using public funding and is made available to the public for free or at low cost. The remote-sensing-based data may be augmented with field assessment, especially to assess the progress of regeneration in different stands.

5.2   Forest-Planning Challenges of Continuous Cover Forestry

Forest planning allows decision makers the opportunity to explore possible outcomes from a wide range of silvicultural methods. Silvicultural treatments that follow the concepts of continuous cover forestry (CCF) and rotation forestry (RF) can be simulated for the forest, and the choice of how the forest is managed should be based on the goals of the decision maker. For example, in thinning a near-mature even-aged stand, the understorey can be spared. The decision on how to continue can be made later, when the decision maker sees what kind of understorey has been created. A slightly different approach is any-aged forestry, where the management schedules are never classified to represent either CCF or RF (Pukkala 2020).

The choice between CCF and RF can also be made as a categorical decision. In this case, the selection of the silvicultural method itself does not require more detailed calculations: when preparing the forest plan, only treatment options that relate to the chosen silvicultural method are simulated. Such a decision can also take into account how well the consequences of the decision are currently thought to be known. RF has prevailed in the Nordic countries for the last 70 years, so its productivity and risks are quite well known, at least under past conditions. There is less empirical information about the productivity and risks of CCF. However, some risks (e.g. disturbances under a changed climate) may be large and are poorly known for both forest management regimes.

In Chap. 2, the CCF management approaches were classified into selection systems, shelterwood systems and group systems. Because all these approaches may require natural regeneration, a realistic description of ingrowth is essential for planning in all cases. In addition, the selection and shelterwood systems require implementation of the specific thinning operations and a description of their effects on the growth of remaining trees. The group systems can be implemented by treating the gaps as separate management units when simulating growth and harvests. Furthermore, specific models for describing forest-stand structure, such as diameter distributions and height-diameter curves, are needed for the selection system. The challenges set by CCF to different parts of forest planning are discussed in detail in the next subsections.

5.2.1 Challenges for Field Data Collection

The canopy structure of stands managed according to selection systems is different compared to RF (Bianchi et al. 2020). However, in the practical implementation of ABA for forest inventories, only a small share of training plots typically originates from forests managed using selection systems. Therefore, remote sensing may not provide very good estimates of stand characteristics for such forests (Maltamo et al. 2000). However, currently it is even more important to be able to detect forests with an understorey that allows switching to CCF. Such forests do exist in the training data, but the estimation methods may not be sufficiently well optimised for detecting them. Jarron et al. (2020) were able to predict the structure of understorey trees with reasonable accuracy for a study area in British Columbia. Bollandsås et al. (2008a) also obtained similar results from uneven-aged forests in Norway. In Finland, understorey characteristics could be estimated from laser scanning data, but the accuracy of the assessment depended on the density of the dominant tree layer (Maltamo et al. 2005). In general, forest planning for CCF requires either information on the diameter distribution in the form of several differently defined diameters (i.e. moments or percentiles) or information by tree storey. Neither is well supported by the currently implemented ABAs; field-measured forest resource information is needed. The situation might change over time as methods for the aerial inventories develop.

Attempts are being made to efficiently obtain tree-level forest inventory data for planning purposes. The potential methodologies include smartphone applications (e.g. Trestima, Katam) and individual-tree detection from remote sensing data (e.g. Kostensalo et al. 2023). These methods produce either a tree list (diameter distribution) or true tree-level data where the location of each tree in the stand is known. The latter case makes it possible to take the spatial distribution of trees into account in growth prediction and harvest planning. It is also possible to optimise harvest decisions at the tree level, at least for the next cutting.

5.2.2 Challenges Describing Stand Structure

Predicting diameter distributions and height-diameter curves is often based on regression models from existing datasets which are mainly based on even-aged forest management. The diameter distribution is usually narrower and the height-diameter curve flatter compared to uneven-aged stands that include several cohorts (Siipilehto et al. 2023). Therefore, using the models developed for even-aged stands in CCF leads to insufficient variation in tree diameter and height.

A theoretically sound way to predict the diameter distribution is using parameter recovery methods based solely on the mathematical relationships of the stand characteristics (Siipilehto and Mehtätalo 2013, Mehtätalo and Lappi 2020; Chap. 11), or calibrating the models using field-measured tree samples or diameter quantiles (Mehtätalo et al. 2006). These methods require additional information besides mean diameter. If this information is not available, but characteristics measured by tree strata in the field are available, the models developed for even-aged stands can be used for each stratum separately. In both cases, the diameter distribution can be further rescaled based on other field-measured quantiles such as the maximum and minimum diameters and the Gini coefficient derived from them (Valbuena et al. 2017). The height model can also be calibrated separately for each canopy layer using the layers' field-assessed mean diameters and heights. Siipilehto et al. (2023) developed separate models based on uneven-aged stands. They also found that stand-level calibration of the height-diameter curves based on even-aged stand data will work if sufficiently many height-sample trees are used for calibration.

The stand characteristics commonly used in RF poorly describe the condition of the uneven-aged stands generated by selection systems, but work well for forests under transition. They give means and totals of (more or less) even-aged cohorts. In selection systems, characteristics that describe means, variance and distribution of uneven-aged tree stocks may be of interest, as well as methods using such information. One essential difference is that the age of a CCF forest cannot be defined. In this respect, the field inventory is simpler than in RF forests, but on the other hand, stand age cannot be used in models of stand structure and development. Other commonly used stand-level characteristics, such as mean diameter or height, also become less meaningful because they do not vary much between truly uneven-aged stands. Therefore, in addition to the traditional mean variables, characteristics that can describe differences between the stands, for example the variation in diameter, may be useful in CCF stands. If both the basal area and number of stems are known, the mean height and diameter combined with this information might describe the trees well enough (Mehtätalo et al. 2007). The advantage of such sample-tree measurements is that they can be chosen based on the desired accuracy of the stand description.

5.2.3 Challenges for Predicting Stand Dynamics

Technically, the growth models used in forest planning should be suitable for both continuous cover and rotation forestry, as discussed in Chap. 4. In CCF, it is essential to be able to predict the establishment of new trees. Since the survival of small seedlings is very uncertain, usually only seedlings that have exceeded the set minimum height or diameter (known as ingrowth) are considered. Ingrowth modelling has proven very challenging. Correctly predicting average ingrowth is not enough. The model must be able to describe (1) whether ingrowth occurs, (2) how much ingrowth occurs, if it does occur, and (3) whether ingrowth is distributed sufficiently evenly over the forest (Lappi and Pukkala 2020).

It is also worth distinguishing whether the model realistically describes these three aspects of ingrowth regionally, or whether it can also reliably assign forecasts to different target stands. The latter is hardly possible without field checks even though laser scanning gives reasonably good information about the existence of undergrowth. In large-scale scenario analyses, treatment schedules are not allocated to specific stands. For planning calculations, it is therefore sufficient to have a model that accurately predicts the proportion of forest stands where sufficient ingrowth takes place, by site condition, age group, and region, with continuous cultivation in mind.

Even though the models can technically be used in both forest management systems, it is possible that not all factors affecting the differences between the even-aged and uneven-aged forestry are appropriately taken into account. For example, information describing the health status of the trees and the condition of the canopy is not currently used in the growth models. Therefore, there may be systematic differences in these factors favouring one of the silvicultural methods.

In general, ideal data for modelling stand structure and dynamics in continuous cover forests is not available. Stands in Nordic countries have high variety in tree size, and there is plenty of empirical information on the growth of both suppressed and dominant trees in them. However, there is not enough knowledge on the growth reaction of trees after a strong thinning treatment. Existing experimental datasets are highly valuable for that purpose, even though they do not fully represent all spatial distributions, sites and species. However, extensive highly representative data will not be available in the near future, especially for growth modelling that requires repeated measurements. Therefore, it seems that forest planning in CCF requires more field information than RF, and would benefit greatly from approaches combining remote sensing data and predictions with local field data (e.g., Myllymäki et al. 2024).

5.2.4 Challenges for Optimisation

Optimisation does not differ between CCF and RF if all relevant factors affecting growth and yield are described with a simulator suitable to describe both silvicultural methods. However, if the simulator systematically over- or underestimates the development of one of the methods or response to a specific treatment, the problem will be further exacerbated in the optimisation. This is because optimisation, by its very nature, chooses extreme solutions, and even the smallest advantage in favour of one method can produce a plan in which treatments according to this method are chosen significantly more often than others (e.g., Kangas and Kangas 1999).

An optimisation method that recognises stochasticity is needed to consider the different uncertainties between uneven-aged and even-aged forestry (Eyvindson and Kangas 2014; Pukkala 2015; Malo et al. 2022). The stochastic method can take into account, for example, the methods' different risks for regeneration. Other damage risks that may differ between the methods (such as wind, bark beetles or root rot) should be accounted for in a similar way. If different risks are not considered, the planning system inevitably favours the silvicultural system that produces a better result on average, regardless of the differences in risks.

5.3 Continuous Cover Forestry in Currently Available Decision Support Systems

5.3.1 Tools in Sweden

Heureka is a forest decision support system widely used in Sweden with different modules applicable for analysis at stand, estate, landscape and national levels (Lämås et al. 2023). The system includes both a simulator for generating treatment schedules using tree- and plot-level growth and yield models and an optimisation tool based on linear and integer programming. Heureka can generate treatment schedules with uneven-aged forestry. However, the growth functions have not been fully designed or properly validated for uneven-aged forestry. Therefore, results for uneven-aged forestry are less certain compared to even-aged forestry. A serious drawback of the current growth function for uneven-aged forestry is that it includes age as variable. There are indications that the current growth function underestimates long-term growth in uneven-aged forestry and thus introduces a bias in the simulation and optimisation results (Fagerberg et al. 2022).

While the empirical experience from uneven-aged forestry in Sweden is limited, long-term field experiments do exist and more data is now available compared to 15 years ago, when Heureka was initially developed. There are ongoing research efforts to validate and improve or replace the current growth models for uneven-aged forestry in Heureka.

Heureka includes functionality for patch cutting. However, no edge effects are included, which means that the new stand in the gaps is not affected by competition from the adjacent mature forest. The remaining mature forest is also unaffected by release from the gap felling. The models for shelterwood management in Heureka also need to be validated and improved.

5.3.2   Tools in Finland

MELA and MOTTI are area-level and stand-level decision-support systems developed at Luke for assessing the effects of alternative forest management practices on growth, yield and profitability (Salminen et al. 2005). Previously, MOTTI was used mainly as a test platform for new growth and yield models as well as for formulating and comparing stand-level forest management alternatives to develop silvicultural guidelines. MELA has been used for large-scale analyses since the 1980s (Siitonen 1983). These systems currently include the same growth and yield models which can predict growth responses to silvicultural practices. Both simulators produce growth and yield information covering the development and structure of growing stock and deadwood by tree species, the amount and structure of cutting removals and natural mortality, and predictions of the biomass dynamics of different tree strata. The main difference between the systems is currently in the simulation of treatment schedules: MELA automatically simulates all treatment options that are acceptable according to the silvicultural guidelines, while in MOTTI treatments tailored specifically for a problem at hand can be simulated. In the future, the best properties of both systems will be merged to form a next-generation planning system.

In both simulators, optimisation can be done using JLP, a system to handle large-area calculations using linear programming (Lappi 1992). Spatial description of the forested area and spatial optimisation are not available in JLP.

Development of the young even-aged stands is based on stand-level models until a stand reaches an 8-m dominant height. Thereafter, tree-level models are used, which are fitted separately for mineral soils (Hynynen et al. 2014) and peatlands (Repola et al. 2018). For uneven-aged stands and CCF, models by Eerikäinen et al. (2007) are implemented based on the permanent experimental ERIKA plots which represent single-tree selection in Norway-spruce-dominated uneven-aged stands. Models cover regeneration establishment, and height development of the established seedlings. For larger trees, calibration functions for diameter and height growth for CCF are under development (Lee et al. 2024).

The Monsu forest planning system has been developed for multi-objective forest planning at the University of Eastern Finland. It has been widely used in education and research for more than 20 years, and is also used in practical forest planning. Monsu allows both automatic and manual simulation of management alternatives. Optimisation is based on multi-attribute utility functions using heuristic methods. Spatial optimisation is also possible, which could be useful in CCF based on a group system. CCF using selection systems became possible in Monsu already in 2014 when growth models that allow both RF and CCF were implemented (Pukkala 2014). Several growth models are available for the simulation, including those of Pukkala et al. (2021). The management schedules can be simulated according to both RF and CCF, and the optimisation selects the one that best meets the management objectives. Alternatively, the simulation can be restricted to RF, CCF, or any-aged management where treatment schedules cannot be classified as RF or CCF (Pukkala 2020). Beyond the models that describe the dynamics of living trees, Monsu includes models for the dynamics of deadwood, soil carbon, wood products, and biodiversity which enable additional analyses. These include the effects of forest management on forest carbon balance and wood-based products (Heinonen et al. 2018), on water systems (Nieminen et al. 2023), and a responsibility report of the applied forest management (Pukkala 2022).

5.3.3   Tools in Norway

Norway has long traditions of developing decision-support systems for forest management, but they have mainly focused on RF in even-aged stands and used area-based growth models (e.g., Eid and Hobbelstad 2000; Strimbu et al. 2023). A few attempts have been made to adapt the area-based growth models for selective cutting in uneven-aged stands, but these solutions have been based on rather subjective considerations due to a lack of empirical data from experimental research plots (Eid and Hobbelstad 2005).

The only decision-support system developed for Norwegian conditions based on individual tree-growth models aiming for selective cutting in uneven-aged stands is T (Gobakken et al. 2008). T is a bio-economic forest simulator that can be applied to stand-level analyses. The simulator produces alternative treatment schedules with all feasible combinations of user-defined treatment and regeneration options, which were supposed to cater for both even- and uneven-aged forestry. The system has no functionality related to large-area (forest holding) analyses. The simulator relied on distance-independent diameter growth and mortality models, and area-based recruitment models (Bollandsås et al. 2008b; Bollandsås and Næsset 2009) calibrated to Norwegian NFI plots. The economic features of the simulator include timber values and cost calculations (harvesting, forwarding) for different harvesting methods (clearcutting, seed tree establishment and cutting, and selective cutting) based on individual trees. The simulator also provides costs related to silvicultural treatments such as planting and young growth tending, and calculates a net present value (NPV) for all treatment schedules over an infinite planning horizon.

Gobakken et al. (2008) evaluated and demonstrated the performance of the simulator. At the time of development, the simulator was judged to work appropriately, meaning vital silvicultural treatments under different conditions were handled reasonably, yielding logical and consistent biological and economic results, in accordance with previous experience and research. Fifteen years later, it has become clear that the biological and economic empirical data the simulator relies on is both uncertain and outdated.

In Norway we have seen two recent forest-simulator initiatives facilitating analyses of selective cutting through SiTree (Antón-Fernández and Astrup 2022) and TreeSim (Nabhani and Sjølie 2022). These initiatives are interesting as frameworks and technical solutions, but the simulators, as they are now, still rely on old biological models (Bollandsås et al. 2008b), and incorporate no or very limited economic features. An ideal decision-support system covering both even-aged and uneven-aged forestry developed for Norwegian conditions will require significant long-term research. Among other features, such a system should probably include a simulator based on distance-dependent tree growth (diameter and height) and mortality models for a more accurate description of horizontal and vertical variations in tree growth (Sharma and Brunner 2017). Such models could possibly be calibrated based on a combination of NFI plots and research plots from experiments where selective cuttings have been followed over time.

5.4 Conclusions

From a technical point of view, forest planning does not differ between CCF and RF. Forest planning has similar stages of data collection, simulation and optimisation, and also similar information needs about decision makers, forest resources and their dynamics under different treatments. For forest planning, a strict difference between RF and CCF is not very useful. Treatment schedules can be simulated and selected that best serve the management objectives of the decision maker. These may include a variety of management alternatives, including CCF, RF, extended rotations and other intermediate approaches.

Incorporation of CCF into forest planning involves a variety of challenges to the specific application of models. These challenges involve natural regeneration, description of forest structure, and reactions of forests to types of forest management operations that have not been extensively applied during the last 70 years.

Some tools for planning CCF are already available in the Nordic countries, but further development is still needed for better decision support. Fulfilling some of the requirements will take time because long-term datasets on CCF are scarce. The planning of CCF management might benefit from approaches combining model predictions with local field data from stands of interest.

The development of remote sensing methods for tree-level mapping of forests opens exciting possibilities for forest planning. First, models are no longer needed to generate the initial tree list for simulation. In addition, a spatial description of the stands could be used, allowing distance-dependent growth models and simulating tree extraction in a spatially and temporally explicit manner. These new possibilities will be especially useful under forest management like CCF that does not follow previously used RF practices.

References

Antón-Fernández C, Astrup R (2022) SiTree: a framework to implement single-tree simulators. SoftwareX 18:100925. https://doi.org/10.1016/j.softx.2021.100925

Bianchi S, Siipilehto J, Hynynen J (2020) How structural diversity affects Norway spruce crown characteristics. For Ecol Manag 461:117932. https://doi.org/10.1016/j.foreco.2020.117932

Bollandsås OM, Buongiorno J, Gobakken T (2008b) Predicting the growth of stands of trees of mixed species and size: a matrix model for Norway. Scand J For Res 23:167–178. https://doi.org/10.1080/02827580801995315

Bollandsås OM, Hanssen KH, Marthiniussen S, Næsset E (2008a) Measures of spatial forest structure derived from airborne laser data are associated with natural regeneration patterns in an uneven-aged spruce forest. For Ecol Manag 255(3):953–961. https://doi.org/10.1016/j.foreco.2007.10.017

Bollandsås OM, Næsset E (2009) Weibull models for single-tree increment of Norway spruce, scots pine, birch and other broadleaves in Norway. Scand J For Res 24(1):54–66. https://doi.org/10.1080/02827580802477875

Eerikäinen K, Miina J, Valkonen S (2007) Models for the regeneration establishment and the development of established seedlings in uneven-aged, Norway spruce dominated forest stands of southern Finland. For Ecol Manag 242(2–3):444–461. https://doi.org/10.1016/j.foreco.2007.01.078

Eid T, Hobbelstad K (2000) AVVIRK-2000 - a large scale forestry scenario model fo– long-term investment, income and harvest analyses. Scand J For Res 15:472–482. https://doi.org/10.1080/028275800750172736

Eid T, Hobbelstad K (2005) Long-term investment, harvest and income analyses using Avvirk-2000 (in Norweigian). Aktuelt fra skogforskningen 2(05):1–29

Eyvindson K, Hartikainen M, Miettinen K, Kangas A (2018) Integrating risk management tools for regional forest planning: an interactive multiobjective value-at-risk approach. Can J For Res 48(7):766–773. https://doi.org/10.1139/cjfr-2017-0365

Eyvindson K, Kangas A (2014) Stochastic goal programming in forest planning. Can J For Res 44:1274–1280. https://doi.org/10.1139/cjfr-2014-0170

Fagerberg N, Lohmander P, Eriksson O et al (2022) Evaluation of individual-tree growth models for *Picea abies* based on a case study of an uneven-sized stand in southern Sweden. Scand J For Res 37(1):45–58. https://doi.org/10.1080/02827581.2022.2037700

Gobakken T, Lexerød N, Eid T (2008) T—a forest simulator for bioeconomic analyses based on models for individual trees. Scand J For Res 23:250–265. https://doi.org/10.1080/02827580802050722

Heinonen T, Pukkala T, Asikainen A et al (2018) Scenario analyses on the effects of fertilization, improved regeneration material and ditch network maintenance on timber production of Finnish forests. Eur J For Res 37:93–107. https://doi.org/10.1007/s10342-017-1093-9

Hoganson HM, Rose DW (1984) A simulation approach for optimal timber management scheduling. For Sci 30(1):220–238

Hynynen J, Salminen H, Ahtikoski A et al (2014) Scenario analysis for the biomass supply potential and the future development of Finnish forest resources. Working Paper of the Finnish Forest Research Institute 302

Jarron L, Coops N, MacKenzie W et al (2020) Detection of sub-canopy forest structure using airborne LiDAR. Remote Sens Environ 244:111770. https://doi.org/10.1016/j.rse.2020.111770

Kangas A, Kangas J (1999) Optimization bias in forest management planning solutions due to errors in forest variables. Silva Fenn 33:303–315

Kangas A, Kurttila M, Hujala T et al (2015) Decision support for forest management, 2nd edn. Springer, Berlin. https://doi.org/10.1007/978-3-319-23522-6

Kangas J, Pukkala T (1996) Operationalization of biological diversity as a decision objective in tactical forest planning. Can J For Res 26:103–111

Kilpeläinen H, Miina J, Store R et al (2016) Evaluation of bilberry and cowberry yield models by comparing model predictions with field measurements from North Karelia, Finland. For Ecol Manag 363:120–129. https://doi.org/10.1016/j.foreco.2015.12.034

Kostensalo J, Mehtätalo L, Tuominen S et al (2023) Recreating structurally realistic tree maps with airborne laser scanning and ground measurements. Remote Sens Environ 298:113782. https://doi.org/10.1016/j.rse.2023.113782

Lämås T, Sängstuvall L, Öhman K et al (2023) The multi-faceted Swedish Heureka forest decision support system: context, functionality, design, and 10 years experiences of its use. Front For Global Change 6. https://doi.org/10.3389/ffgc.2023.1163105

Lappi J (1992) JLP: A linear programming package for management planning. Finnish Forest Research Institute, Research Notes 414

Lappi J, Pukkala T (2020) Analyzing ingrowth using zero-inflated negative binomial models. Silva Fenn 54(4):10370. https://doi.org/10.14214/sf.10370

Lee D, Repola J, Bianchi S et al (2024) Calibration models for diameter and height growth of Norway Spruce in uneven-aged stands in Finland. For Ecol Manag. (in press)

Malo P, Tahvonen O, Suominen A et al (2022) Reinforcement learning in optimizing forest management. Can J For Res 51(10):1393–1409. https://doi.org/10.1139/cjfr-2020-0447

Maltamo M, Kangas A, Uuttera J et al (2000) Comparison of percentile based prediction methods and the Weibull distribution in describing the diameter distribution of heterogeneous scots pine stands. For Ecol Manag 133:263–274

Maltamo M, Packalén P, Yu X et al (2005) Identifying and quantifying structural characteristics of heterogeneous boreal forests using laser scanner data. For Ecol Manag 216(1):41–50. https://doi.org/10.1016/j.foreco.2005.05.034

Mehtätalo L, Kangas A (2005) An approach to optimizing data collection in an inventory by compartments. Can J For Res 35(1):100–112. https://doi.org/10.1139/x04-139

Mehtätalo L, Lappi J (2020) Biometry for forestry and environmental data: with examples in R. Chapman and Hall/CRC, New York. https://doi.org/10.1201/9780429173462

Mehtätalo L, Maltamo M, Kangas A (2006) The use of quantile trees in prediction of the diameter distribution of a stand. Silva Fenn 40:501–516. https://doi.org/10.14214/sf.333

Mehtätalo L, Maltamo M, Packalén P (2007) Recovering plot-specific diameter distribution and height-diameter curve using ALS based stand characteristics. International archives of photogrammetry, remote sensing and spatial information sciences 36(3/W52): 288–293. Proceedings of the ISPRS Workshop 'Laser Scanning 2007 and SilviLaser 2007' Espoo, September 12–14, 2007, Finland

Myllymäki M, Kuronen M, Bianchi S et al (2024) A Bayesian approach to projecting forest dynamics and related uncertainty: an application to continuous cover forests. Ecol Model 491:110669. https://doi.org/10.1016/j.ecolmodel.2024.110669

Nabhani N, Sjølie HK (2022) TreeSim: an object-oriented individual tree simulator and 3D visualization tool in python. SoftwareX 20:101221. https://doi.org/10.1016/j.softx.2022.101221

Næsset E, Gobakken T, Holmgren J et al (2004) Laser scanning of forest resources: the Nordic experience. Scand J For Res 19(6):482–499. https://doi.org/10.1080/02827580410019553

Nieminen M, Pukkala T, Stenberg L et al (2023) The effect of continuous-cover forestry and rotational forestry on the water stress on forested waterbodies in Finland (in Finnish). Metsätieteen aikakauskirja 2023:22001. https://doi.org/10.14214/ma.22001

Pukkala T (2014) Does biofuel harvesting and continuous cover management increase carbon sequestration? For Pol Econ 43:41–50. https://doi.org/10.1016/j.forpol.2014.03.004

Pukkala T (2015) Optimizing continuous cover management of boreal forest when timber prices and tree growth are stochastic. For Ecosyst 2:6. https://doi.org/10.1186/s40663-015-0028-5

Pukkala T (2020) Instructions for optimal any-aged forestry. Forestry 91(5):563–574. https://doi.org/10.1093/forestry/cpy015

Pukkala T (2022) Assessing the externalities of timber production. For Pol Econ 135:102606. https://doi.org/10.1016/j.forpol.2021.102646

Pukkala T, Heinonen T, Kurttila M (2009) An application of a reduced cost approach to spatial forest planning. For Sci 55(1):13–22. https://doi.org/10.1093/forestscience/55.1.13

Pukkala T, Vauhkonen J, Korhonen KT et al (2021) Self-learning growth simulator for modelling forest stand dynamics in changing conditions. Forestry 94(3):333–346. https://doi.org/10.1093/forestry/cpab008

Repola J, Hökkä H, Salminen H (2018) Models for the diameter and height growth of scots pine, Norway spruce and pubescent birch in drained peatland sites in Finland. Silva Fenn 52(5):10055. https://doi.org/10.14214/sf.10055

Salminen H, Lehtonen M, Hynynen J (2005) Reusing legacy FORTRAN in the MOTTI growth and yield simulator. Comput Electron Agr 49(1):103–113. https://doi.org/10.1016/j.compag.2005.02.005

Sharma RP, Brunner A (2017) Modeling individual tree height growth of Norway spruce and scots pine from national forest inventory data in Norway. Scand J For Res 32:501–514. https://doi.org/10.1080/02827581.2016.1269944

Siipilehto J, Mehtätalo L (2013) Parameter recovery vs. parameter prediction for the Weibull distribution validated for scots pine stands in Finland. Silva Fenn 47(4):1–22. https://doi.org/10.14214/sf.1057

Siipilehto J, Sarkkola S, Nuutinen Y et al (2023) Predicting height-diameter relationship in uneven-aged stands in Finland. For Ecol Manag 549:121486. https://doi.org/10.1016/j.foreco.2023.121486

Siitonen M (1983) A long-term forestry planning system based on data from the Finnish national forest inventory. Proceedings of the IUFRO subject group 4.02 meeting in Finland, September 5–9, 1983. Univ of Helsinki, Dept of Forest Mensuration and Management, Res Notes 17: 195–207

Strimbu VF, Eid T, Gobakken T (2023) A stand level scenario model for the Norwegian forestry—a case study on forest management under climate change. Silva Fenn 57(2):23019. https://doi.org/10.14214/sf.23019

Tahvanainen V, Miina J, Pukkala T et al (2018) Optimizing the joint production of timber and marketed mushrooms in Picea abies stands in eastern Finland. J For Econ 32(1):34–41. https://doi.org/10.1016/j.jfe.2018.04.002

Tikkanen O-P, Heinonen T, Kouki J et al (2007) Habitat suitability models of saproxylic red-listed boreal forest species in long-term matrix management: cost-effective measures for multi-species conservation. Biol Conserv 140(3–4):359–372. https://doi.org/10.1016/j.biocon.2007.08.020

Valbuena R, Maltamo M, Mehtätalo L et al (2017) Key structural features of boreal forests may be detected directly using L-moments from airborne lidar data. Remote Sens Environ 194:437–446. https://doi.org/10.1016/j.rse.2016.10.024

Wikberg PE (2004) Occurrence, morphology and growth of understory saplings in Swedish forests. Acta Universitatis Agriculturae Sueciae. Silvestria 322. Dept of Silviculture, Swedish Univ. of Agricultural Sciences (dissertation)

Chapter 6
Harvesting of Continuous Cover Forests

Heikki Korpunen, Yrjö Nuutinen, Paula Jylhä, Lars Eliasson, Aksel Granhus, Juha Laitila, Stephan Hoffmann, and Timo Muhonen

Abstract

- Overall forest management objectives and stand properties set the requirements and possibilities for harvesting in continuous cover forestry (CCF).
- Harvester and forwarder operators play a key role in successful CCF harvesting, as both productivity and quality of work are essential factors in harvesting operations.
- Optimal stand conditions improve work productivity on selection harvesting sites; harvested stem volume correlates well with work productivity in cutting, and density of remaining trees does not significantly reduce work productivity in forwarding.
- Carefully executed group cutting and shelterwood harvesting can reduce the number of damaged remaining trees, which is beneficial for future tree generations.
- Research-based information is needed about work productivity in harvesting, damage caused by harvesting, and optimisation of strip road and forest road networks for CCF.

Keywords Harvesting · Cutting · Forwarding · Productivity · Forest operation management

H. Korpunen (✉) · A. Granhus · S. Hoffmann
Norwegian Institute of Bioeconomy Research (NIBIO), Ås, Norway
e-mail: heikki.korpunen@nibio.no; aksel.granhus@nibio.no; stephan.hoffmann@nibio.no

Y. Nuutinen · J. Laitila · T. Muhonen
Natural Resources Institute Finland (Luke), Joensuu, Finland
e-mail: yrjo.nuutinen@luke.fi; juha.laitila@luke.fi; timo.muhonen@luke.fi

P. Jylhä
Natural Resources Institute Finland (Luke), Kokkola, Finland
e-mail: paula.jylha@luke.fi

L. Eliasson
Skogforsk, Uppsala Science Park, Uppsala, Sweden
e-mail: lars.eliasson@skogforsk.se

© The Author(s) 2025
P. Rautio et al. (eds.), *Continuous Cover Forestry in Boreal Nordic Countries*,
Managing Forest Ecosystems 45, https://doi.org/10.1007/978-3-031-70484-0_6

6.1 Introduction

Harvesting operations in Fennoscandia are characterised by highly productive, fully-mechanised harvester-forwarder systems using the cut-to-length (CTL) method (Uusitalo 2010). The forest machines are flexible and can be used for all current cutting directives. Apart from occasional cable yarding operations in Norway, motor-manual tree felling and processing has been replaced by the fully-mechanised CTL system in Fennoscandian commercial forestry, accounting for up to 95% of harvests in Finland, Sweden and Norway (Lundbäck et al. 2021). This has shifted the sector from labour intensive to capital intensive, where high levels of machine utilisation and throughput volumes have reduced harvesting costs and increased work safety (Axelsson 1998). A shift towards CCF will significantly affect current harvesting operations, by increasing demands on planning and operator skills (Uusitalo 2010).

Technological advances in machinery and other equipment can also facilitate more precise and efficient harvesting operations, supporting the implementation of CCF. For example, modern harvesters and forwarders equipped with satellite positioning systems (GNSS) and advanced boom controls can accurately fell and extract selected trees, while minimising damage to the remaining trees. In Fennoscandian countries, it is the responsibility of the harvester operator to select trees, which further increases the mental workload during logging operations. However, improvements in remote sensing and forest inventories are improving strategic and operative decision-making in CCF harvesting.

In Fennoscandian CCF, the harvester operator is responsible not only for selecting and harvesting individual trees but also for maintaining optimal forest health for future activities. This requires a high level of skill and precision in identifying and targeting specific trees for removal, while leaving remaining trees intact. The operator must assess the quality, size, position, and growth potential of individual trees when making selection decisions during harvesting. In Fennoscandia, the working environments can vary significantly, from peatland soils with poor bearing capacity in Finland and Sweden to steep rocky terrains in Norway. Operators must be able to adapt to various conditions when logging, so they play a key role in successful execution of CCF harvesting. Training and skills maintenance, and continuous learning, are extremely important.

There are currently several national and international ongoing research and development projects, aimed at improving efficiency of forest operations in CCF harvesting. This is achieved by improving working methods or by using decision-support tools when planning the harvesting activities.

The aim of this chapter is to provide information about planning and execution of selection, group, and shelterwood harvesting in CCF, focusing on stand-level decision-making. The chapter considers first the planning phase of the logging activities, and then key features, such as expected work productivity or harvesting damage, associated with each harvesting method.

6.2   Planning Harvesting

Once the forest owner has made the decision to apply CCF harvesting in a selected stand, operational planning begins. Harvesting planning is usually carried out by a forestry professional, representing either the forest industry, the local forest owners' association, or a forestry service provider. In the planning phase, the forest owner chooses the desired and most suitable harvesting method—selection, shelterwood, or group cutting. The method selected depends on access, stand properties, management guidelines, regulations, and the forest owner's management objectives. The expected volumes of different timber assortments influence the requirements for landing sites and logistics.

Planning of harvesting (cutting and extraction) of a stand begins with evaluating accessibility. Accessibility depends on forest road quality and network density, and stand accessibility (topography). These are affected by weather conditions, which also impact logging operations. CCF is particularly affected by accessibility of forest roads, since the frequency of harvesting operations in a single stand is assumed to be higher compared with rotation forestry (RF). For example, Finnish forestry management guidelines (e.g. Metsänhoidon suositukset Tapio 2023) state that selection harvesting in spruce stands in southern Finland should take place after approximately 10–20 years, whereas in RF, thinning can be carried out after around 30–40 years and final felling at 60–100 years, depending on forest management plans. The more frequent harvesting in CCF favours a permanent forest road network, but this is an additional cost factor due to construction and maintenance work. In individual cases where volume of timber from harvests is expected to be low over longer periods, a focus on efficient machine trails might be more relevant to keep road costs low, despite higher extraction costs associated with the longer routes. Innovative planning aids, such as depth-to-water maps (Hoffmann et al. 2022) to identify suitable and trafficable machine routes for ground-based equipment, and software tools like Seilaplan (Bont et al. 2022) for cable yarding layouts, could be used in such situations to ensure low impact and efficient timber extraction.

Throughout Fennoscandia, current planning methods for harvesting from stump to roadside storage were mainly developed for traditional RF. In CCF harvesting operations, it is crucial that strip roads are placed according to log concentration, so that the work can be carried out efficiently with minimum damage to soil and remaining growing stock. Moist peatlands with low bearing capacity should be harvested in winter when the soil is frozen, whereas stands on mineral soils with good bearing capacity can be harvested at any time of the year. Compared to selection cutting, strip cutting, which can be considered a variant of group cutting, allows greater freedom in the location of on-site forwarding routes, as well as in organising route schedules. This is particularly beneficial in peatland forests (Laitila et al. 2020).

Stand properties set the framework for the work environment, harvesting productivity, costs, and quality of silvicultural outcome. The work environment includes technical harvesting factors: (1) stand structure before cutting, (2) amount of removal (m^3/ha), (3) stem volume (m^3/stem), (4) size distribution of removed trees, and (5) terrain characteristics (soil bearing capacity, slope, terrain roughness, and potential ditch network). These technical factors influence the productivity and costs of harvesting, and affect the accessibility of stands from and to the roadside. A recent study by Manner et al. (2023) supports earlier studies showing that the stem volume of removed trees is the most significant variable affecting productivity in selection cutting (Fjeld 1994; McNeel and Rutherford 1994; Suadicani and Fjeld 2001).

Stands on steep terrain comprise challenging environments for harvesting, mainly in Norway, but also in Sweden to some extent. Harvesting operations in steep terrain are expensive due to restricted access, the high level of planning required, use of specialised work systems with adapted equipment, and commonly lower productivity rates (Ghaffariyan et al. 2010; Böhm and Kanzian 2023). Lundbäck et al. (2021) note that, globally, mechanisation level falls as terrain steepness increases, although in Fennoscandia, the level of mechanisation used in steep terrain is higher than the worldwide average. Specialised machinery and equipment designed for steep terrain is required, such as ground-based machines with traction winches and steep-slope cable yarders that improve worker safety and operational efficiency (Holzfeind et al. 2020).

Planning of harvesting in steep terrain must also include protecting forest land from erosion as much as possible. Cable yarding has low environmental impact and is suitable for application in complex alpine silvicultural systems (Spinelli et al. 2015), but is only used on a limited scale in Fennoscandia. In Norway, less than 30,000 m^3 timber is produced annually through cable yarder operations, with no indication of expansion due to unavailability of a suitable workforce (Ottaviani et al. 2011).

CCF operations are not excluded from steep terrain, but they require well-conducted and adapted mechanised operations. Suadicani and Fjeld (2001) proved that carefully-planned mechanised selective harvesting in steep terrain can be productive, especially with larger stem volumes. The suitability of traction winches for overcoming technical terrain limitations and mitigating site impacts will be of particular interest in upcoming research, and projects are ongoing in Norway. Future technical developments must determine whether fully mechanised operations can be used in terrain conditions previously only harvestable using motor-manual felling and yarder extraction. One consideration is whether harvesting operations need to be conducted in all terrain conditions in the forested landscape, or whether other management objectives should be given priority in less-accessible areas.

Forest management guidelines, successful planning of harvesting, and appropriate technologies and cutting methods all contribute to the profitability of timber production, forest recreational values, and high biodiversity levels. However, the stand structure, i.e. distribution of suitable seedlings, can vary considerably within an individual cutting area (Fig. 6.1), which is why cutting may not always follow local forest management guidelines (e.g. Metsänhoidon suositukset Tapio 2023).

Professionalism and expertise of the harvester operator are particularly important in CCF, where the work environment, work planning, feasibility of the work method, and avoidance of damage to remaining trees ultimately determine the

Fig. 6.1 The thinning treatment and its timing are crucial in the success of CCF. Left: an even-structured stand. Right: a more irregular-structured stand. Photos: Erkki Oksanen/Luke

success of harvester work. The impact of the harvester operators' skills has been identified as a major factor in work efficiency (Sirén 1998; Ryynänen and Rönkkö 2001; Väätäinen et al. 2005; Kariniemi 2006; Palander et al. 2012; Purfürst and Erler 2011; Liski et al. 2020). Differences in operator skills increase as harvesting conditions become more difficult, and work planning and skills of the harvester operator significantly affect the number of damaged seedlings. In a simulation study (Miettinen 2005), when trees away from the strip road were felled, 45% of seedlings were exposed to damage. When trees were felled on the strip road where possible, 38% of seedlings were exposed. Use of the same felling directions and strip roads requires the harvester to have sufficient power and good control over the tree during felling. According to a pilot study by Manner and Ersson (2023), forwarding productivity in selection harvesting was found to be dependent on log concentration on the strip roads, while the density of remaining trees had very little effect. These issues must be carefully considered when planning and selecting a suitable CCF harvesting method for each stand.

Especially in CCF, where logging activities are frequent, the quality of remaining trees after each harvesting is crucial for maximising the future harvesting potential. However, the amount of information available is rather limited. Generally, selection cutting is perhaps the most demanding CCF harvesting method, while for example in group cutting, the proportion of damaged remaining neighbouring trees has been observed to be small.

Silvicultural outcome refers to the structure and quality of the residual trees, and soil damage and root system breakage in strip roads after cutting and forwarding

(Surakka and Sirén (2007). Harvesting maps predicting the bearing capacity of the terrain are available for most of Finland through the Finnish Forestry Center (Peuhkurinen 2017). In Sweden (Mohtashami et al. 2022) and in Norway (Heppelmann et al. 2022) there are hydrological models that can be used for predicting the risks of track rutting when planning CCF logging operations. The controller area network (CAN) of the harvester's on-board production statistics system makes it possible to compile a map of strip roads for forwarding (Ala-Ilomäki et al. 2020). With better pre-harvesting information from remote sensing and other sources such as soil maps, harvesting activities can be planned to improve cost-efficiency and minimise damage.

6.3   Selection Cutting

In Fennoscandia, selection cutting is a method used in CCF management in forest dominated by Norway spruce (*Picea abies* (L) Karst.; Lähde et al. 1999). It is mostly the largest trees that are felled to vacate growth space for the remaining trees, and dense groups of smaller trees are thinned (Surakka and Sirén 2007; Puettmann et al. 2015; Sirén et al. 2015, Lundqvist 2017).

There are challenges in the actual practice of selection cutting. One is the constant care needed to avoid damaging standing trees, which hampers harvester crane movements (Surakka and Sirén 2007). There is also a risk of damage to smaller trees in the lower canopy layers (Fjeld and Granhus 1998; Hämäläinen 2014; Sirén et al. 2015; Nyman 2016).

Finnish management recommendations for CCF (Metsänhoidon suositukset Tapio 2023) suggest no pre-clearing of undergrowth before selection cutting. However, dense undergrowth reduces visibility, and disrupts harvester head operation. In forests harvested by selection cutting, the location and quality of log bunches are not always as good for forwarding as in traditional thinning. This slows the loading work phase in the CCF stand.

The structure of timber assortments and volume of removal in the first CCF selection cutting (the conversion of even-aged stands into continuous cover stands) is generally positive for harvesting productivity. Laamanen (2014) explored the structure of eight CCF logging sites. The basal area of the sites before cutting was approximately 19–30 m^2/ha and the volume 157–285 m^3/ha. After cutting, the basal area was 6.6–14.3 m^2/ha and the volume 46–121 m^3/ha. The harvested volume was 110–231 m^3/ha, and average stem volume of the removed trees varied between 0.251 and 0.410 m^3. It should be noted that volumes from conversion harvestings may not fully correlate with future harvesting; in a long perspective, the volumes would vary according to variables like tree species, stand structure, and site index.

Several studies on work processes (e.g. Suadicani and Fjeld 2001; Manner et al. 2023) indicate that the harvester's processing time consumption (seconds per stem) in CCF selection cutting does not vary much from that in clearcutting, given trees of similar size. However, overall harvester productivity (m^3/hour) in selection cutting may differ slightly from final felling in RF, because selection cutting removes mostly the largest stems in the stand (Fjeld 1994; Lilleberg 1998; Suadicani and Fjeld 2001; Andreassen and Øyen 2002; Hämäläinen 2014).

When harvesting costs of CCF are compared with those of RF, calculations must include all harvesting treatments over the entire rotation period. However, forest owners and forest managers are more interested in revenues (timber sales) and harvesting costs (cutting, forwarding and relocation costs) when assessing feasibility in the next management decision. Lilleberg (1998) and Imponen et al. (2003) reported that the lower productivity of cutting and forwarding increased the cost of selection thinning by about 10% compared to clearcutting. However, the experimental sites were RF spruce stands without undergrowth. In Sweden, Jonsson (2015) found that the cost of CCF harvesting over the entire rotation cycle was 28% higher and machinery fuel consumption 21% higher than in RF. In Norway, Andreassen and Øyen (2002) compared selection thinning and clearcutting in uneven-aged spruce stands, with an average stem volume of 0.6 m^3 in selection thinning and 0.3 m^3 in clearcutting. The harvesting cost of selection cutting was about 10% higher than in clearcutting.

The structure and condition of standing trees after selection cutting determine the development of the forest and the timing of the next cut, and therefore the silvicultural outcome of harvesting is crucial (Sirén et al. 2015). The most important feature for near-term timber production and harvesting opportunities is trees of height 5–15 m. Studies report that 10–20% of these trees are damaged (Fig. 6.2) in

Fig. 6.2 Breakage of tops and damage to bark are typical examples of damage caused by selection cutting. Photos: Erkki Oksanen/Luke

the first mechanised selection cutting (Fjeld and Granhus 1998; Hämäläinen 2014; Sirén et al. 2015; Nyman 2016).

In CCF, the emergence of new seedlings and survival of existing ones are crucial for the long-term stand development. According to Hagström (1994), Granhus and Fjeld (2001), Vanha-Majamaa et al. (2002), Hanssen et al. (2007), and Surakka et al. (2011a, b), the damage rate of saplings of 0.5–3.0 m was between 2 and 61% in mechanised selection cutting. Laitila and Repola (2023) found that 2–4% of remaining trees were damaged after selection cutting of two Scots pine stands. The greater the volume extracted (m³/ha) and the closer the seedlings are to the strip road, the greater the proportion of damage. There is also a large variation in the proportion of damage (Sirén et al. 2015). Similar observations in selection harvesting were found by Metslaid et al. (2018). Generally, the highest damage rates may be expected at high harvest levels in densely-stocked stands, since these conditions leave little room for the harvester operator to ensure the felling direction is kept away from residual trees and advance regeneration (Fjeld and Granhus 1998).

Weather conditions may also affect levels of damage in the residual stand. While snow cover may offer some protection to young seedlings, severe frost makes the shoots and stems of seedlings and saplings increasingly brittle and prone to breakage (Eliasson et al. 2003). Severe frost also increases risk of breakage of the tops among the intermediate canopy layer trees. However, more important than the damage proportions is the quantity, condition, and uniform spatial distribution of the undamaged standing seedlings.

In selection cutting, it can be challenging to obtain enough logging residue to create an adequate brush mat for protecting the roots of residual trees (Fig. 6.3). Availability of brush material is determined by the amount and distribution of trees to be removed, and their size and species (Surakka et al. 2011a, b). According to Sirén et al. (2013a, b), in RMF spruce stands, 15–20 kg/m² spruce and pine logging residues can be generated by processing as many of the trees on the strip road as possible.

Digital operator-assistance systems will be important in the challenging working environments of selection cutting (Ylimäki et al. 2012; Väätäinen et al. 2013). Pre-information about the stand and terrain is important in selection cutting, and this can

Fig. 6.3 Uneven-structured forests after selection cutting during the previous winter. Left: peatland. Right: mineral soil. Photos: Erkki Oksanen/Luke

be obtained through laser scanning. However, detection of undergrowth shorter than 3–4 m may be poor (Hovi 2011).

6.4 Group Cutting and Shelterwood Cutting

Knowledge about mechanised group cutting is based on a small set of studies, e.g. Fjeld (1994) and Eliasson et al. (2020). The knowledge base is greater for shelterwood cutting, with studies of shelterwood establishment (Eliasson et al. 1999; Eliasson 2000), shelterwood thinning (Eliasson 2000; Hånell et al. 2000), and final overstorey removal (Glöde 1999, 2001; Glöde and Sikström 2001). Harvesting of groups or gaps has been found to be more expensive than final fellings due to reduced extraction productivity and, to some extent, lower harvester productivity. More studies are needed of group and patch cutting, not only to identify efficient work methods but also to find suitable group designs that enable efficient harvesting operations. In the establishment and thinning phases of a shelterwood, the smaller size of the harvested trees and the restrictions caused by remaining trees both contribute to lower cutting productivity than in final felling. In the final overstorey removal of a shelterwood, the trees, on average, are larger than in clearcutting and thinning but harvesting profitability is lower, due to both the low removal volume per hectare (Mäkelä 1992; Glöde 2001) and the care needed to avoid damage to the regeneration (Glöde 2001; Glöde and Sikström 2001).

Group cutting and shelterwood cutting are hampered by the need for seedlings to become established, so strip roads must be planned well and their density minimised. In stands with established regeneration, the strip roads should be located, where possible, by making use of natural and harvested groups or gaps. In shelterwood stands with a dense understorey at the sapling stage, the strip roads should be systematically opened as in traditional thinning stands; when possible, existing strip roads should be used. When trees are being felled in small groups, care should be taken to avoid damaging the trees at the edge of the group (Isomäki and Niemistö 1990; Mäkelä 1992). In shelterwood cutting, the distance between strip roads can be extended by using a chain saw towards the end of the harvester's boom reach to fell inaccessible trees, which will be processed and bunched by the harvester. However, organising the work safely can be problematic, and the harvesting cost is higher than in simple harvester work (Mäkelä 1992).

The shelterwood is usually removed all at once to minimise harvesting costs and reduce damage to growing stock and soil (Hyppönen and Niemistö 1998), although there may be variation in the number of shelterwood removals depending on tree species. In seedling stands where maximum height is 0.5 m, it is best to remove the overstorey trees in winter when a thick snow cover protects the seedlings (Maukonen 1987). In both shelterwood cutting and gap cutting, seedlings protruding from the snow surface are especially susceptible to harvesting damage during frost (Roiko-Jokela 1983), so harvesting should take place in frost-free periods. In harvester work, the accumulation of logging residues can be controlled by adjusting working technique without reducing productivity (Nurmi 1994). The emergence of new seedlings in group cutting can be promoted by leaving bare soil, and seedling damage can be reduced by not processing logging residues on them (Fig. 6.4).

In shelterwood cutting, seedlings are damaged both during felling and processing of trees, but a solution can be to cover them with logging residues or log piles. Forwarder wheels or tracks can damage seedlings close to strip roads (Maukonen 1987; Niemistö 1995; Hyppönen 1996). The amount of damage is mostly affected by the removal volume and density of seedlings, and the strip road system (Peltoniemi 1991; Sikström and Glöde 2000). Hyppönen and Niemistö (1998) investigated the impact of shelterwood harvesting on the development of a pine sapling stand. Almost one-fifth of the seedlings were damaged, with the damage mostly associated with the removed volume per hectare and the density of the strip road network. Forwarding caused 60% of the damage. More important, however, is the number of undamaged seedlings remaining, and Sikström and Glöde (2000) reported that the number remaining was satisfactory. Niemistö et al. (2012) reported that careful releasing of spruce undergrowth in a shelterwood stand decreased the productivity of cutting by 11–17% compared to clearcutting. However, after harvesting, sufficient vigorous spruces were detected. According to Tamminen (1985) and Piri (1996), root damage to the remaining undergrowth in removal of shelterwood may be a greater risk than damage to seedlings themselves. This is because root rot infecting the root system spreads over a long period of time.

Fig. 6.4 Group cutting with logging residues piled at the edge. Photo: Yrjö Nuutinen/Luke

6.5   Conclusions and Further Research Needs

The overall impact of CCF harvesting, and especially selection cutting, will depend on its contribution at regional and national levels, because a possible increasing prevalence of CCF harvesting has a significant impact on harvesting areas, logistics, costs, and technology. According to Hannerz (2017), harvesting costs would increase and harvesting removals would be significantly reduced if clearcutting were to be stopped completely.

In the future, the key research and development area will be adaptation of harvesting technology and working methods to the work environment; this need was noted by Bianchi et al. (2023). The stand structure mainly influences the harvester's work method and choice of machine type. Knowledge of the working environment is necessary (i.e. stand structure and trafficability of the terrain) to see the impact of different CCF methods on timber production and future harvesting conditions. More frequent timber harvesting, with long-distance transportation logistics, will most likely have effects on soil and groundwater levels. These effects need to be understood better to plan the operations in the most effective and least damaging way.

Further research on CCF harvesting is needed to complement existing knowledge. In Fennoscandia, a large proportion of forests with a quite regular stand structure need a preliminary conversion thinning to start creating a CCF structure. A small number of forests are already unevenly structured or already managed using CCF methods. One important aim will be to develop logging activities through improved planning and working methods to reduce damage to remaining trees and the environment. The forest management recommendations may require revision from a methodology and technology perspective. For example, according to the Finnish recommendations (Metsänhoidon suositukset Tapio 2023), uncommercial small trees should not be cleared in selection cutting. Long-term follow-up information is needed from monitored stands about, for example, whether a permanent strip-road network should be used or whether a network should be designed separately for each thinning. Previous study results vary greatly on the productivity of CCF harvester cutting.

Since CCF is relatively new in Fennoscandia, there is a need to evaluate the use of different methods. Most of the stands to be harvested using the selection cutting method are in the early transition phase from RF towards CCF, so there is an interest in completely new harvesting methods to produce an uneven-aged structure. There are many stands where the density is relatively high due to a lack of treatment when the stand was young. One possible cutting method is boom-corridor thinning (BCT), which is a single-grip harvester's working method for young dense stands. In BCT all trees from specific areas are harvested in one crane movement cycle, and adjacent areas will be untreated, producing strand structures that are more heterogeneous than those created by uniform thinning of the stand from below or above (Nuutinen et al. 2021a, b; Bergström et al. 2022; Nuutinen and Miina 2023). The boom-corridors are about 1–2 m wide and 10 m deep, and are systematically placed on both sides of the strip road.

The development of CCF harvesting requires cooperation between forestry companies and contractors, machine operators, forestry authorities, and research and teaching organisations (Engeström 1987; Nuutinen 2013). The development of CCF training for harvester operators and the transfer and processing of new research knowledge into training programmes is important. The digital devices and methods that already exist provide good opportunities for this.

References

Ala-Ilomäki J, Salmivaara A, Launiainen S et al (2020) Assessing extraction trail trafficability using harvester CAN-bus data. Int J Forest Eng 31:138–145. https://doi.org/10.1080/1494211 9.2020.1748958

Andreassen K, Øyen BH (2002) Economic consequences of three silvicultural methods in uneven-aged mature coastal spruce forests of Central Norway. Forestry 75:483–488

Axelsson S-Ã (1998) The mechanization of logging operations in Sweden and its effect on occupational safety and health. J For Eng 9:25–31

Bergström D, Fernandez-Lacruz R, De La Fuente T et al (2022) Effects of boom-corridor thinning on harvester productivity and residual stand structure. Int J Forest Eng:1–17. https://doi.org/1 0.1080/14942119.2022.2058258

Bianchi S, Ahtikoski A, Muhonen T et al (2023) Evaluation of operating cost management models for selection cutting in Scandinavian continuous cover forestry. iForest 16:218–225. https://doi.org/10.3832/ifor4204-016

Böhm S, Kanzian C (2023) A review on cable yarding operation performance and its assessment. Int J Forest Eng 34:229–253

Bont LG, Moll PE, Ramstein L et al (2022) SEILAPLAN, a QGIS plugin for cable road layout design. Croat J For Eng 43(2):241–255. https://doi.org/10.5552/crojfe.2022.1824

Eliasson L (2000) Effects of establishment and thinning of shelterwoods on harvester performance. J For Eng 11(1):21–27

Eliasson L, Bengtsson J, Cedergren J et al (1999) Comparison of single-grip harvester productivity in clear- and shelterwood cutting. J For Eng 10(1):43–48

Eliasson L, Lageson H, Valinger E (2003) Influence of sapling height and temperature on damage to advance regeneration. For Ecol Manag 175:217–222

Eliasson L, Grönlund Ö, Lundström H et al (2020) Harvester and forwarder productivity and net revenues in patch cutting. Int J For Eng 32(1):3–10

Engeström Y (1987) Learning by expanding: an activity-theoretical approach to developmental research. Orienta-Konsultit Oy, Helsinki. ISBN: 951-95933-2-2

Fjeld D (1994) Time consumption for selection and patch cutting with a one-grip harvester. Commun Skogforsk 47(4):1–27

Fjeld D, Granhus A (1998) Injuries after selection harvesting in multi-storied spruce stands—the influence of operating systems and harvest intensity. J For Eng 9(2):33–40

Ghaffariyan MR, Stampfer K, Sessions J (2010) Optimal road spacing of cable yarding using a tower yarder in southern Austria. Eur J For Res 129:409–416. https://doi.org/10.1007/s10342-009-0346-7

Glöde D (1999) Single- and double-grip harvesters - productivity measurements in final cutting of shelterwood. J For Eng 10(2):63–74

Glöde D (2001) Final cutting of shelterwood. Harvesting techniques and effect on the Picea abies regeneration. Swedish University of Agricultural Sciences, Faculty of Forestry, Umeå. Ph.D. Thesis, Silvestria 179

Glöde D, Sikström U (2001) Two felling methods in final cutting of shelterwood, single-grip harvester productivity and damage to the regeneration. Silva Fenn 35(1):71–83

Granhus A, Fjeld D (2001) Spatial distribution of injuries to Norway spruce advance growth after selection harvesting. Can J For Res 31:1903–1913

Hagström S (1994) En studie av avverkningsskador på inväxningsbeståndet vid blädning. (A study of logging damage to the regeneration stand during thinning) SLU, inst. för skogsteknik, Studentuppsatser 2 [SLU institution for forest operations, studentwork 2], p. 16

Hämäläinen J (2014) Poimintahakkuun nykykäytännöt: työohjeistus, ajanmenekki ja korjuujälki (Current practices in selective cutting: work instructions, time consumption, and harvesting quality). University of Helsinki. Faculty of Forestry and Agriculture. Department of Forest Sciences. M.Sc. thesis, p. 103

Hånell B, Nordfjell T, Eliasson L (2000) Productivity and costs in shelterwood harvesting. Scand J For Res 15:561–569

Hannerz M (2017) Hyggesfritt. Det handlar om procenten. (No clear cuttings. It is about the percentage.). Skogforsk Vision 3:18–21

Hanssen KH, Granhus A, Brean R (2007) Vitalitet, avgang og skader på foryngelsen ved selektiv hogst. (Vitality, mortality, and damage to regeneration in selective logging). Forskning fra Skog og landskap 3:11–16

Heppelmann JB, Talbot B, Antón Fernández C et al (2022) Depth-to-water maps as predictors of rut severity in fully mechanized harvesting operations. Int J For Eng 33(2):108–118. https://doi.org/10.1080/14942119.2022.2044724

Hoffmann S, Schönauer M, Heppelmann J et al (2022) Trafficability prediction using depth to water maps: the status of application in northern and central European forestry. Curr For Rep 8:55–71. https://doi.org/10.1007/s40725-021-00153-8

Holzfeind T, Visser R, Chung W et al (2020) Development and benefits of winch-assist harvesting. Curr For Rep 6:201–209. https://doi.org/10.1007/s40725-020-00121-8

Hovi A (2011) Alikasvoksen mittaus ja kartoitus laserkeilauksella (Measurement and mapping of understory with laser scanning). M.Sc. thesis. University of Helsinki. Faculty of Forestry and Agriculture. Department of Forest Sciences, p. 74

Hyppönen M (1996) Ylispuiden korjuun vaikutus mäntytaimikoiden kasvatuskelpoisuuteen ja arvoon Lapissa (The effect of overstory logging on the suitability and value of pine seedlings in Lapland). Lisensiaattityö MML-tutkintoa varten (Licentiate's thesis). University of Joensuu. p. 43

Hyppönen M, Niemistö P (1998) Ylispuuhakkuut ja taimikkovauriot. (Undergrowth in forest regeneration. In: Moilanen M, Saksa T (eds) From shade to light. Pihlaja Series 3

Imponen V, Keskinen S, Linkosalo T (2003) Monimuotoisuus talousmetsän uudistamisessa— kuusikoiden käsittelyvaihtoehtojen vaikutukset puuntuotannon ja -hankinnan talouteen (Diversity in economic forest regeneration - effects of spruce stand management options on timber production and procurement economy). Metsätehon raportti [Forest Technology Report] 163, p. 22

Isomäki A, Niemistö P (1990) Ajourien vaikutus puuston kasvuun Etelä-Suomen nuorissa kuusikoissa. (effect of strip roads on the growth and yield of young spruce stands in southern Finland.). Folia Forestalia 756:40

Jonsson R (2015) Prestation och kostnader i blädning med skördare och skotare [Performance and costs in felling with harvester and forwarder]. Arbetsrapport från Skogforsk nr 863–2015 [Working report from Skogforsk], p. 28

Kariniemi A (2006) Kuljettajakeskeinen hakkuukonetyönmalli—työn suorituksen kognitiivinen tarkastelu (Operator-centered logging machine model - cognitive examination of work performance). Helsingin yliopiston Metsävarojen käytön laitoksen julkaisuja [Publications of the Department of Forest Resources Use at the University of Helsinki] 38. University of Helsinki, p. 131

Laamanen V (2014) Poimintahakkuukohteiden puuston rakenne, korjuutekniset olosuhteet, korjuukustannukset ja korjuujälki. (Structure of selection cutting stands, logging conditions, logging costs and logging damages.) University of Helsinki. Faculty of Forestry and Agriculture. Department of Forest Sciences. M.Sc. thesis, p. 85

Lähde E, Laiho O, Norokorpi Y (1999) Diversity-oriented silviculture in the boreal zone of Europe. For Ecol Manag 118:223–243. https://doi.org/10.1016/S0378-1127(98)00504-0

Laitila J, Repola J (2023) Korjuukustannukset Lapin poimintahakkuukohteissa (Harvesting costs of selective harvesting in Lapland). Working report. Luonnonvara- ja biotalouden tutkimus. Luonnonvarakeskus [Natural resources and bioeconomy research. Natural Resources Institute Finland], p. 60

Laitila J, Väätäinen K, Kilpeläinen H (2020) Integrated harvesting of industrial roundwood and energy wood from clearcutting of a scots pine-dominated peatland forest. Int J For Eng 31(1):19–28. https://doi.org/10.1080/14942119.2020.1672462

Lilleberg R (1998) Puunkorjuun tuotos ja kustannukset. (Output and costs of timber harvesting). In: Kaila S (ed) Monimuotoisuus talousmetsän uudistamisessa -hankkeen väliraportit (MONTA-hanke). (Economic differences in forest regeneration (MONTA project)). Metsätehon raportti 62:29–32

Liski E, Jounela P, Korpunen H et al (2020) Modeling the productivity of mechanised CTL harvesting with statistical machine learning methods. Int J For Eng 31(3):253–262. https://doi.org/10.1080/14942119.2020.1820750

Lundbäck M, Häggström C, Nordfjell T (2021) Worldwide trends in methods for harvesting and extracting industrial roundwood. Int J For Eng 32(3):202–215

Lundqvist L (2017) Tamm review: selection system reduces long-term volume growth in Fennoscandic uneven-aged Norway spruce forests. For Ecol Manag 391:362–375. https://doi.org/10.1016/j.foreco.2017.02.011

Mäkelä M (1992) Ylispuiden poistohakkuiden korjuutekniikka (the harvesting technique of over-story removal cuts). Metsätehon katsaus [Forest Technology Review] 6:8

Manner J, Ersson BT (2023) A pilot study of continuous cover forestry in boreal forests: do remaining trees affect forwarder productivity? J For Sci 69(4):317–323

Manner J, Karlsen T, Ersson BT (2023) A pilot study of continuous cover forestry in boreal forests: decreasing the harvest intensity during selection cutting increases piece size, which in turn increases harvester productivity. J For Sci 69(4):172–177

Maukonen A (1987) Ylispuuhakkuun taimikolle aiheuttamat vauriot. (damages caused by cutting of hold-over trees to the regeneration stand). Metsäntutkimuslaitoksen tiedonantoja 244:30

McNeel JF, Rutherford D (1994) Modelling harvester-forwarder system performance in a selection harvest. J For Eng 6(1):7–14

Metsänhoidon suositukset, Tapio (2023) The Best Practice Guidelines for Sustainable Forest Management. Online version accessed 4.9.2023. (metsanhoidonsuositukset.fi)

Metslaid M, Granhus A, Scholten J et al (2018) Long-term effects of single-tree selection on the frequency and population structure of root and butt rot in uneven-sized Norway spruce stands. For Ecol Manag 409:509–517

Miettinen A (2005) Paikkatietoanalyysien soveltaminen eri-ikäisrakenteisten metsien hakkuiden tutkimuksessa. Päättötyö, paikkatiedonhallinnan erikoistumisopinnot (Application of spatial data analyses in the study of harvesting in forests of different age structures. Final work, specialization studies in spatial data management). Hämeen ammattikorkeakoulu [Häme University of Applied Sciences], p. 10

Mohtashami S, Thierfelder T, Eliasson L et al (2022) Use of hydrological models to predict risk for rutting in logging operations. Forests 13(6):901. https://doi.org/10.3390/f13060901

Niemistö P (1995) Turvemaan hieskoivikon tiheyden vaikutus alikasvoskuusikon tiheyteen. (The effect of the density of drained peatland birch stands on the density of undergrowth spruce stands) In: Poikolainen J, Väärä T (eds) Metsäntutkimuspäivä Kuusamossa 1994. Metsäntutkimuslaitoksen tiedonantoja 552:87–103

Niemistö P, Korpunen H, Laurén A et al (2012) Impact and productivity of harvesting while retaining young understorey spruces in final cutting of downy birch (Betula pubescens). Silva Fennica 46(1):81–97. article id 67

Nurmi J (1994). Työtavan vaikutus hakkuukoneen tuotokseen ja hakkuutähteen kasautumiseen (The impact of work method on harvester productivity and harvest residue accumulation). Metsätieteen aikakauskirja 2(1994):113–121

Nuutinen Y (2013) Possibilities to use automatic and manual timing in time studies on harvester operations. Dissertationes Forestales 156, p. 68

Nuutinen Y, Miina J (2023) Effect of boom corridor and selective thinning on the post-treatment growth of young scots pine and birch stands. Silva Fennica 57(2) article id 23017:11. https://doi.org/10.14214/sf.23017

Nuutinen Y, Miina J, Saksa T et al (2021a) Comparing the characteristics of boom-corridor and selectively thinned stands of scots pine and birch. Silva Fennica 55(3):22

Nuutinen Y, Miina J, Saksa T et al (2021b) Hakkuukoneella tehtävän väyläharvennuksen vaikutus harvennuskertymään ja kasvatettavaan puustoon nuorissa metsissä. (The effect of boom-corridor thinning with a harvester on thinning yield and growing stock in young forests.) Metsätieteen aikakauskirja 2021–10623. Tutkimusseloste, p. 5

Nyman D (2016) Kalhyggesfritt skogsbruk—lätt avverkning eller hård gallring? En översikt av vår nygamla skogsbruksmetod. (Uneven-aged forestry—light logging or heavy thinning? A review of our new-old silvicultural system.) Yrkeshögskolan NOVIA. Examensarbete för skogsbruksingenjör (YH-examen). Utbildningprogrammet för naturbruk. (University of applied sciences NOVIA. Thesis for forest engineering (Vocational degree). Educational program for agriculture.) Raseborg, p. 60

Ottaviani G, Talbot B, Nitteberg M et al (2011) Workload benefits of using a synthetic rope Strawline in cable yarder rigging in Norway. Croat J For Eng 32(2):561–569. https://doi.org/10.5552/crojfe.2011.013

Palander T, Ovaskainen H, Tikkanen L (2012) An adaptive work study method for identifying the human factors that influence the performance of a human–machine system. For Sci 58(4):377–389. https://doi.org/10.5849/forsci.11-081

Peltoniemi T (1991) Ylispuiden poisto konetyönä, miestyönä ja niiden yhdistelmänä. (removal of dominant trees with machine work, manual work, and their combination.). Metsätehon katsaus [Forest Technology Review] 18:5

Peuhkurinen J (2017) Staattinen maastomalli—päivitetyn version ominaisuudet ja implementointi käytäntöön. (Static terrain model—characteristics and implementation) MEOLO/Asiantuntijaryhmän kokous [Expert group meeting]. 17.2.2017. Arbonaut

Piri T (1996) The spreading of the S type of Heterobasidion annosum from Norway spruce stumps to the subsequent tree stand. Eur J Plant Pathol 26:193–204

Puettmann KJ, Wilson SM, Baker SC et al (2015) Silvicultural alternatives to conventional even-aged forest management—what limits global adoption? For Ecosyst 2:8. https://doi.org/10.1186/s40663-015-0031-x

Purfürst FT, Erler J (2011) The human influence on productivity in harvester operations. Int J For Eng 22(2):15–22. https://doi.org/10.1080/14942119.2011.10702606

Roiko-Jokela P (1983) Taimikoiden kunto ylispuun poiston jälkeen. (the condition of seedlings after the removal of overstory). In: Metsäntutkimuspäivät Rovaniemellä [Forest research days in Rovaniemi] 1983. Metsäntutkimuslaitoksen tiedonantoja 105:72–82

Ryynänen S, Rönkkö E (2001) Harvennusharvestereiden tuottavuus ja kustannukset. (Productivity and expenses associated with thinning harvesters) Helsinki. Työtehoseuran julkaisuja 381:67

Sikström U, Glöde D (2000) Damage to Picea abies regeneration after final cutting of Shelterwood with single- and double-grip harvester systems. Scand J For Res 15:274–283

Sirén M (1998) Hakkuukonetyö, sen korjuujälki ja puustovaurioiden ennustaminen (One-grip harvester operation, its silvicultural result and possibilities to predict tree damage). Doctoral thesis. Finnish Forest Research Institute. Research papers 697, p. 179

Sirén M, Ala-Ilomäki J, Mäkinen H et al (2013a) Harvesting damage caused by thinning of Norway spruce in unfrozen soil. Int J For Eng 24(1):60–75

Sirén M, Hytönen J, Ala-Ilomäki J et al (2013b) Integroitu aines- ja energiapuun korjuu turve-maalla sulan maan aikana - korjuujälki ja ravinnetalous. (Integrated harvesting of industrial and energy wood on peatlands under unfrozen period). Working Papers of the Finnish Forest Research Institute 256, p. 24

Sirén M, Hyvönen J, Surakka H (2015) Tree damage in mechanized uneven-aged selection cut-tings. Croat J For Eng J Theory Appl For Eng 36:33–42

Spinelli R, Maganotti N, Visser R (2015) Productivity models for cable Yarding in alpine forests. Eur J For Eng 1(1):9–14

Suadicani K, Fjeld D (2001) Single-tree and group selection in montane Norway spruce stands: factors influencing operational efficiency. Scand J For Res 16:79–87. https://doi.org/10.108 0/028275801300004433a

Surakka H, Sirén M (2007) Poimintahakkuiden puunkorjuun nykytietämys ja tutkimustarpeet. (Current knowledge and research needs in selective logging). Metsätieteen aikakauskirja 4(2007):373–390

Surakka H, Sirén M, Heikkinen J, Valkonen S (2011a) Damage to saplings in mechanized selec-tion cutting in uneven-aged Norway spruce stands. Scand J For Res 26(3):232–244. https://doi. org/10.1080/02827581.2011.552518

Surakka H, Sirén M, Heikkinen J, Valkonen S (2011b) Damage to saplings in mechanized selec-tion cutting in uneven-aged Norway spruce stands. Scand J For Res 26:232–244

Tamminen P (1985) Butt-rot in Norway spruce in southern Finland. Commun Instituti Forestalis Fenniae 127:1–52

Uusitalo J (2010) Introduction to Forest operations and technology. JVP Forest Systems Oy, Hämeenlinna, p 287

Väätäinen K, Ovaskainen H, Ranta P, Ala-Fossi A (2005) Hakkuukoneenkuljettajan hiljaisen tie-don merkitys hakkuutulokseen työpistetasolla (The significance of harvester operator's tacit knowledge on cutting with a single-grip harvester). Metlan tiedonantoja 937(9):100

Väätäinen K, Lamminen S, Ala-Ilomäki J et al (2013) Kuljettajaa opastavat järjestelmät koneel-lisessa puunkorjuussa - kooste hankkeen avaintuloksista. (Operator assistance systems in mer-chanized harvesting). Working Papers of the Finnish Forest Research Institute 279, p. 24

Vanha-Majamaa I, Jalonen J, Hautala H (2002) Puusto, taimettuminen ja muu kasvillisuus (Stand, regeneration and other vegetation). Monta-tulosseminaari (MONTA Result seminar) 17.5.2002. Helsinki, p. 9

Ylimäki R, Väätäinen K, Lamminen S et al (2012) Kuljettajaa opastavien järjestelmien tarve ja hyötypotentiaali koneellisessa puunkorjuussa. (The need and potential for benefit of operator-assisting systems in mechanized wood harvesting). Working Papers of the Finnish Forest Research Institute 224, p. 70

Chapter 7
Genetic Effects

Katri Kärkkäinen, Sonja T. Kujala, Rosario Garcia-Gil, Arne Steffenrem, Johan Sonesson, Liina Hoikkala, Harri Mäkinen, and Sauli Valkonen

Abstract

- Genetic effects of continuous cover forestry (CCF) are not well known. We need more research, especially on the genetics of spruce-dominated CCF sites. Levels of relatedness are of interest, as are estimates of safe limits for the intensity and duration of CCF practices that secure genetic potential for good growth and quality.
- With even-aged forestry, genetically improved regeneration material can be used to mitigate climate change-related risks through breeding and deployment recommendations. In CCF, currently based on natural regeneration, we assume that enough seedlings establish, and that sites contain enough genetic variation to enable natural selection and evolutionary processes.
- Based on research in other regions, the number of reproducing trees must be kept large to avoid excessive levels of relatedness and inbreeding and to maintain sufficient levels of genetic diversity.

K. Kärkkäinen (✉) · S. T. Kujala · L. Hoikkala
Natural Resources Institute Finland (Luke), Oulu, Finland
e-mail: katri.karkkainen@luke.fi; sonja.kujala@luke.fi

R. Garcia-Gil
Department of Forest Genetics and Plant Physiology, Umeå Plant Science Centre, Swedish University of Agricultural Sciences, Umeå, Sweden
e-mail: m.rosario.garcia@slu.se

A. Steffenrem
Norwegian Institute of Bioeconomy Research, Steinkjer, Norway
e-mail: arne.steffenrem@nibio.no

J. Sonesson
The Forestry Research Institute of Sweden (Skogforsk), Uppsala, Sweden
e-mail: johan.sonesson@skogforsk.se

H. Mäkinen · S. Valkonen
Natural Resources Institute Finland (Luke), Helsinki, Finland
e-mail: harri.makinen@luke.fi; sauli.valkonen@luke.fi

© The Author(s) 2025
P. Rautio et al. (eds.), *Continuous Cover Forestry in Boreal Nordic Countries*,
Managing Forest Ecosystems 45, https://doi.org/10.1007/978-3-031-70484-0_7

- In some well-documented long-term experiments in other regions, intensive high-grading has led to slower growth rates, which could partly be due to genetic degradation of the stand. If contemporary recommendations for selection cutting are followed, negative genetic effects should be unlikely.

Keywords Continuous cover forestry · Genetic diversity · Inbreeding · Selection · Growth potential

7.1   Research on Genetic Effects of CCF Is Difficult But Important

There is a clear knowledge gap about the genetic effects of CCF in northern, conifer-dominated forests. This issue is further complicated by the diversity of CCF management methods (Fady et al. 2016). Under the umbrella of CCF methods, trees can be harvested by selection cutting, gap/group cutting, or shelterwood cutting (see Chap. 2). The genetic effects of seed-tree cutting with Scots pine and other light-demanding species have been well studied (e.g., Savolainen and Kärkkäinen 1992; García Gil et al. 2015). However, the reproduction patterns and genetic effects of harvesting in spruce-dominated uneven-aged stands in the Nordic region have not received much attention.

Heavy high-grading (see Chap. 2, Sect. 2.2.3), commonly practised in the past, involved harvesting all the large and high-quality trees without considering the condition or future of the remaining stand. Although high-grading was common in Nordic forests until the early 1900s, there has been no genetic research on the effects of these exploitative practices. Some observers noted the degraded state of heavily high-graded sites in Finland and Sweden (e.g., Lindqvist 1946). CCF is currently uncommon in Nordic forests (see Fig. 7.1, Mason et al. 2022) and genetic studies on CCF study sites have only recently begun. Therefore, our evaluation of the genetic effects of CCF is based only on research from regions where CCF has been practised for a long time, but where species and management methods differ from those in Nordic forests.

From a human perspective, the genetic effects of forest management methods manifest very slowly, making them hard to study. It is fairly easy to assess the rate of matings between related trees that can lead to inbreeding depression or reduced molecular genetic diversity, by using contemporary DNA-marker or sequencing techniques (e.g., García Gil et al. 2015). It is more difficult to estimate whether repeated selection cutting diminishes the growth potential of the stand. Multiple generations of growth and genetic data would be needed to fully understand the dynamics. To some extent this can be estimated with simulations.

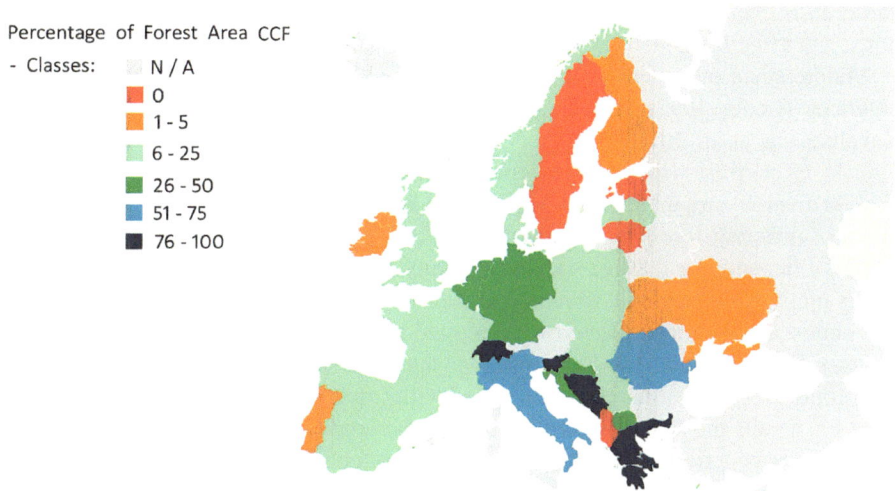

Fig. 7.1 The percentage of total forest area managed under continuous cover forestry (CCF) in different European countries (Fig. 1 in Mason et al. 2022; CCBY 4.0, link to the license: creative-commons.org/licenses/by/4.0/)

7.2   Does CCF Affect Genetic Diversity, Levels of Inbreeding and Adaptive Potential of Forests?

Genetic variation in essential traits of trees is a prerequisite for adapting to climate change. The amount of genetic variation in traits like growth, quality, cold tolerance, and pest/pathogen resistance can be estimated in specific trials. This is, however, rather laborious, and it is difficult to know exactly which traits are most important for adapting to future biotic threats and climate (e.g. new pests and pathogens, drought). Therefore, levels of genetic variation are often approximated by assessing overall genetic diversity using DNA markers or sequence data. These markers or sequenced genome areas do not necessarily control adaptive trait variation, but can be used to estimate within- and between-population genetic diversity, and how they are affected by forest management practices.

Typically, northern pine and spruce forests show abundant molecular genetic diversity. This variation is mostly shared between populations, meaning that at the overall genomic level the forest sites are not differentiated (Tollefsrud et al. 2009; Tyrmi et al. 2020). In many adaptive traits, however, the populations differ from each other. As an example, northern trees stop their yearly growth earlier than southern trees when tested under common conditions (e.g. Savolainen et al. 2007), triggered by the local temperature, day length (e.g. Heide 1974) and light quality (e.g. Ranade and García-Gil 2021). Although clearly different on average, the populations seem to maintain plenty of within-population variation that can facilitate adapting to different conditions (Savolainen et al. 2011). The maintenance of genetic variation in Nordic conifer species is due to their large populations, vast distributions, and effective gene flow through wind pollination (Savolainen et al. 2007).

Maintenance of genetic diversity depends on the choice of forest management practices (Ledig 1992; Savolainen and Kärkkäinen 1992; Ratnam et al. 2014; Kavaliauskas et al. 2018; Aravanopoulos 2018). The number of reproducing individuals is central; with a large reproductive population, good levels of overall genetic diversity are maintained. However, it is hard to conclude generally about the effect of different forest management practices on the levels of variation, since the response depends on the tree species' biology and the forest stand's history. These issues must be studied in Nordic boreal forests.

Fennoscandian forests were overexploited in the eighteenth and nineteenth centuries, followed by extensive clearcuttings and even-aged forestry, with an increasing proportion of commercial plantations with improved material after World War II. As a result, most Nordic forests have experienced periods of human activity, although the genetic effects of forest management on genetic diversity or inbreeding have so far been quite small (Savolainen and Kärkkäinen 1992; Aravanopoulos 2018).

Some have suggested that CCF could induce more family structures (increased frequency of genetically related trees) and therefore increase the rate of inbreeding among close relatives (Finkeldey et al. 2002). However, many such studies are done with tropical tree species experiencing selective cutting, and are therefore not applicable to boreal forests due to differences in the trees' biology (animal-pollinated tropical trees with small, sparse populations vs. wind-pollinated boreal trees with large, continuous populations; Ratnam et al. 2014). In boreal forests, Boyle et al. (1990) detected more relatedness in an uneven-aged black spruce site than in an even-aged stand naturally regenerated after a forest fire. According to Sagnard et al. (2011), spatial clustering and distances between reproducing trees and seedlings within sites influences the level of relatedness. In theory, high relatedness can lead to more inbreeding, in turn reducing growth and quality. However, pollen arriving from outside the stand reduces the frequency of between-relative mating in wind-pollinated species. Inbreeding depression in conifers is strong, as seen in the high mortality of self-pollinated embryos during seed development (Savolainen and Kärkkäinen 1992; Kärkkäinen and Savolainen 1993) and early seedling stages (Koelewijn et al. 1999). Milder levels of inbreeding could, however, result in non-mortal defects, such as reduced growth (Williams and Savolainen 1996; Wu et al. 1998).

Rapidly progressing climate warming creates many opportunities and challenges for our forests. Growing seasons lengthen, but frost damage due to temperature swings may increase, new pests and pathogens will emerge, and droughts will become more frequent and severe. How will CCF forest sites react to these changes compared to sites regenerated with genetically improved material from forest breeding? The possibility of assisted adaptation by regenerating sites with improved seed or seedling material with genetic properties likely to match future needs (e.g. Koralewski et al. 2015; Gómez-Pineda et al. 2021) is currently being studied.

In the Nordic countries climate change is taken into account in tree breeding and deployment by considering observed climate changes (Berlin et al. 2016) and climate forecasts (Hallingbäck et al. 2021). In joint analyses between Finland and Sweden, national deployment area recommendations have already been updated concerning climate warming (https://metsainfo.luke.fi/fi/vilpas). This kind of assisted adaptation is not possible in strictly naturally regenerated CCF sites.

Naturally regenerated forests may contain abundant genetic variation that can be used for adaptation, provided that the number of reproductive individuals is kept high and effective seedling establishment is secured (Brang et al. 2014; Fady et al. 2016). The adaptive potential of CCF sites depends on existing genetic variation and variation arriving by gene flow, mainly through pollination. It is also crucial that the new, better-adapted trait combinations can take hold in the forests; the speed of adaptation is also affected by the available space for seedling establishment (Savolainen et al. 2004). In CCF sites, the reproducing trees and some of the under-storey trees established decades ago, and the climate has already changed signifi-cantly since then. For long-term adaptive potential, it is important to consider the role of low-frequency variants in populations (e.g., Kastally et al. 2022), as some of them can be useful in the future climate. Concerns about losing low-frequency vari-ants in CCF were raised by Finkeldey and Ziehe (2004), especially where variants are correlated with growth rate. The loss of low-frequency variants can be mitigated to some extent by keeping the number of reproducing individuals high, both in natu-rally regenerating forests and in the breeding population.

7.3   Does Selection Cutting Lower Genetically Controlled Growth Potential?

The concern about CCF practices lowering genetic growth potential arises from the history of heavy high-grading practices in Finland, which were detrimental to for-ests. All big trees qualifying as logs were harvested, leaving only small trees of poor quality, without considering the stand's future development. This kind of harvesting was practised outside Finland, too, and its genetic effects are unclear.

In theory, genetic changes in growth rates could be induced by high-grading or, in broader terms, selecting trees that are left to grow and reproduce. The reverse is also possible, in which the fastest-growing and best-quality trees are selected as plus trees. These plus trees are grafted into seed orchards to breed improved forest regen-eration material. This material has been shown to have enhanced growth and quality compared with unimproved plants (Haapanen et al. 2016, 2020). This shows that selection of reproducing individuals can make significant changes in the genetic growth potential in only one generation. This genetic change has been due to high-intensity selection—only a very small fraction of trees qualified as plus trees (Oskarsson 1972). Tree breeding in the Nordic countries has improved growth con-siderably, with ~20–37% gains in spruce volume growth (Haapanen 2020; Liziniewicz et al. 2019), and over 20% better mean annual yield of pine stands (Haapanen et al. 2016). These estimates of the genetic gains from breeding are based on rather young experiments (20–30 years). However, the genetic gains in older experiments abroad have remained similar through later life stages, for example radiata pine experiments in New Zealand (Kimberley et al. 2015). Note that these estimates reflect the genetic potential for better growth in even-aged forestry and cannot be used as reverse estimates under uneven-aged CCF practices. Nevertheless, they clearly show that selection based on tree phenotypes can affect forest growth.

One example of the impact of harvesting method was seen in a long-term red spruce experiment in Maine, USA (Sokol et al. 2004). Over a period of 50 years, fixed diameter-limit harvesting (FDL), removing all trees above ~23 cm in diameter, was repeated twice on one site. On a second site, nine selection cuttings every 5 years were performed without diameter-limit prescriptions. The diameter-limit harvested site had poorer growth in the residual trees compared to the selection-cutting site. However, this experimental setup may be problematic, since management caused variation in residual stand density, which can lead to biased results if not considered in analyses. When estimating genetic effects of CCF practices, intensity and duration (in generations) of phenotypic selection are important factors. Ledig (1992) describes two harvesting scenarios in the history of North American forest usage: harvesting the finest and largest white pines for masts, versus harvesting nearly all trees for buildings and fuel, leaving only the very poorest trees to re-establish the stands. According to Ledig (1992), the latter practice harmed the growth potential of the remaining forest more than the selective logging. The severity of the effect depends mostly on the selection intensity (low intensity in the selective logging as most trees were spared vs. high intensity when leaving only a small proportion of poor trees alive). By quantitative genetic assessments (e.g. Cornelius et al. 2005) or simulating effects of forest management (e.g. Schaberg et al. 2008) we may assess the possibility of unintentional selection on growth or quality in CCF.

7.4 Conclusions and Knowledge Gaps

In the case of the CCF methods applicable to Scots pine (gap and shelterwood cutting), it is not likely that genetic growth potential is decreased by forestry practices. In those cases, there is no artificial selection in regeneration but big canopy trees next to the gaps are reproducing. In spruce-dominated forests, selection cutting of big and good-quality trees is used, but some of these large trees should be left to ensure efficient regeneration, coupled with careful maintenance of the remaining forest. Furthermore, if natural regeneration is limited in the stand and relies on a few reproducing trees, additional artificial regeneration via seeding or planting with genetically improved regeneration material could not only produce sufficient and spatially even seedling establishment, but also mitigate possible genetic worries. Overall, the lessons learned from the cases described above in other countries suggest that CCF practices are unlikely to lower the genetic growth potential, at least not during the first or second rounds of harvesting (Ledig 1992; Cornelius et al.

2005; Ratnam et al. 2014). It is, however, important to pay attention to the regenerative potential of the tree stands.

References

Aravanopoulos FA (2018) Do silviculture and forest management affect the genetic diversity and structure of long-impacted forest tree populations? Forests 9(6):355. https://doi.org/10.3390/f9060355

Berlin M, Persson T, Jansson G, Haapanen M, Ruotsalainen S, Bärring L, Andersson Gull B (2016) Scots pine transfer effect models for growth and survival in Sweden and Finland. Silva Fenn 50(3). https://doi.org/10.14214/sf.1562

Boyle T, Liengsiri C, Piewluang C (1990) Genetic structure of black spruce on two contrasting sites. Heredity 65(3):393–399

Brang P, Spathelf P, Larsen JB, Bauhus J, Boncčina A, Chauvin C, Svoboda M (2014) Suitability of close-to-nature silviculture for adapting temperate European forests to climate change. Forestry 87(4):492–503

Cornelius JP, Navarro CM, Wightman KE, Ward SE (2005) Is mahogany dysgenically selected? Environ Conserv 32(2):129–139

Fady B, Cottrell J, Ackzell L, Alía R, Muys B, Prada A, González-Martínez SC (2016) Forests and global change: what can genetics contribute to the major forest management and policy challenges of the twenty-first century? Reg Environ Chang 16(4):927–939

Finkeldey R, von Gadow K, Nagel J, Saborowski J (2002) Continuous Cover Forestry Reproduction in Continuous Cover Forests — The Geneticist's Perspective. Springer Netherlands Dordrecht, pp 67–79

Finkeldey R, Ziehe M (2004) Genetic implications of silvicultural regimes. Forest Ecol Manag 197(1–3):231–244

García Gil MR, Floran V, Östlund L et al (2015) Genetic diversity and inbreeding in natural and managed populations of scots pine. Tree Genetics Genomes 11:28

Gómez-Pineda E, Blanco-García A, Lindig-Cisneros R, O'Neill GA, Lopez-Toledo L, Sáenz-Romero C (2021) Pinus pseudostrobus assisted migration trial with rain exclusion: maintaining Monarch Butterfly Biosphere Reserve forest cover in an environment affected by climate change. New For 52(6):995–1010. https://doi.org/10.1007/s11056-021-09838-1

Haapanen M (2020) Performance of genetically improved Norway spruce in one-third rotation-aged progeny trials in southern Finland. Scand J For Res 35(5–6):221–226

Haapanen M, Hynynen J, Ruotsalainen S et al (2016) Realised and projected gains in growth, quality and simulated yield of genetically improved scots pine in southern Finland. Eur J Forest Res 135:997–1009

Hallingbäck HR, Burton V, Vizcaíno-Palomar N, Trotter F, Liziniewicz M, Marchi M, Berlin M, Ray D, Benito Garzón M (2021) Managing uncertainty in scots pine range-wide adaptation under climate change. Front Ecol Evol 9:724051

Heide OM (1974) Growth and dormancy in Norway spruce ecotypes (*Picea abies*). I, interaction of photoperiod and temperature. Physiol Plant 30:1–12

Kärkkäinen K, Savolainen O (1993) The degree of early inbreeding depression determines the selfing rate at the seed stage: model and results from *Pinus sylvestris* (Scots pine). Heredity 71(2):160–166

Kastally C, Niskanen AK, Perry A et al (2022) Taming the massive genome of scots pine with PiSy50k, a new genotyping array for conifer research. Plant J 109(5):1337–1350

Kavaliauskas D, Fussi B, Westergren M, Aravanopoulos F, Finzgar D, Baier R, Kraigher H (2018) The interplay between forest management practices, genetic monitoring, and other long-term monitoring systems. Forests 9(3):133

Kimberley MO, Moore JR, Dungey HS (2015) Quantification of realised genetic gain in radiata pine and its incorporation into growth and yield modelling systems. Can J For Res 45(12):1676–1687. https://doi.org/10.1139/cjfr-2015-0191

Koelewijn HP, Koski V, Savolainen O (1999) Magnitude and timing of inbreeding depression in scots pine (*Pinus sylvestris* L.). Evolution 53(3):758–768

Koralewski TE, Wang HH, Grant WE, Byram TD (2015) Plants on the move: assisted migration of forest trees in the face of climate change. For Ecol Manag 344:30–37

Ledig FT (1992) Human impacts on genetic diversity in forest ecosystems. Oikos:87–108

Lindqvist B (1946) Den skogliga rasforskningen och praktiken. Svenska Skogsvårdsföreningens förlag, Stockholm, p 176

Liziniewicz M, Karlsson B, Helmersson A (2019) Improved varieties perform well in realized genetic gain trials with Norway spruce seed sources in southern Sweden. Scand J For Res 34(6):409–416. https://doi.org/10.1080/02827581.2019.1622035

Mason WL, Diaci J, Carvalho J, Valkonen S (2022) Continuous cover forestry in Europe: usage and the knowledge gaps and challenges to wider adoption. Forestry 95(1):1–12

Oskarsson O (1972) Finnish plus trees of scots pine and Norway spruce. Folia For 150:1–138. http://urn.fi/URN:ISBN:951-40-0028-5

Ranade SS, García-Gil MR (2021) Molecular signatures of local adaptation to light in Norway spruce. Planta 253:53. https://doi.org/10.1007/s00425-020-03517-9

Ratnam W, Rajora OP, Finkeldey R et al (2014) Genetic effects of forest management practices: global synthesis and perspectives. For Ecol Manag 333:52–65

Sagnard F, Oddou-Muratorio S, Pichot C, Vendramin GG, Fady B (2011) Effects of seed dispersal, adult tree and seedling density on the spatial genetic structure of regeneration at fine temporal and spatial scales. Tree Genetics Genomes 7(1):37–48

Savolainen O, Bokma F, Garcia-Gil R, Komulainen P, Repo T (2004) Genetic variation in cessation of growth and frost hardiness and consequences for adaptation of *Pinus sylvestris* to climatic changes. For Ecol Manag 197(1–3):79–89

Savolainen O, Kärkkäinen K (1992) Effect of forest management on gene pools. In: Adams WT, Strauss SH, Copes DL, Griffin AR (eds) Population genetics of Forest trees. Forestry sciences, vol 42. Springer, Dordrecht. https://doi.org/10.1007/978-94-011-2815-5_17

Savolainen O, Kujala ST, Sokol C, Pyhäjärvi T, Avia K, Kn ürr T, Kärkkäinen K, Hicks S (2011) Adaptive potential of northernmost tree populations to climate change with emphasis on Scots pine (Pinus sylvestris L.) J Hered102(5) 526-536 https://doi.org/10.1093/jhered/esr056

Savolainen O, Pyhäjärvi T, Knürr T (2007) Gene flow and local adaptation in trees. Annu Rev Ecol Evol Syst 38(1):595–619. https://doi.org/10.1146/annurev.ecolsys.38.091206.095646

Schaberg PG, DeHayes DH, Hawley GJ, Nijensohn SE (2008) Anthropogenic alterations of genetic diversity within tree populations: implications for forest ecosystem resilience. For Ecol Manag 256(5):855–862

Sokol KA, Greenwood MS, Livingston WH (2004) Impacts of long-term diameter-limit harvesting on residual stands of red spruce in Maine. North J Appl For 21(2):69–73

Tollefsrud MM, Sønstebø JH, Brochmann C, Johnsen Ø, Skrøppa T, Vendramin GG (2009) Combined analysis of nuclear and mitochondrial markers provide new insight into the genetic structure of north European *Picea abies*. Heredity 102(6):549–562

Tyrmi JS, Vuosku J, Acosta JJ et al (2020) Genomics of clinal local adaptation in *Pinus sylvestris* under continuous environmental and spatial genetic setting. G3: Genes, Genomes, Genetics 10(8):2683–2696

Williams CG, Savolainen O (1996) Inbreeding depression in conifers: implications for breeding strategy. For Sci 42(1):102–117

Wu H, Matheson A, Spencer D (1998) Inbreeding in *Pinus radiata*. I. The effect of inbreeding on growth, survival and variance. Theor Appl Genet 97:1256–1268

Chapter 8
Financial Performance

Anssi Ahtikoski, Peichen Gong, Per Kr Rørstad, Esa-Jussi Viitala, and Jaakko Repola

Abstract

- Financial comparisons between rotation forestry (RF) and continuous cover forestry (CCF) are based on simulations in which the growth and yield of trees is estimated using a growth simulator. These often include an optimisation tool to present the maximum value of the objective function (usually the present value of net income).
- Studies have shown that the profitability of CCF depends on the initial state of a stand, especially the diameter distribution of the trees. The effect of interest discount rate also depends on the initial state.
- As a rule, it is safe to say that the more the forest structure resembles the target diameter distribution of the trees in CCF (i.e., a forest with heterogeneous structures), the more profitable it is to shift from RF to CCF.
- The more heterogeneous the tree structure on mineral soil, the higher the applied interest rate, the higher the forest establishment costs (soil preparation and culti-

A. Ahtikoski (✉)
Natural Resources Institute Finland (Luke), Tampere, Finland
e-mail: anssi.ahtikoski@luke.fi

P. Gong
Department of Forest Economics, Swedish University of Agricultural Sciences (SLU), Umeå, Sweden
e-mail: peichen.gong@slu.se

P. K. Rørstad
Faculty of Environmental Sciences and Natural Resource Management, Norwegian University of Life Sciences (NMBU), Ås, Norway
e-mail: per.kristian.rorstad@nmbu.no

E.-J. Viitala
Natural Resources Institute Finland (Luke), Helsinki, Finland
e-mail: esa-jussi.viitala@luke.fi

J. Repola
Natural Resources Institute Finland (Luke), Rovaniemi, Finland
e-mail: jaakko.repola@luke.fi

© The Author(s) 2025
P. Rautio et al. (eds.), *Continuous Cover Forestry in Boreal Nordic Countries*,
Managing Forest Ecosystems 45, https://doi.org/10.1007/978-3-031-70484-0_8

vation), and the poorer the growth conditions (site type and temperature sum), the more profitable CCF is. Few studies have been found that focus on peatland forests.

• Future financial studies should also consider risks associated with wind, harvesting, and insect and fungus damage as well as carbon payments and nontimber benefits

Keywords Stand-level optimisation · Rotation forestry · Continuous cover forestry · Profitability · Economic-ecological model

8.1   Recent Financial Studies

The profitability of continuous cover forestry (CCF) has been studied in Finland (e.g., Tahvonen 2009, 2011; Pukkala et al. 2010; Tahvonen et al. 2010; Rämö and Tahvonen 2014, 2015, 2017; Tahvonen and Rämö 2016; Juutinen et al. 2018a; Assmuth and Tahvonen 2018; Assmuth et al. 2018, 2021; Parkatti et al. 2019; Parkatti and Tahvonen 2021). These studies have focused on the stand level and compared the financial outcome of rotation forestry, RF (i.e., a management system based on clearcuttings and artificial regeneration) with those of CCF or have examined how an even-aged stand could optimally be converted into a CCF stand (Rämö and Tahvonen 2017). Recent studies have also examined the impact of carbon payments on optimal forest management (Tahvonen 2022). These studies share a significant multidisciplinary approach in which ecological models are combined with an economic calculation framework, thereby integrating algorithms of numerical optimisation into the economic-ecological model. Tahvonen's review (2022) is a modern synthesis of economic studies focusing on CCF. Common to all these studies (apart from Juutinen et al. 2018a) is that they are based on numerical optimisation, which enables a transparent and approved calculation method based on economic theory (e.g., Amacher et al. 2009; Tahvonen 2022). Another common factor is that forest growth is estimated by simulating the development of individual trees or tree size classes.

The simulation of tree growth introduces a need for caution. Historically, in Fennoscandia, tree growth modelling has been mainly based on field measurements representing RF (see, e.g., Bianchi et al. 2020). Consequently, growth models for RF have been based on considerably larger and longer field measurements than the corresponding models for CCF. A recent comprehensive study (Hynynen et al. 2019) suggests that RF and CCF stands respond differently to cutting treatments, and there are likely differences between RF and CCF stands with respect to the risk of abiotic and biotic damage (Nevalainen 2017). The effects of damage on stand growth have not been considered in detail in financial analyses.

8.2 Profitability Comparisons

The financial outcome of forest management is calculated by comparing revenues with expenses. Revenues mainly derive from the sale of trees (usually at stumpage), and expenses arise from investments, i.e., silvicultural measures. The time period between investments and return on investments can be very long in forestry—the period between stand establishment and clearcutting may be 50–100 years, in northernmost boreal biome even longer. Capital may also have other uses (opportunities), and borrowed capital also comes with a price, so the profitability of forestry is usually examined by incorporating commensurate revenues and expenses occurring at different times.

The profitability of forestry can be affected by length of rotation period, stand density and management intensity. Traditionally, no silvicultural expenses arise from CCF, since it relies on natural regeneration (i.e., trees are assumed to regenerate naturally), nor is there need for sapling stand management.

The most robust comparison method involving economic theory is based on calculating the net present value (NPV) of timber production (Haight 1985; Tahvonen and Viitala 2006; Tahvonen 2011). Calculating NPV considers all revenues deriving from forest management and all expenses arising from the present into the infinite future. Revenues and expenses generated at different times are discounted into present values, and the difference between them calculated to obtain the NPV. For RF starting from stand establishment at time zero, the NPV of the first and all future rotations can be calculated using a simple formula called the Faustmann model (Faustmann 1849):

$$V(T) = \left(1 - e^{-rT}\right)^{-1} \left[\sum_{t=1}^{T} R(t) e^{-rt} - c \right] \qquad (8.1)$$

where $V(T)$ = present value of net incomes (EUR/ha), r = continuous time interest rate, T = rotation period (years), e^{-rt} = continuous discount factor, $R(t)$ = net revenue or silvicultural cost at time t (EUR/ha), and c = stand regeneration cost (EUR/ha). [Note that for each continuous time interest rate r there is a corresponding discrete time interest rate i, $i = e^{r} - 1$]. Revenues are calculated by multiplying timber assortment volumes by corresponding unit prices. Net revenues are obtained by deducting harvesting costs from timber revenues. In Eq. 8.1, the numerator represents the NPV during a rotation starting from bare land, and the denominator is used to repeat the NPV to infinity (Amacher et al. 2009). When $V \geq 0$ using the chosen interest rate, forest management is financially profitable. When an existing stand is managed according to an RF regime, the present value of the net revenues in current and all future time periods is

$$V_{RF}(s) = \sum_{t=0}^{T_r} R(s,t) e^{-rt} + V(T) e^{-rT_r} \qquad (8.2)$$

where s = the initial state of the stand, T_r = the time when the existing stand is harvested and regenerated, $R(s, t)$ = net revenues at time t for the existing stand (EUR/ha), $V(T)$ = the bare land value determined according to Eq. 8.1. If the stand is to be managed according to a CCF regime (through selective harvesting), clearcutting will never be carried out in the stand, which means that $T_r = \infty$, and the NPV of the stand becomes

$$V_{CCF} = \sum_{t=0}^{\infty} R_s\left(s,t\right)e^{-rt} \tag{8.3}$$

where $R_s(s,t)$ = net revenues from the existing stand (EUR/ha) at time t. Assuming that timber prices are constant over time, and ignoring uncertainty in stand growth as well as risks of forest damage, optimal selective harvesting will lead to a cyclical steady state of the stand and a constant periodic net revenue (Rämö and Tahvonen 2017; Parkatti and Tahvonen 2020). Let T_c denote the time when the stand reaches a steady state, $\bar{R}$ the steady state periodic net revenue, and T_h the length of the cutting cycle. Eq. 8.3 can be rewritten as

$$V_{CCF} = \sum_{t=0}^{T_c} R_s\left(s,t\right)e^{-rt} + \bar{R}e^{-rT_c}\left(e^{rT_h} - 1\right)^{-1} \tag{8.4}$$

By comparing the NPVs associated with RF (Eqs. 8.1 and 8.2) and CCF (Eqs. 8.3 or 8.4), we can determine which method is financially more advantageous in each situation.

8.3 Optimisation Method

Forest management is primarily a goal-driven activity, requiring the best possible solution within the given limits. In economics, the best solution can be achieved by seeking an optimal solution in which the objective function is at its maximum (Intriligator 2002). Finding an optimal solution in RF requires the identification of thinning timings and intensities and the length of a rotation period. These are based on growth and yield models (in short, growth models) that produce estimates for tree growth. Because growth models are usually included in simulators consisting of several subprograms (e.g., diameter growth model, survival model, and height model), the financial assessment becomes especially complex, and it is practically impossible to solve them analytically (Pukkala 2009). In CCF, the problems are similar in principle, while the rotation period does not need to be defined separately, unless the technical calculation framework has not been built to enable solutions for both RF and CCF (Tahvonen 2015a, b; Tahvonen and Rämö 2016; Parkatti et al. 2019). Due to the complexity of the calculation framework, financial analyses at stand level are often based on numerical optimisation, in which the maximum value of the objective function (usually the NPV) is determined numerically using mathematical calculation algorithms that have been integrated with tree growth models (Niinimäki et al. 2012).

The optimisation methods on which calculation algorithms are based can be roughly divided into two groups: gradient-based and derivative-free methods (Cao 2010; Pyy 2021). Which optimisation method is most suitable depends on the convexity of the target function, the dimensions of the variables, and how growth models have been technically implemented in the simulator (Valsta 1992; Cao 2010). In turn, growth models can be based on stand-level models, process-based models, size distribution transition matrix models, size class matrix models, or growth models representing individual tree growth dynamics (Parkatti 2021).

Research based on the stand-level financial optimisation of CCF can be considered to have started systematically in the 1970s (Adams and Ek 1974), continued in the 1980s (e.g., Haight 1985; Haight and Getz 1987), and increased significantly in the 2010s, especially in Finland (e.g. Tahvonen et al. 2010; Pukkala et al. 2010; Tahvonen 2015a, b; Rämö and Tahvonen 2017; Parkatti et al. 2019). At the same time, financial studies on CCF have undergone a methodological shift, from static to dynamic models, so that recent studies have even been able to simultaneously examine the respective profitability of RF and CCF using the same initial stands. Methods applied to optimisation have also developed, so that recent financial studies (e.g., Parkatti et al. 2019; Parkatti and Tahvonen 2021) have used hierarchical multilevel optimisation, in which the rotation period usually represents the highest-level problem, and the thinning intensity the lowest level. Conceptually, this is a mixed-integer non-linear optimisation problem, in which some variables are integers and some are continuous real numbers. Technically, solving such an optimisation task requires a separate calculation algorithm, which is linked to growth models in the simulator software. In practice, solving a single optimisation task numerically may take several hours, even with the most powerful computers, while the calculation can be accelerated using machine learning. As a result, models can also include randomness related to tree growth, as well as natural damage (Malo et al. 2021). Recently, reinforcement learning (RL) algorithms have been applied in stand-level optimisation to capture stochasticity related to stand growth and prices (Tahvonen et al. 2022).

8.4 Profitability Comparisons in Different Situations

The table below presents the situations (site type, tree species, geographical location, and initial state) in which CCF on mineral soils is a financially better option than RF. In the literature, initial stands have often been divided into (1) bare land, (2) young, and (3) mature stands (ready for clearcutting). The structural features of stands have been described to make it easier to interpret the results. According to recent studies, it seems that when starting the financial analysis from bare land, spruce forests and spruce-dominated mixed forests (<40% deciduous trees) are more profitable when using CCF than RF, at least in southern and central Finland

and when applying an interest rate of at least 3%. The situation is less straightforward in pure pine forests. If stand establishment costs can be kept below EUR 1000/ha, RF could be a better option financially than CCF in dryish sites in central Finland, for example. For young even-aged spruce forests, it is probable that CCF would become more profitable than RF, at least in mesic forests in southern Finland. For mature stands, CCF is categorically more profitable than RF, at least in spruce forests, provided that the site type is not too fertile. The financial outcome depends on the initial stand structure (diameter distribution and stand age): the more the structural features resemble the target stand structure, i.e., uneven-aged forest, the more profitable CCF is compared to RF with clearcutting and artificial regeneration. Profitability of CCF also seems to improve in spruce forests further north compared with RF (e.g., Tahvonen 2011).

The Table 8.1 below presents the key results of financial studies divided according to the initial state of a stand, tree species, site type and location. The profitability of CCF is compared with RF. Results apply for mineral soils only.

Only a few studies have compared NPV of CCF and RF in Sweden. In contrast to the results from the Finnish studies, Swedish studies showed that, in most cases, CCF is less profitable than RF and that the difference in NPV between the two regimes is greater when the interest rate is higher (Granath and Söderström 2022; Nordström et al. 2013; Sonesson 2017; Wikström 2000, 2008).

There are several possible reasons behind the Swedish conclusion that selection cutting would result in a lower NPV than RF. First, a majority of the cases examined are old multi-storeyed stands with a large growing stock of timber, and the harvest intensity in the selective harvesting regime is strictly constrained (Granath and Söderström 2022; Sonesson 2017; Wikström 2008). In such cases, the stands should be harvested and regenerated immediately or in the near future under a RF regime, whereas a selective harvesting regime (CCF) implies that harvest of much of the growing stock of timber should be postponed, which means the retained timber capital would incur a high opportunity cost. This also explains why selective harvesting (CCF) becomes even less profitable than RF when the interest rate is high. Relaxing the constraints on harvest intensity would increase the NPV associated with selective harvesting (Sonesson 2017; Wikström 2008), but the Swedish studies did not consider the high harvesting intensities allowed in Finnish studies that showed CCF to be more profitable.

Secondly, in two of the studies, simulated timber growth was significantly lower under selective harvesting compared with the RF alternative (Sonesson 2017; Wikström 2000). The growth simulators used in these studies are not necessarily suitable for predicting the growth of the stand and future timber yield under selective harvesting management. Thirdly, in most of the studies, selective harvesting plans (cutting cycle and harvest intensity) were determined based on some simple rules, whereas the management programme under RF was optimised to maximise the NPV. This may have led to underestimation of the NPV associated with the CCF regime. Although Wikström (2000) determined the maximum NPV under both management regimes by optimising harvest plans, the growth simulator used in the

Table 8.1 Background information, key results and limitations related to financial studies on CCF-RF comparison

Initial state	Tree species	Location/Site type	Key results	Specifications and limitations
Bare land	Spruce	Central Finland/ Mesic forest (H_{100} 24 m)	If forest regeneration costs are at least EUR 1000/ha, and the interest rate is at least 3%, CCF is more profitable[a]	The initial state comprised a 20-year-old seedling stand, in which the tree density is 1750 or 2250 seedlings/ha, but the results can be generalised for bare land
	Spruce	Southern Finland/Mesic forest	With an interest rate of 3%, RF is more profitable[b] than CCF. The cutting cycle ranged from 10 to 30 years, and the basal area after cutting was 4–16 m²/ha	The study was not based on optimisation, as it identified alternative cutting cycles and intensities and compared results with RF
	Spruce, birch, pine (mixed forest)	Central Finland/ Mesic forest	With an interest rate of 3% and a deciduous tree coverage of 40%, CCF is more profitable than RF in mixed forests[c]	The initial state comprised a 20-year-old seedling stand with 1750 spruce seedlings/ha and 250 natural seedlings of other tree species (deciduous trees), but the results can be generalised for bare land
	Spruce, pine	Central Finland/ relatively sub-xeric (+) forest	For spruce, CCF is more profitable[d] if the interest rate is 3%, and forest regeneration costs are more than EUR 0/ha. For pine, RF is more profitable when the Bollandsås model is applied, regeneration costs are no more than EUR 1000/ha, and the calculated interest rate is 3%. When the Pukkala model is applied, CCF is more profitable[d] when the interest rate is 3%, and regeneration costs range from EUR 0 to 2000/ha	The results were calculated according to two different ecological models (Bollandsås and Pukkala) using two calculated interest rates (1% and 3%) and different regeneration costs: EUR 0, 500, 1000, 1500 and 2000/ha

(continued)

Table 8.1 (continued)

Initial state	Tree species	Location/Site type	Key results	Specifications and limitations
Young forest	Spruce	Southern and Central Finland/ fresh herb-rich forest heath	CCF is categorically more profitable[c] than RF when the interest rate is 1–5%	The financial examination involved three different initial states, one of which represented an optimal steady-state tree structure, one a young even-aged forest, and the third a mature stand
	Spruce	Southern Finland (Juupajoki) / Mesic forest	RF is more profitable than CCF[f] when using the interest rate of 3%. The cutting cycle ranged from 10–30 years, and the basal area after felling was 4–16 m²/ha	The study was not based on optimisation, as it identified alternative cutting cycles and intensities, and compared results with RF
Mature stand	Spruce	Southern Finland/Mesic forest	CCF is more profitable[g], provided that the cutting cycle ranges from 10 to 20 years, and the cutting intensity is 4–16 m³/ha (area after cutting), when using the calculated interest rate of 3%	Stand-level optimisation was not applied
	Spruce	Southern and Central Finland/ fresh herb-rich forest heath	When the interest rate is 3% or higher, CCF is more profitable than clearcutting and artificial regeneration, i.e., RF[h]	The financial examination involved three different initial states, one of which represents an optimal steady-state tree structure, one a young even-aged forest, and the third an even-aged mature stand

The initial state categorized to bare land, young forest and mature stand
[a] Tahvonen and Rämö (2016), Tables 5 and 6
[b] Juutinen et al. (2018a), Fig. 3a (note: forest-level optimisation not used in the study)
[c] Parkatti and Tahvonen (2020), Figs. 4c, 4 g, 4i and Table 5
[d] Parkatti et al. (2019), Table 4
[e] Tahvonen et al. (2010), Fig. 7
[f] Juutinen et al. (2018a), Fig. 3b
[g] Juutinen et al. (2018a), Fig. 4
[h] Tahvonen et al. (2010), Fig. 8

optimisation underestimated the growth of the stands when they were managed through selective harvesting.

Research on the economics of CCF is also limited in Norway. In a series of field trials on 12 mature fields in southeast Norway it was found that selection harvest increased both timber price and harvesting cost slightly compared to RF. The difference in gross margin ranged from −2% to 9% (Glommen Skogeierforening and Mjøsen Skogeierforening 2005). The reported increase in harvesting costs was about one-third of the figure reported from Sweden by Jonsson (2015). Økseter and Myrbakken (2005) calculated the NPV of CCF and RF for the same trials. Their main result is that profitability on average is roughly the same for both regimes and is independent of the interest rate (2.5–4%). There is a large spread between the individual trials. At a 2.5% discount rate, CCF yielded a NVP from 90 to 120% of that for RF, varying according to the field characteristics and assumptions about growth response to selective harvest.

Granhus et al. (2022) studied 19 Norway spruce fields, comparing CCF with RF. The stands cover a large range of different forest conditions and CCF harvest types, ranging from shelterwood cutting to multi-layer CCF. Thinning quotient ranged from 0.77 to 1.69. The results from the economic analysis revealed that, for about half of the stands, CCF resulted in a higher NPV than RF. Since the stands were not selected systematically, it is hard to generalise the results.

A matrix model for economic evaluation of forestry including CCF under Norwegian conditions—the T model (Gobakken et al. 2008)—was developed more than a decade ago. The underlying growth model (Bollandsås et al. 2008) has been used in Finland by, e.g., Tahvonen and Rämö (2016), but is hardly used in Norwegian studies. Gobakken et al. (2008) estimated the optimal choice of management system (CCF vs. RF) in terms of NPV for a set of regenerated spruce stands. These had the same basal area (20 m²/ha), but different diameter distributions. CCF yielded the largest NPV for the stands with a clear inverse J-shaped tree diameter distribution. For uniformly and normally distributed stands, RF with natural regeneration was most profitable.

8.5 Conclusions and Further Measures

Based on recent studies, CCF is more profitable than RF in spruce forests and spruce-dominated mixed forests on mineral soils when the interest rate is at least 3%, and the site type is not too fertile (e.g., Parkatti et al. 2019; Parkatti and Tahvonen 2020; Tahvonen 2022). The initial state (bare land, young forest, or forest mature for regeneration felling) is a less significant factor than the stand structure in the initial situation (size class distribution). However, estimates of financial profitability involve various uncertainties (Juutinen et al. 2020), and existing growth models need to be constantly improved to better describe tree growth dynamics (Kuuluvainen et al. 2012). Processes needing better models include natural regeneration (e.g., Hokkanen 2001; Eerikäinen et al. 2014; Saksa and Nerg 2008), the growth of large trees, higher than 1.3 m (e.g., Lundqvist 2017; Hynynen et al. 2019; Bianchi et al. 2020), harvesting costs (e.g. Granhus and Fjeld 2001; Surakka and Sirén 2007; Laitila and Repola 2023), and damage risks (e.g., Piri and Valkonen 2013; Hanewinkel et al. 2014; Pukkala et al. 2016; Nevalainen 2017; Nevalainen and Piri 2020). To date, these processes have not been taken into account in detail in financial analyses. This also applies to how stumpage prices develop, including the prices of different assortments relative to one another.

Financial studies of CCF have focused on mineral soils and only one study has considered Finnish peatland forests (Juutinen et al. 2021). The financial performance of CCF should also be studied further from the perspective of climate change mitigation. New studies will also be required to identify the financially reasonable growing stock in CCF on peatlands, while taking into account the ability of trees to maintain a favourable water level for growth through evaporation without the need for remedial ditching (Juutinen et al. 2021; Shanin et al. 2021).

Carbon sequestration should also be considered with regard to CCF. According to preliminary research results, if landowners were paid for the carbon sequestered in trees, CCF would be even more profitable than RF, but this depends on the carbon price applied (Parkatti et al. 2023). A higher carbon-emission trading price seems to extend the cutting cycle in CCF and increase the average tree capital (Assmuth et al. 2018, 2021; Parkatti and Tahvonen 2021). This also seems to apply to RF. In pine forests where RF is applied, a carbon price of EUR 20–60/tCO_2 extends the cycle by 10–40 years and increases carbon stocks per hectare by 27–80% (Niinimäki et al. 2013; Pihlainen et al. 2014).

Some studies (e.g., Assmuth et al. 2018; Rørstad 2022) suggest that the impact of carbon payments is more significant at higher interest rates. When low interest rates are used, the tree capital is usually higher, the cutting cycle longer, and carbon sequestration higher compared to when high interest rates are applied. A high interest rate favours early revenues, considering the carbon payments relative to cutting revenues. However, the results partly depend on what proportion of a forest's carbon stocks is included in calculations, and what carbon sequestration payment scheme is assumed (Juutinen et al. 2018b).

In the light of recent studies, CCF is an optimal solution in northern Lapland when the interest rate is 3% or higher (Parkatti and Tahvonen 2021). If the price of carbon is between €EUR 60-€EUR 100 tCO_2^{-1}, harvesting any old pine forests in the region could never be profitable. If an old-growth forest is, however, managed according to RF the carbon choke price would be €EUR 29 (interest rate 1%) or €EUR 14 tCO_2^{-1} (interest rate 3%), when the benefits of multiple uses (reindeer husbandry) are considered.

In conclusion, the reliability of estimates of tree development has a significant impact on the results of profitability comparisons between CCF and RF. The development dynamics of even-aged forests are fairly well known, and their growth and regeneration models are based on extensive long-term empirical studies on mineral soils and peatlands. However, the impact of CCF on growth and profitability is not yet well known, and reliable tree growth and regeneration models are currently unavailable. Regeneration, initial tree development and the revival of suppressed trees are key to assessing the long-term growth and yield impact and cutting possibilities of CCF, but little research data is yet available about harvesting costs and damage. No harvesting cost models based on empirical data have been published, and the costs presented are often based on models derived from even-aged forests.

Take Home Messages

When deciding whether to apply CCF or RF the structure and volume of the initial stand define the elements for success. As a rule of a thumb: on mineral soils low interest rates (<3%) favour RF, while CCF outperforms RF when high interest rates are applied. Then including carbon sequestration might change the financial ranking between RF and CCF, but the outcome depends greatly on the carbon price applied and how carbon is included in the models. Also benefits of multiple uses (e.g., reindeer husbandry) need to be taken into account which might alter results, too. To date, few results are available for peatlands—more studies are needed to enable financial comparison between CCF and RF.

References

Adams DM, Ek AR (1974) Optimizing the management of uneven-aged forest stands. Can J For Res 4(3):274–287

Amacher GS, Ollikainen M, Koskela E (2009) Economics of Forest resources. The MIT Press, Cambridge

Assmuth A, Rämö J, Tahvonen O (2018) Economics of size-structured forestry with carbon storage. Can J For Res 48(1):11–22

Assmuth A, Rämö J, Tahvonen O (2021) Optimal carbon storage in mixed-species size-structured forests. Environ Res Econ 79(2):249–275

Assmuth A, Tahvonen O (2018) Optimal carbon storage in even- and uneven-aged forestry. Forest Policy Econ 87:93–100

Bianchi S, Huuskonen S, Siipilehto J, Hynynen J (2020) Differences in tree growth of Norway spruce under rotation forestry and continuous cover forestry. Forest Ecol Manag 458:117689

Bollandsås OM, Buongiorno J, Gobaken T (2008) Predicting the growth of stands of trees of mixed species and size: a matrix model for Norway. Scand J Forest Res 23:167–178

Cao T (2010) Silvicultural decisions based on simulation-optimization systems. Dissertation, University of Helsinki

Eerikäinen K, Valkonen S, Saksa T (2014) Ingrowth, survival and height growth of small trees in uneven-aged Picea abies stands in southern Finland. Forest Ecosystems 1(5):10

Faustmann M (1849) Berechnung des Werthes, welchen Waldboden, sowie noch nicht haubare Holzbestände für die Waldwirthschaft besitzen. [Calculation of the value which forest land and immature stands possess for forestry]. Allgemeine Forst- und Jagd-Zeitung 25:441–455

Glommen Skogeierforening & Mjøsen Skogeierforening (2005) Prosjekt KONTUS—Sluttrapport (Project KONTUS—final report) Report: Glommen Skogeierforening, Mjøsen Skogeierforening (Glommen forest owner association, Mjøsen forest owner association)

Gobakken T, Lexerød NL, Eid T (2008) A forest simulator for bioeconomic analyses based on models for individual trees. Scand J Forest Res 23(3):250–265

Granath J, Söderström M (2022) Hyggesfritt skogsbruk - Ekonomisk inverkan på skogsbruket (Non-clearcut forestry—economic effects on the forest sector/forest industry). Bachelor thesis, Department of Forest Economics, Swedish University of Agricultural Sciences

Granhus A, Fjeld D (2001) Spatial distribution of injuries to Norway spruce advance growth after selection harvesting. Can J For Res 31(11):1903–1913

Granhus A, Ødegård E, Bergseng E, Bergsaker E (2022) Lukkede hogster - produksjon, foryngelse og økonomi (Closed cover forestry - production, regeneration and economics) Rapport 7/148/2021: Norsk institutt for bioøkonomi. (Norwegian Institute of Bioeconomy Research)

Haight RG (1985) A comparison of dynamic and static economic models of uneven-aged stand management. For Sci 31(4):957–974

Haight RG, Getz WM (1987) Fixed and equilibrium endpoint problems in uneven-aged stand management. For Sci 33(4):908–931

Hanewinkel M, Kuhn T, Bugmann H, Lanz A, Brang P (2014) Vulnerability of uneven-aged forest to storm damage. Forestry 87(4):525–534

Hokkanen T (2001) Siemenet ja siemensato. In: Valkonen S, Ruuska J, Kolström T, Kubin E, Saarinen M (eds) Onnistunut metsänuudistaminen (Succesful forest regeneration) Metsäntutkimuslaitos. Kustannusosakeyhtiö Metsälehti, pp 69–79

Hynynen J, Eerikäinen K, Mäkinen H, Valkonen S (2019) Growth response to cuttings in Norway spruce stands under even-aged and uneven-aged management. Forest Ecol Manag 437:314–323

Intriligator MD (2002) Mathematical optimization and economic theory. Classics in Applied Mathematics 39, Philadelphia SIAM

Jonsson R (2015) Prestation och kostnader i blädning med skördare och skotare (Performance and costs in selective harvesting with harvester and forwarder) Arbetsrapport från Skogforsk nr. 863–2015: Skogforsk (Working paper from the Forestry Research Institute of Sweden)

Juutinen A, Ahtikoski A, Lehtonen M, Mäkipää R, Ollikainen M (2018b) The impact of a short-term carbon payment scheme on forest management. Forest Policy Econ 90:115–127

Juutinen A, Ahtikoski A, Mäkipää R, Shanin V (2018a) Effect of harvest interval and intensity on the profitability of uneven-aged management of Norway spruce stands. Forestry 91:589–602

Juutinen A, Ahtikoski A, Rämö J (2020) Puuntuotannon kannattavuuteen vaikuttavat tekijät jatku-vapeitteisessä metsänkasvatuksessa (Factors affecting profitability associated with continuous cover forestry) Metsätieteen Aikakauskirja, Katsaus. 10313

Juutinen A, Shanin V, Ahtikoski A, Rämö J, Mäkipää R, Laiho R, Sarkkola S, Lauren A, Penttilä T, Hökkä H, Saarinen M (2021) Profitability of continuous cover forestry in Norway spruce-dominated peatland forest and the role of water table. Can J For Res 51(6):859–870

Kuuluvainen T, Tahvonen O, Aakala T (2012) Even-aged and uneven-aged forest management in Boreal Fennoscandia: a review. Ambio 41(7):720–737

Laitila J, Repola J (2023) Korjuukustannukset Lapin poimintahakkuukohteissa (Harvesting costs in selection cuttings in the Finnish Lapland). Luonnonvara- ja biotalouden tutkimus 45/2023. Luonnonvarakeskus, p 58

Lundqvist L (2017) Tamm review: selection system reduces long-term volume growth in Fennoscandic uneven-aged Norway spruce forests. Forest Ecol Manag 391:362–375

Malo P, Tahvonen O, Suominen A, Back P, Viitasaari L (2021) Reinforcement learning in optimizing forest management. Can J For Res 51:1393–1409

Nevalainen S (2017) Comparison of damage risks in even- and uneven-aged forestry in Finland. Silva Fenn 51(3):1741

Nevalainen S, Piri T (2020) Metsätuhoriskit tasa- ja eri-ikäismetsätaloudessa. Metsätieteen Aikakauskirja; Tieteen Tori, artikkeli id 10310

Niinimäki S, Tahvonen O, Mäkelä A (2012) Applying a process-based model in Norway spruce management. Forest Ecol Manag 265:102–115

Niinimäki S, Tahvonen O, Mäkelä A, Linkosalo T (2013) On the economics of Norway spruce stands and carbon storage. Can J For Res 43:637–648

Nordström EM, Holmström H, Öhman K (2013) Evaluating continuous cover forestry based on the forest owner's objectives by combining scenario analysis and multiple criteria decision analysis. Silva Fenn 47(4)

Økseter P, Myrbakken S (2005) Økonomi og planlegging ved forvaltningsprinsippet KONTUS sammenlignet med flatehogst (Economics and planning of the management system KONTUS compared to clearcut forestry). Rapport nr. 13–2005: Høgskolen i Hedmark (Hedmark regional College)

Parkatti VP (2021) On the economics of continuous and rotation forestry. Academic Dissertation, Dissertationes Forestales 312. University of Helsinki. https://doi.org/10.14214/df.312

Parkatti VP, Assmuth A, Rämö J, Tahvonen O (2019) Economics of boreal conifer species in continuous cover and rotation forestry. Forest Policy Econ 100:55–67

Parkatti V-P, Tahvonen O (2020) Optimizing continuous cover and rotation forestry in mixed-species boreal forests. Can J For Res 50:1138–1151

Parkatti V-P, Tahvonen O (2021) Economics of multifunctional forestry in the Sámi people homeland region. J Environ Econ Manag 110:102542

Parkatti V-P, Tahvonen O, Viskari T, Liski J (2023) Including soil alters the optimization of forestry with carbon sinks. Can J For Res:1–14. https://doi.org/10.1139/cjfr-2022-0226

Pihlainen S, Tahvonen O, Niinimäki S (2014) The economics of timber and bioenergy production and carbon storage in Scots pine stands. Can J For Res 44(9):1091–1102, https://doi.org/10.1139/cjfr-2013-0475

Piri T, Valkonen S (2013) Incidence and spread of Heterobasidion root rot in uneven-aged Norway spruce stand. Can J For Res 43(9):872–877

Pukkala T (2009) Population-based methods in the optimization of stand management. Silva Fenn 43(2):261–274

Pukkala T, Lähde E, Laiho O (2010) Optimizing the structure and management of uneven-sized stands in Finland. Forestry 82:129–142

Pukkala T, Laiho O, Lähde E (2016) Continuous cover forestry reduces wind damage. Forest Ecol Manag 372:120–127

Pyy J (2021) Forest management optimization according to nonlinear partial differential equation (PDE) and gradient-based optimization algorithm. Dissertation, University of Oulu

Rämö J, Tahvonen O (2014) Economics of harvesting uneven-aged forest stands in Fennoscandia. Scand J Forest Res 29:777–792

Rämö J, Tahvonen O (2015) Economics of harvesting boreal uneven-aged mixed-species forests. Can J For Res 45:1102–1112

Rämö J, Tahvonen O (2017) Optimizing the harvest timing in continuous cover forestry. Environ Resour Econ 67:853–868

Rørstad PK (2022) Payment for CO2 sequestration affects the Faustmann rotation period in Norway more than albedo payment does. Ecol Econ 199:107492. https://doi.org/10.1016/j.ecolecon.2022.107492

Saksa T, Nerg P (2008) Kuusen istutus, luontainen uudistaminen ja näiden yhdistelmät kuusen uudistamisessa. Metsätieteen aikakauskirja 4(2008):255–267

Shanin V, Juutinen A, Ahtikoski A, Frolov P, Chertov O, Rämö J, Lehtonen A, Laiho R, Mäkiranta P, Nieminen M, Lauren A, Sarkkola S, Penttilä T, Tupek B, Mäkipää R (2021) Simulation modelling of greenhouse gas balance in continuous-cover forestry of Norway spruce stands on nutrient-rich drained peatlands. Forest Ecol Manag 496:119479

Sonesson J (2017) Virkesproduktion och ekonomi för olika exempel på hyggesfri skogsskötsel (timber production and economics of different cases of non-clearcut forestry). In: Hannerz M, Nordin A, Saksa T (eds) Hyggesfritt skogsbruk. Erfarenheter från Sverige och Finland Nono-clearcut forestry—Experiences from Sweden and Finland. Future Forests Rapportserie 2017:1. Sveriges lantbruksuniversitet, Umeå. Swedish University of Agricultural Sciences, Umeå

Surakka H, Sirén M (2007) Poimintahakkuiden puunkorjuun nykytietämys ja tutkimustarpeet (Current knowledge on selection cuttings and research needs). Metsätieteen aikakauskirja 4(2007):373–390

Tahvonen O (2009) Optimal choice between even- and uneven-aged forestry. Nat Resour Model 22:289–321

Tahvonen O (2011) Optimal structure and development of uneven-aged Norway spruce forests. Canadian J Forest Research 41:2389–2402

Tahvonen O (2015a) Economics of naturally regenerating, heterogeneous forests. J Assoc Environ Resour Econ 2(2):309–337

Tahvonen O (2015b) Economics of rotation and thinning revisited: the optimality of clearcuts versus continuous cover forestry. Forest Policy Econ 62:88–94

Tahvonen O (2022) Metsien hoito jatkuvapeitteisinä: katsaus taloudelliseen tutkimukseen. Suomen Luontopaneelin julkaisuja 1C/2022. (In Finnish). [Managing forests according to continuous cover forestry, a review of forest economics literature]

Tahvonen O, Pukkala T, Laiho O, Lähde E, Niinimäki S (2010) Optimal management of uneven-aged Norway spruce stands. Forest Ecol Manag 260:106–115

Tahvonen O, Rämö J (2016) Optimality of continuous cover vs. clearcut regimes in managing forest resources. Can J For Res 46:1–11

Tahvonen O, Suominen A, Malo P, Viitasaari L, Parkatti V-P (2022) Optimizing high-dimensional stochastic forestry via reinforcement learning. J Econ Dyn Control 145:104553. https://doi.org/10.1016/j.jedc.2022.104553

Tahvonen O, Viitala EJ (2006) Does Faustmann rotation apply to fully regulated forests? For Sci 52:23–30

Valsta L (1992) An optimization model for Norway spruce management based on individual-tree growth models. Acta Forestalia Fennica 232. The Society of Forestry in Finland, Helsinki

Wikström P (2000) A solution method for uneven-aged management applied to Norway spruce. For Sci 46(3):452–463

Wikström P (2008) Jämförelse av ekonomi och produktion mellan trakthyggesbruk och blädning i skiktad granskog (Comparison of economics and production between rotation forestry and closed cover forestry in multistory spruce forests). Skogsstyrelsen (The Swedish Forest Agency). Rapport, 24

Chapter 9
Wood Properties and Quality

Riikka Piispanen, Jiri Pyörälä, and Sauli Valkonen

Abstract

- Trees in continuous cover forestry (CCF) typically form very narrow rings at young ages in suppressed positions but can grow very quickly at older ages in dominant positions, maintaining long-lived crowns.
- CCF trees have slightly higher mean wood density in stems than rotation forestry (RF) trees.
- CCF trees have better fiber properties for pulp than RF trees.
- CCF trees have a relatively short stem section of small dead knots in sawn timber.
- CCF trees have a relatively long section of large green knots in sawn timber.
- Sawn goods produced from spruce logs yielded with the selection system do not differ markedly from those from RF.
- The application of a shelterwood system with overstorey retention for Scots pine facilitates the production of very high-grade timber.

Keywords Wood quality · Wood density · Fiber dimensions · Timber quality · Juvenile wood

9.1 Introduction

Wood properties and quality in uneven-aged stands are key focus areas in the debate that has taken place in the Nordic region since the 1980s about the pros and cons of continuous cover forestry (CCF). Expectations regarding the production of potentially high-quality sawn timber have been one of the arguments in favour of

R. Piispanen · S. Valkonen (✉)
Natural Resources Institute Finland (Luke), Helsinki, Finland
e-mail: sauli.valkonen@luke.fi

J. Pyörälä
Department of Forest Sciences, University of Helsinki, Helsinki, Finland
e-mail: jiri.pyorala@helsinki.fi

© The Author(s) 2025   149
P. Rautio et al. (eds.), *Continuous Cover Forestry in Boreal Nordic Countries*,
Managing Forest Ecosystems 45, https://doi.org/10.1007/978-3-031-70484-0_9

CCF, as most of the wood harvested from CCF stands is sawlogs. However, in addition to sawlogs and sawn goods, attention should be paid to properties that impact pulp making. A large portion of the harvested volume is used for making pulp from culled logs that are unsuitable for producing lumber, and from chips produced from the outside of the logs during lumber manufacturing (slabs, chips, and sawdust).

Scientific knowledge about the effects of CCF on wood properties and quality in both sawn timber and pulp is limited, and arguments have been based on general knowledge regarding wood quality in even-aged forests. Very few studies have directly focused on wood properties in trees grown under any of the CCF regimes in the Nordic countries. The global situation is not much better. However, studies conducted in stands at various stages of transformation from rotation forestry (RF) to CCF are more common and very relevant (e.g., Pape 1999; Seeling 2001; Macdonald et al. 2010, see review by Pretzsch and Rais 2016).

There is one rather unique Norwegian study on wood quality in Norway spruce, based on 35 mature (d > 28 cm) sample trees in seven experimental selection stands. The study focused on the ring profile and density at different heights along the stems, mechanical, technical, and physical wood properties measured on small immaculate samples, and timber strength of standard structural sizes (Eikenes et al. 1995). An earlier Norwegian review focused on the basic quality traits, and factors influencing them, of wood from the selection and shelterwood methods (Vadla 1992).

Fagerberg et al. (2023) produced a simulation framework for modeling knot size for uneven-sized Norway spruce stands with the competition-dependent approach in Sweden. At this stage, the data has only originated from one stand, but provides some important perspective to the topic.

Most comprehensive information on wood properties and quality in CCF trees comes from a series of intensive studies conducted in Finland in stands that are part of the ERIKA experimental set on single-tree selection in Norway spruce stands in southern Finland (Pyörälä 2013; Piispanen et al. 2014; Kumpu et al. 2020; Piispanen et al. 2020; Pyörälä et al. 2022. In these experimental stands, a single-tree selection system was adopted in the 1980s for the maintenance or establishment of typical complex stand structures to resemble the classic "Plenterwald" system. Selection harvests were carried out 3–4 times before the data was collected in 2007–2011 from nine stands at three sites (Lapinjärvi, Vesijako, Suonenjoki). All stands at that time could be characterized as truly multi-aged (with tree ages up to 170 years) and most were also full-storied (in the sense of Ahlström and Lundqvist 2015). In connection with a single-tree selection harvest, a total of 156 sample trees were acquired on the plots. The trees were felled and measured on site and in the lab for a number of wood quality variables at fiber, annual ring, stem disc, log, stem, and branch levels.

Based on these studies, this chapter largely concentrates on the following wood properties and quality traits in Norway spruce grown under the selection system, with reference to the influences of stand structure and tree dominance:

- Growth rates and patterns (ring profiles and branching patterns)
- Wood density
- Fiber dimensions

- Quality of sawn timber (twist and other distortions, knottiness, other quality traits, and strength and quality grading results)

At the end of the chapter, we briefly consider timber quality in two-storied Scots pine stands originating from shelterwood cuttings with subsequent partial overstorey retention. Some basic considerations but few research results are available in the Nordic area for Scots pine (review by Vadla 1992), while much more work has been done in Germany on scientific and practical aspects. These are briefly reviewed.

9.2 Selection Silviculture with Norway Spruce

9.2.1 Tree Growth Patterns

Silvicultural practices are used to control the environmental factors that influence tree growth and wood formation and properties (Wimmer and Downes 2003). Based on this general knowledge, wood properties can be expected to differ between CCF and RF, as the typical growth rhythms of individual trees are distinctively different.

RF is characterized by uniform stands with a single dominant species, and with homogeneous spacing, stem diameter, height, and canopy structure. After clearcutting and reforestation, the emerging seedlings and young trees have ample resources and growing space at their disposal and, for a long period, they grow virtually free of competition from their neighboring trees. Consequently, their initial growth is very rapid. Later, the canopy closes, and competition gradually intensifies, suppressing tree diameter growth. Thinning will temporarily boost the diameter growth of the remaining trees. Growth is consistently relatively rapid, but with a slow decline from the initial maximum.

In contrast, the single-tree selection system in CCF is associated with a great degree of structural heterogeneity within and between stands. Stands are made up of trees of multiple ages and sizes, mixed at small spatial scales, resulting in complex competitive interactions between the trees. Trees typically experience consecutive suppression and release phases, especially at early ages (Schütz 2001; Eerikäinen et al. 2014). Seedlings emerge from natural regeneration among the matrix of trees of all sizes. They are subjected to intensive competition, and their height and diameter growth is slow. The duration of this undergrowth stage varies considerably within and between stands. Many undergrowth trees must endure such unfavorable conditions for several decades. Others develop more rapidly while occupying growing spaces with lower stand density or in the presence of shade-intolerant tree species with a lower shading intensity. Seedlings and saplings that survive this stage will gradually be able to accelerate their growth as larger trees around them are gradually removed in repeated harvesting (Fig. 9.1). Growth will further accelerate when the tree has finally achieved a better competitive position in the intermediate and, especially, in the codominant and dominant layers. Diameter growth of the

Fig. 9.1 Slow growth at the undergrowth stage is shown as very thin annual rings near the pith at the tree base. Photo from the ERIKA experiment. Photo: Riikka Piispanen

large trees in a selection stand can be faster than that of their counterparts in an even-aged stand.

These differences have been documented in literature, for example based on the ring samples collected from the 96 Norway spruces in the ERIKA experiment (Pyörälä 2013; Piispanen et al. 2014, 2020) and 35 Norway spruces in the Norwegian study (Eikenes et al. 1995). These materials often contained very narrow rings near the pith covering a variable number of years up to several decades, depending on the length of the suppression period (Fig. 9.2). Trees that had obtained dominant canopy position by the time of sampling showed a rather stable ring width, with a modest initial increase and a gentle decrease in the later stages (Fig. 9.2). For the intermediate trees, ring width was smaller but increased throughout, while the suppressed trees displayed an increasing ring width at the later stages only. It should be noted that, with constant radial growth, tree basal area growth is amplified by default, as the annual growth ring (of equal width) becomes longer as it is added to a greater girth.

The typical growth pattern is most pronounced at the base of the tree where the rings have been formed during the lowest canopy position of the tree, i.e., in suppressed position. Further up the stems of large trees, the zone of narrow rings fades away as the canopy position improves, and the growth ring pattern becomes increasingly like that in trees from RF, where ring width is greatest near the pith and decreases steadily towards the bark (Eikenes et al. 1995; Pyörälä 2013; Piispanen et al. 2020). The results confirm that the characteristic slow initial growth and higher age in selection stands will not impede tree growth when the tree has risen into a

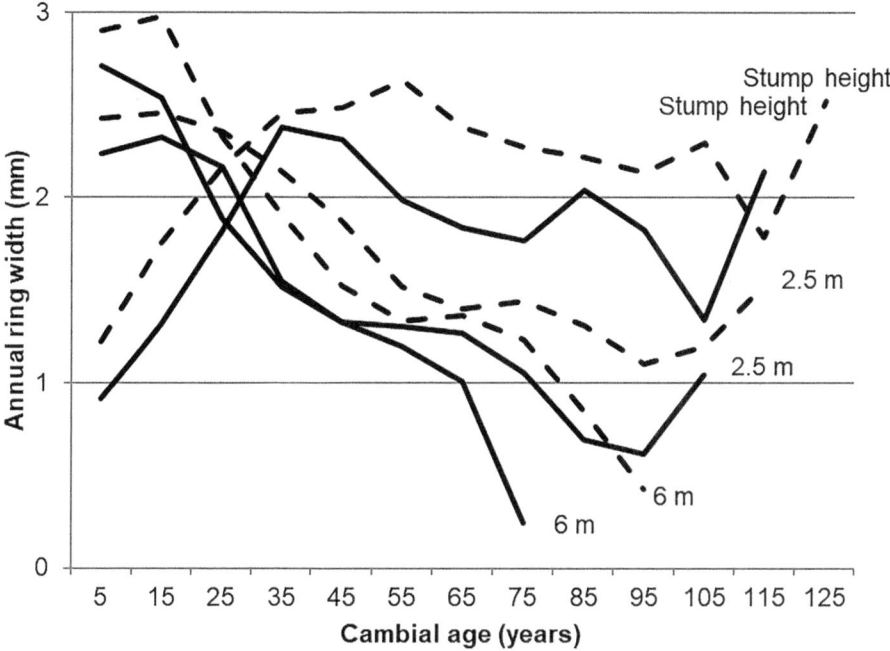

Fig. 9.2 Average annual ring width in Norway-spruce stands managed with single-tree selection by cambial age and by two tree size classes (solid line = 20–29.9 cm, dashed line = equal to or greater than 30 cm) at different heights along the stem (stump height, 2.5 m, and 6 m). Values are averages for the diameter classes of 10 rings, shown at class midpoints. Data from Piispanen et al. (2020)

dominant canopy position. The oldest trees in the ERIKA data were 170 years old, with about half of the period representing slow growth. However, they had grown quite normally during their last decades and had eventually become large (d = 40–65 cm).

An important consequence of the tree growth pattern in selection stands is that the trees tend to have many branches in their base section, but these remain thin and are shed rapidly (Fig. 9.3), appearing as small knots in the wood near the pith (Piispanen et al. 2020). Another feature arising from the differences in growth patterns and inter-tree competition is that large trees in selection stands have longer crowns (of living branches) than those in even-aged stands (Kumpu et al. 2020; Hasenauer and Monserud 1996). They have consistently more growing space and fewer strong competitors in their vicinity than in RF, where stands tend to become much denser towards the end of the rotation. Under intense competition pressure and mutual shading, branches in the lower canopy die off, and length of the live crown contracts. The branches remain attached even when dead, and are visible as loose knots in the wood. Consequently, the zone of live knots is longer in selection stands, but the zone of dead knots is shorter, and the wood has smaller knots than in RF. However, the inner sections near the pith at the tree base tend to have more small dead knots. In contrast, Fagerberg et al. (2023) concluded from simulations

Fig. 9.3 Trees in selection stands tend to have many thin branches at the undergrowth stage, which will remain as small knots in the wood near the pith in the base section of the tree. Photo Luke/ Erkki Oksanen

with a novel competition-dependent model that the impact from local competition on knot size in Norway spruce trees was rather limited. The study was conducted in one uneven-aged stand treated with selective cuttings in Sweden, and this conclusion must be considered rather tentative.

In addition to the obvious differences in the ring profiles and branching patterns, many differences in wood properties and quality between CCF and RF regimes arise from the respective stand dynamics. Shifting canopy positions can be reflected in the wood properties of trees grown in selection stands. One of the most fundamental factors that contributes to these differences is the smaller share of juvenile wood in the mature selection trees (Eikenes et al. 1995; Piispanen et al. 2014). Juvenile wood (or corewood) is associated with poor quality traits like lower density, thinner cell walls, lower length and width, greater variation in fiber dimensions, greater fiber angle in the S2 layer of the secondary cell wall, lower latewood proportion, and lower strength (Vadla 1992). Juvenile wood generally encompasses the first 5–20 annual rings in Norway spruce, as defined by the degree of microfibril angle (Vadla 1992). In the ERIKA experiment, the juvenile wood region at the base of the stem extended 20 rings in the uneven-aged stands, and 10 rings in the even-aged comparison stands (Piispanen et al. 2014). These juvenile rings, very narrow in selection stands, extended 2–3 cm from the pith and constituted a minuscule proportion of the stem volume in the mature trees, when compared to planted trees in even-aged stands (e.g., Downes et al. 2002; Pyörälä et al. 2022), where initial growth is intentionally maximized. Greater ring width can also amplify the poor properties of the juvenile wood (Piispanen et al. 2014; Pyörälä et al. 2022).

9.2.2   Wood Density

Wood density is one of the most important and most frequently studied wood properties due to its effect on timber strength, pulp yield, and the biomass accumulation and carbon storage of the wood. Wood density is closely related to growth rates, and in selection stands, the tree growth patterns significantly affect wood density ring by ring. During the initial period of suppressed growth at the stem base, the first, very narrow annual rings tend to have very high wood density, with a subsequent decrease towards the bark beyond around 20 rings (Piispanen et al. 2020; Piispanen et al. 2014; see Fig. 9.4). Density then increases and finally levels off in the outermost part. A large earlywood content in the mature outerwood of the CCF spruces was also observed, but this was not significantly associated with low overall wood density, due to the relatively high latewood density in the old rings (Piispanen et al. 2014).

In RF, wood density is lowest close to the pith, reflecting rapid juvenile growth, and increases evenly towards the bark. Density profiles in the upper parts of the stems from selection stands resemble those from RF (Eikenes et al. 1995; Kumpu et al. 2020). Consequently, at the whole-tree level, wood density in selection stands is slightly higher than in RF stands (Kumpu et al. 2020).

In the ERIKA studies, the highest wood densities in the narrowest rings near the pith were over 600 kg/m^3, and around 450 kg/m^3 in the mature wood near the bark (Piispanen et al. 2014). In the Norwegian study on mature CCF spruces near the timberline (Eikenes et al. 1995), basic density at stump height declined from an

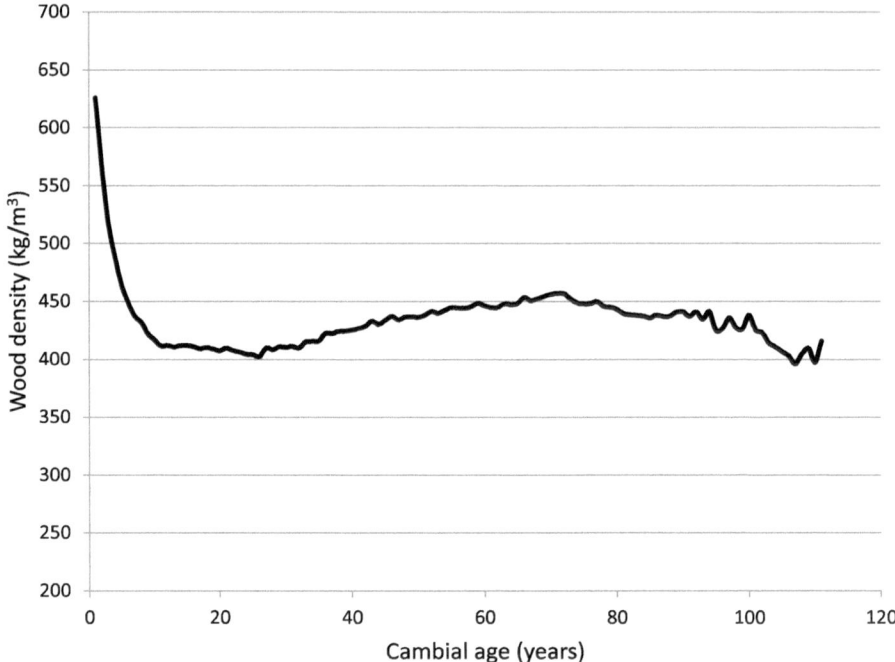

Fig. 9.4 Average wood density in Norway spruce trees in ERIKA selection stands at a height of 0.6 m from tree base, calculated with the model presented in Piispanen et al. (2014), applied to dominant trees (d/D$_{dom}$ = 1), and with annual ring widths (ir) representing the average value by distance from the pith

initial average of 450 kg/m³ near the pith to some 350 kg/m³ near the bark. Further up the stem (20–80% tree height) the variations in density from pith to bark were very small.

9.2.3 Fiber Properties

Wood fibers particularly affect the quality of pulp, but also indirectly the strength and stiffness of sawn timber. A considerable proportion of the merchantable timber ends up as pulp, as the surface slabs of the sawlog stems and sawdust are directed to pulping.

The fiber properties in Norway spruce change because of tree growth and aging of the cambium. Both the fiber length and diameter tend to increase in response to the increasing mechanical stress and the hydraulic pressure required to move water to the crown. The analyses of the ERIKA data indicated that different fiber properties have differing responses to tree age and tree size. Pyörälä et al. (2022) found that, with respect to the cambial age, the Norway spruce fibers in the selection

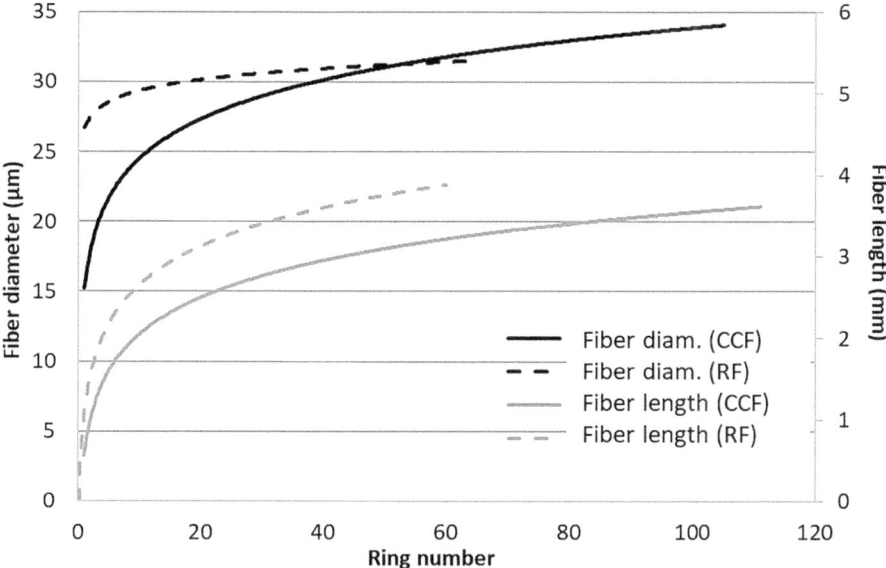

Fig. 9.5 Smoothed trends of fiber dimensions with respect to the age of the cambia in the ERIKA experimental data (CCF), and comparison data from RF materials. Data from Pyörälä et al. (2022)

stands lengthened more rapidly close to the pith than in RF stands, while the diameters of the fibers were more dependent on tree size (Fig. 9.5). As a result, Norway spruce grown under the selection system yielded a greater percentage of large fibers, which are associated with better pulp quality. This, too, was due to their specific growth patterns that lead to dominant mature spruces. These maintain relatively large and long crowns and require large-diameter earlywood fibers at stem base to sustain adequate water uptake, and long, thick-walled latewood fibers that contribute to the mechanical bearing capacity of the stem.

9.2.4 Traits and Quality of Sawn Goods

Expectations are high regarding the possibility of producing high-quality sawn timber in CCF. However, the quality of sawn goods is a result of several different factors. The complexity of the growth patterns in CCF can be expected to induce larger variation in the quality of sawn timber, both between and within individual trees.

In the study based on the ERIKA experiments with single-tree selection, 96 Norway spruces were processed into sawn timber at commercial sawmills, measured for quality traits, and graded according to the Nordic standard (Nordic Timber 1994) to grades ranging from A1 (flawless) down to A2–A4, B, C and D (reject) (Fig 9.6). The factors considered in the grading were the number, diameter and characteristics of knots, deformations (bow, cup, spring, and twist), the number and size of

Fig. 9.6 Demonstration of the quality grading in the ERIKA study. Photo Erkki Oksanen/Luke

pitch pockets and bark pockets, grain angle or distortion, proportion of compression wood, and the presence of resin wood, wetwood, firm rot or blue stain.

In the CCF trees, the most common reasons for quality downgrade in planks and boards of the butt logs were the large number of dead knots, compression wood, loose knots, and large diameter of the knots, which are typical for spruce wood irrespective of source. In the top logs, green knot diameter, wood twist and resin pockets caused the most problems, resulting in poorer quality grades (Pyörälä 2013; Piispanen et al. 2020).

Sawn timber from the selection stands was, on average, of similar quality to the lumber from the RF stands used for comparison, but the variation in quality was greater in the selection system; see Fig. 9.7 (Piispanen et al. 2020). The number of loose knots and the severity of twist often resulted in increased proportions of grade C (36%) compared with grade B (25%). The number of high-quality boards (grade A) was high in the selection trees (20%). Factors that caused boards to be downgraded to the lowest D category (19%) were often the proportion of compression wood and decay, and twist in the first boards from the pith (Piispanen et al. 2020).

In the butt logs of the selection trees, annual rings were very narrow close to the pith, but this was not correlated with deformations of boards (twist, bow, curved edge, and bottom) as previously suspected (Piispanen et al. 2020). On the contrary, the smaller amount of juvenile wood tended to reduce the amount of twist in boards close to the pith. In butt logs, twist was more evenly distributed, when the first boards were compared with the second and third boards from the pith. In even-aged spruces, twist is usually most severe in boards close to the pith. In uneven-aged trees, in the outermost boards, the presence of a high number of pin knots typically increased twist (Fig. 9.8).

Compression wood has also been expected to be a problem in single-tree selection, because the tree canopy positions change repeatedly due to recurring selection

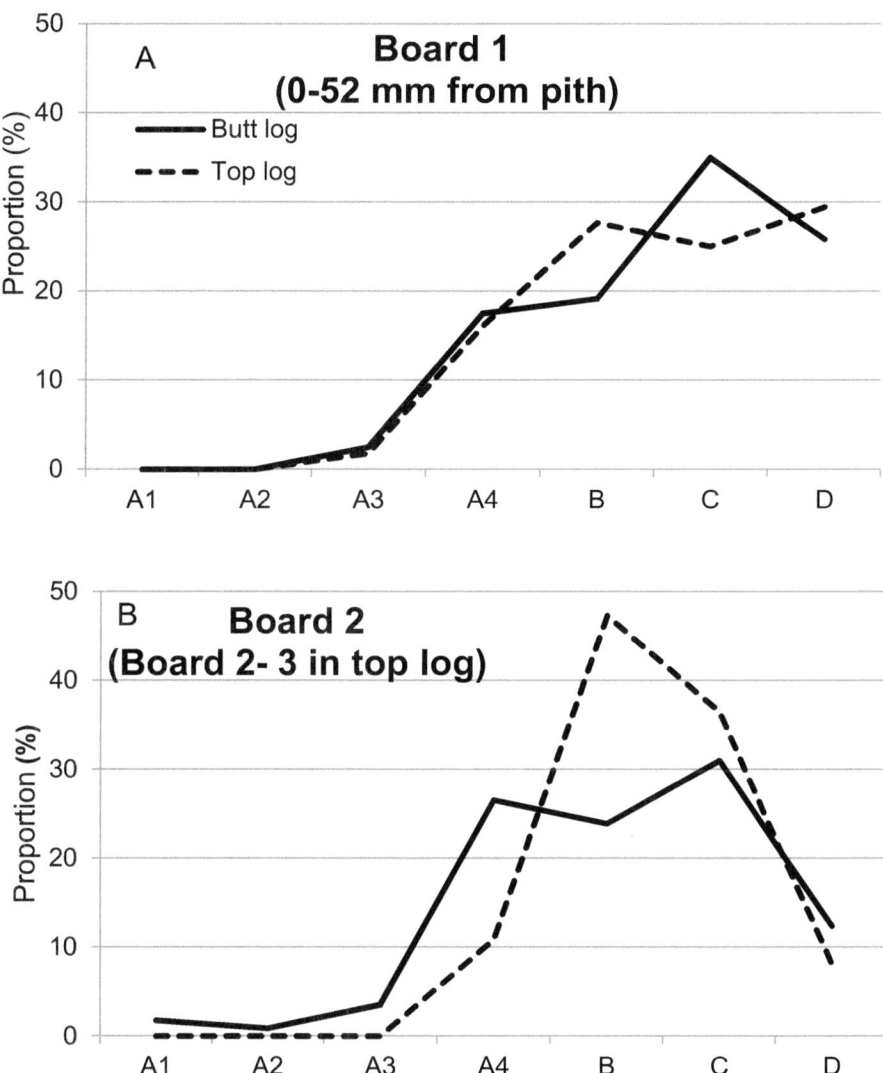

Fig. 9.7 Distribution of grades of the sawn boards (Nordic Timber 1994) obtained from the butt and top logs according to board location from the pith outwards—first board (0–52 mm from pith) to fifth board. Data from Piispanen et al. (2020)

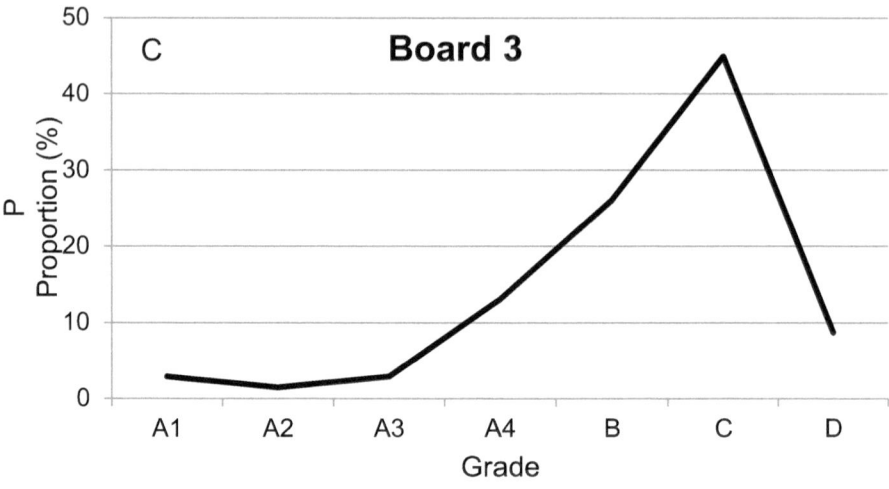

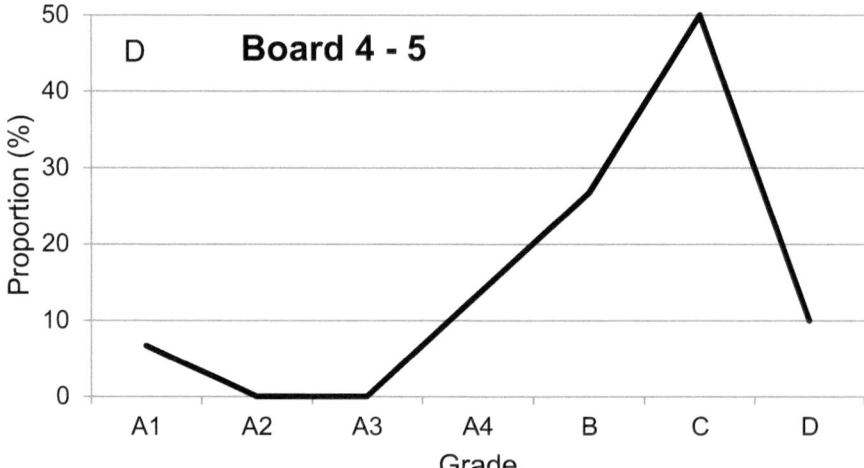

Fig. 9.7 (continued)

cuttings. In the edge trees of tree groups, asymmetrical crowns may develop, and the tensions can result in non-circular or slightly inclined tree stems, after the "supporting" trees are removed. However, the average proportion of compression wood in the ERIKA data was no larger than in southern Finnish lumber in general (Piispanen et al. 2020). The compression wood factor had no significant correlation to twist found in the ERIKA data.

Eikenes et al. (1995) also studied the strength of structural timber in Norway spruce selection stands using stress grading. They reported a mean bending strength of 42 MPa, which was higher than in their even-aged comparison materials. Twenty

Fig. 9.8 Severe twist in the board in the center among boards with little or no twist. Photo Erkki Oksanen/Luke

percent of the boards were visually strength-graded to the highest class T30, and 39% to the second class T24, according to the Norwegian standard NS 3470. Their findings showed that wood with very good strength properties can be produced in selection forests near the treeline in Norway.

9.3 Scots Pine Stands—From Shelterwood to Overstorey Retention

Uneven-aged structures are difficult to create and maintain with the shade-intolerant species of Scots pine, as already demonstrated by the Dauerwald selection system in northern Germany (Wiedemann 1925; Heinsdorf 1994; Helliwell 1997). Instead, a semi-continuous system based on natural regeneration with shelterwood (or seed-tree) cutting with subsequent overstorey retention seems to provide a more viable method for pine (see Chaps. 2 and 3). Part of its attractiveness in the contemporary CCF context could be the possibility to produce high-quality wood connected to a degree of continuity in large trees, with assumed benefits to biodiversity, amenity, and other ecosystem services. The long silvicultural experience with contemporary aspirations to increase structural complexity in pine stands (e.g., Der Wald in Sachsen 2005; Huth et al. 2022) in northern Germany is tangential and useful, together with basic research in Fennoscandia (see review by Vadla 1992 and Valkonen 2000; Valkonen et al. 2002).

Unfortunately, direct research on wood quality in such systems in the Nordic area is virtually lacking, apart from one that is clearly a pilot study (Kulmala 2016).

It seems safe to hypothesize that these kinds of management regimes can facilitate the production of high quality wood and wood products, especially compared to intensively managed pine stands in RF. Serious problems with abundant thick branches and a high proportion of juvenile wood associated with rapid initial tree growth have resulted from the large-scale planting of Scots pine on fertile sites with wide spacing patterns (Varmola 1996; Huuskonen 2008; Huuskonen et al. 2014).

The three principal elements linking the overstorey retention system with wood properties and quality are:

1. Wood quality in overstorey trees
2. Overstorey influence on wood quality in the understorey
3. Management for better wood quality in the understorey

When the best seed bearers or shelterwood trees of the old generation are retained for several decades on top of their previous rotation of 70–100 years, the additional growth that they accumulate tends to develop high-quality traits, producing wood quality in the highest categories. If high external quality is emphasized in the initial selection of trees to be retained, the stems are straight with few defects and free of branches up to a considerable height (Niemistö et al. 1993). The wood added with each annual ring will be knot-free up to a considerable height, which increases with time. The tree ring width tends to be moderate and rather constant, which is generally associated with good wood properties.

If the early development of the trees has been characterized by close spacing and the presence of overstorey trees, suppressing branch and stem diameter growth, wood quality can also be very good in the inner parts (towards the pith), especially in the lower stem parts (butt logs). On the other hand, the crown length in trees retained in relatively solitary positions does not contract, unlike trees in a closed canopy. In southern Finland, seed trees retained for 10–15 years had an average crown ratio of 0.42–0.51, depending on site fertility (Sarvas 1949). Comparable figures would be 0.30–0.33 in old unthinned natural stands and about 0.4 in mature managed stands (Hynynen 1995; Hynynen and Siipilehto 1996). The remaining branches become thicker, and the live crown zone will be associated with poorer quality of wood for sawn goods.

The results of the pilot study in Evo, southern Finland, supported these findings (Kulmala 2016). The study was conducted in one pine stand with an overstorey up to 180 years old and a managed understorey of 40–50 years of age. The overstorey consisted of large trees (diameter = 35–65 cm, dominant height 30 m) with a very high stem density (110 stems/ha). Gradual thinnings had been carried out during the stand's life cycle. The overstorey probably represented dimensions and traits that could be expected at the final stages of overstorey management corresponding to two normal rotations, but its density was far greater than what could be expected (e.g., 10 stems/ha, Valkonen 2020). The understorey had been gradually thinned to its current density of 2000 stems/ha. It had thrived surprisingly well underneath such a dense overstorey, with the largest individuals already achieving diameters of 20–25 cm (max 28 cm). The over- and understoreys were sampled separately, applying stratification by diameter. The main results were obtained from sample logs taken at three heights (butt log 0–4 m, intermediate log 8–12 m, top log downwards from live crown) in the overstorey and at one or two heights in the understorey. The logs were sawn with a regular sawing pattern, and the sawn goods graded according to the Nordic standard (Nordic Timber 1994; see Sect. 9.2.4 for details of the system) and measured for strength according to the CSN EN 338 standard, based on the modulus of elasticity with an acoustic meter.

The results indicated that the sawn goods yielded from the overstorey pines were of exceptionally good quality. No less than 80% of boards sawn from the outer parts of the butt logs were assigned to the best A grade, with 50% in the A1 grade (not a single defect detected on a board). They originated mostly from the knot-free zone with consistent growth and no reaction wood or other traits that could have caused distortions in sawn goods. Their density was high for pine, 500–600 kg/m^3. Planks sawn from the inner parts of butt logs were also high quality, with 70% in the A grade and 30% in A1. The boards were given very high grades in the EN 338 strength classification, with 80% in class 35 or above. Quality decreased upwards on the stem as the knot-free proportion decreased, but the grades were still very good for the intermediate logs.

Intense competition within the understorey and by the overstorey is a key factor influencing wood properties and quality from understorey trees in two-storey pine stands. It tends to reduce the diameter growth in both stems and branches, enhancing wood quality in terms of greater density and a smaller number and diameter of knots (Voegeli 1961; Ackzell and Lindgren 1992; Vadla 1992; Varmola 1996; Agestam et al. 1998). The proportion of poor-quality juvenile wood, which encompasses up to 10–20 first annual rings, becomes smaller with slower growth. The seed-tree and shelterwood methods facilitate the emergence and retention of relatively high stocking levels, with an additional contribution made by the overstorey trees. One of the very few studies addressing this concept in Scots pine stands was conducted by Valkonen et al. (2002) in sapling stage stands (dominant height 1–7 m) with overstoreys of large Scots pines (diameter 30–40 cm) retained for 8–18 years with 30–120 stems/ha. The overall branching (sum of cross-sectional areas of branches in the whorl with the thickest branch of the tree) was clearly suppressed in the vicinity of the overstorey trees (Fig. 9.9). However, there was less impact on the maximum branch diameter, considered a key trait (Varmola 1996). Both benefits were considered much smaller than what may be achieved by a successful site-species match and sufficient stocking level.

The high quality of sawn goods from the inner regions of the overstorey stems and from the understorey stems, and the very thin branches with a high degree of self-pruning in the current understorey trees, were further indications of the quality benefits of this management system.

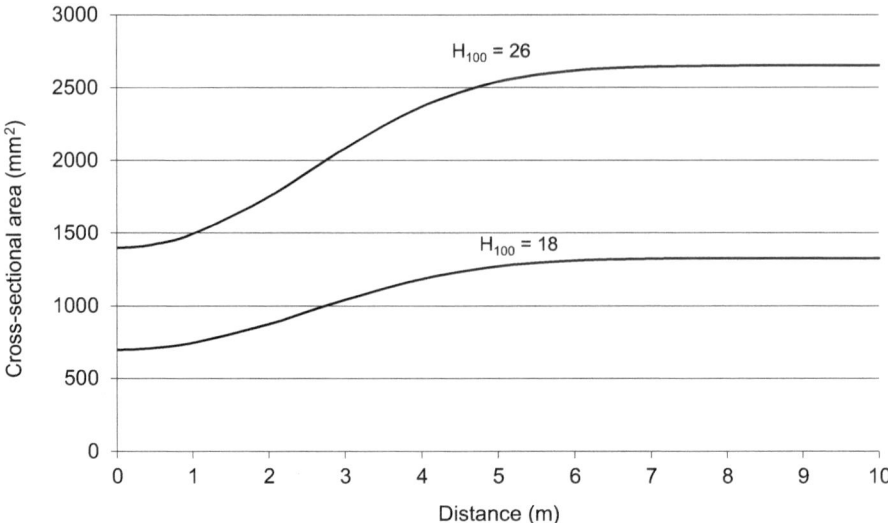

Fig. 9.9 Effect of overstorey trees on Scots pine saplings (height 5 m). Sum of branch cross-sectional area of three main whorls by distance from an overstorey pine tree with a diameter of 40 cm on a fertile (H_{100} = 26 m) and an infertile (H_{100} = 18 m) site. Data from Valkonen et al. (2002)

9.4 Knowledge Gaps

Few studies have considered the wood quality in CCF in Nordic countries. Some key wood properties, such as microfibril angle, remain undescribed in the literature, the geographical representativeness of the materials is still limited, and the length of the experiments still rather short compared to the lifespan of the trees. Few studies directly address either the shelterwood system with other species than Scots pine, or gap cutting for either Norway spruce or Scots pine.

   If such studies are generally sparse, ones that would allow indisputable comparisons of the wood quality between CCF and RF are non-existent. Such comparisons are often problematic due to differences in sampling, measurement methods, and growth conditions. For example, Eikenes et al. (1995) stated that the selection stands used in their study were located near the timberline of spruce at high altitudes in the far north and had very low site indices (11–17 m), making comparisons with the main productive spruce forests with higher site indices rather difficult.

   In the ERIKA studies on single-tree selection, five Swedish and two Finnish middle-aged RF stands were used as comparative data (see Piispanen et al. 2014). The RF material represented sawn timber from operational harvests, where poor-quality logs are usually removed by logistics already in the forest or at least before sawing. In contrast, logs from the ERIKA sample trees were not preselected before sawing, and the boards were graded with no consideration of external characteristics of the log that would have resulted in an advance rejection in an industrial process. Unsurprisingly, direct data-based evaluation was possible for only a very few features, and results from previous studies or literature were used on case-by-case basis.

Despite these challenges, the presented research provides valuable insights about wood properties and quality in forests managed with the CCF regimes, and has revealed the key differences to those managed with RF. This allows logical deductions of some potential outcomes.

9.5 Conclusions

Based on current knowledge, the key determinant of wood properties and quality in spruce selection stands is the extremely slow juvenile growth occurring during the early suppression phase. Accompanied by the relatively rapid growth rates of the mature trees in dominant positions, the proportion of juvenile wood remains relatively small in the stems, which improves the mean density and other traits in the merchantable stems.

The quality of sawn timber in the Norway spruces in the selection system did not differ noticeably from that in RF. However, Eikenes et al. (1995) found very good strength properties in Norway spruce selection forests near the timberline in Norway. The shelterwood method facilitates the production of Scots pine timber with very high quality grades.

The fiber properties in the selection stands were slightly better suited to pulp production than those in RF. The large crowns in dominant trees resulted in long and wide fibers in mature wood, i.e., in the surfaces of sawlogs, much of which ends up in pulping as side-streams of sawing.

Besides the superiority or inferiority of the wood for any particular purpose, a more important observation arising from the wood quality studies in CCF is how the rate and timing of the wood formation processes differ under different regimes. The research results offer further confirmation of the findings that Norway spruce can recover and rejuvenate after decades of suppression, and that the old trees in selection stands can maintain relatively large crowns in the abundant canopy space, boosting the rates of stem diameter growth to levels greater than those in trees with similar ages in the closed-canopy RF. The high growth rates in the old spruces in the selection stands were associated with relatively high ring densities and large tracheids. Similarly, the very old Scots pines in the overstoreys of the shelterwood system (up to 180 years old) retained good volume growth with their large live crowns, and produced high-grade timber in their stem base.

The lifespan of an individual tree in CCF is generally longer than in RF. Slow growth occurs in the suppressed, juvenile trees, but does not necessarily lead to a poor financial outcome. Having assumed an uneven-aged structure, the reserve of regeneration and suppressed trees ensures a steady supply of fast-growing mature trees to the relatively sparse upper canopy when larger trees are repeatedly removed. However, the transition period of an even-aged forest into a balanced uneven-aged structure will cause financial losses, due to the delayed harvests in the upper layers and the establishment of the sapling reserve.

The slower juvenile growth and longer rotation of individual trees is currently not financially compensated for by the increase in wood density, or tracheid quality, due to the volume-focused wood markets that do not consider wood quality. In the future, possible premiums or compensation paid for wood quality or carbon-binding potential could make CCF more attractive on sites with potential for prolonging the lifespans of the dominant trees. The main risk, especially for Norway spruces in CCF, is the root rot caused by *Heterobasidion annosum*, which can develop into a major problem in spruce-dominated forests, causing big financial losses. The main means for mitigating the risks of infection are careful logging and the restriction of harvests to the cool season when there are no spores in the air, a topic discussed in more detail elsewhere in this book.

References

Ackzell L, Lindgren D (1992) Seed-tree stand: threat or protection for artificial regeneration? In: Hagner M (ed) Silvicultural alternatives. An Internordic workshop June 22–25, 1992, Umeå, Sweden. Swedish University of Agricultural Sciences, Department of Silviculture. Rapporter 35, pp 86–95

Agestam E, Ekö P, Johansson U (1998) Timber quality and volume growth in naturally regenerated and planted scots pine stands in S.W. Sweden. Studia For Suecica 204:17

Ahlström M, Lundqvist L (2015) Stand development during 16–57 years in partially harvested sub-alpine uneven-aged Norway spruce stands reconstructed from increment cores. For Ecol Manag 350:81–86

Der Wald in Sachsen (Forests in Sachcen) (2005) Landesforstpräzidium, Freistaat Sachsen. p 106. (In German). https://www.wald.sachsen.de/Der_Wald_in_Sachsen.pdf

Downes GM, Wimmer R, Evans R (2002) Understanding wood formation: gains to commercial forestry through tree-ring research. Dendrochronologia 20(1–2):37–51. https://doi.org/10.107 8/1125-7865-00006

Eerikäinen K, Valkonen S, Saksa T (2014) Ingrowth, survival and height growth of small trees in uneven-aged *Picea abies* stands in southern Finland. For Ecosyst 1(5):10. https://doi.org/1 0.1186/2197-5620-1-5

Eikenes B, Kucera B, Fjærtoft F, Storheim O, Vestøl G (1995) Virkeskvalitet i fleraldret skog. Summary: Wood quality from uneven-aged forest. Rapport XXV fra forskningsprogrammet «Skogøkologi og flersidig skogbruk». Rapport fra Skogforsk 24/95:30. (In Norwegian with English summary)

Fagerberg N, Seifert S, Seifert T, Lohmander P, Alissandrakis A, Magnusson B, Bergh J, Adamopoulos S, Bader M (2023) Prediction of knot size in uneven-sized Norway spruce stands in Sweden. For Ecol Manag 544:121206. https://doi.org/10.1016/j.foreco.2023.121206

Hasenauer H, Monserud RA (1996) A crown ratio model for Austrian forests. For Ecol Manag 84:49–60

Heinsdorf M (1994) Kiefernnatürverjüngung - ein historischer Abriss (natural regeneration with scots pine—a historical outline). Beiträge für Forstwirtschaft und Landschaftsökologie 28(2):62–65. (In German)

Helliwell R (1997) Dauerwald. Forestry 70(4):375–379

Huth F, Wehnert A, Wagner S (2022) Natural regeneration of scots pine requires the application of silvicultural treatments such as overstorey density regulation and soil preparation. Forests 13(6):817. https://doi.org/10.3390/f13060817

Huuskonen S (2008) Nuorten männiköiden kehitys - taimikonhoito ja ensiharvennus. Summary: the development of young scots pine stands – precommercial and first commercial thinning. Dissertationes Forestales 62:61. + 4 appendix papers

Huuskonen S, Hakala S, Mäkinen H, Hynynen J, Varmola M (2014) Factors influencing the branchiness of young scots pine trees. Forestry 87:257–265. https://doi.org/10.1093/forestry/cpt057

Hynynen J (1995) Predicting tree crown ratio for unthinned and thinned scots pine stands. Can J For Res 25(1):57–62

Hynynen J, Siipilehto J (1996) MELA-mallit kasvatusmetsien dynamiikan kuvaajana (prediction of stand dynamics in middle-aged forests with the MELA models). In: Hynynen J, Ojansuu R (eds) Puuston kehityksen ennustaminen - MELA ja vaihtoehtoja, Metsäntutkimuslaitoksen tiedonantoja, vol 612, pp 69–84. (In Finnish)

Kulmala A (2016) Erirakenteisen kasvatustavan vaikutus mäntysahatavaran laatuun (The quality of Scots pine timber in different-aged forest). Ammattikorkeakoulun opinnäytetyö. Hämeen ammattikorkeakolu (HAMK). p 40. (In Finnish with English abstract)

Kumpu A, Piispanen R, Berninger F, Saarinen J, Mäkelä A (2020) Biomass and structure of Norway spruce trees grown in uneven-aged stands in southern Finland. Scand J For Res 35:252–261

Macdonald E, Gardiner B, Mason W (2010) The effects of transformation of even-aged stands to continuous cover forestry on conifer log quality and wood properties in the UK. Forestry 83(1):1–15. https://doi.org/10.1093/forestry/cpp023

Niemistö P, Lappalainen E, Isomäki A (1993) Mäntysiemenpuuston kasvu ja taimikon kehitys pitkitetyn luontaisen uudistamisen aikana. Summary: growth of scots pine seed bearers and the development of seedlings during a protracted regeneration period. Folia Forestalia 826:26

Nordic Timber (1994) Grading rules for pine (Pinus sylvestris) and spruce (Picea abies) sawn timber. Commercial grading based on evaluation of the four sides of sawn timber. Suomen Sahateollisuusmiesten Yhdistys [The Association of Finnish Sawmillmen], Gummerus Kirjapaino Oy, Jyväskylä, Finland. p 80. ISBN 952-90-5750-4. (in Finnish)

Pape R (1999) Influence of thinning and tree diameter class on the development of basic density and annual ring width in Picea abies. Scand J For Res 14(1):27–37. https://doi.org/10.1080/02827589950152269

Piispanen R, Heinonen J, Valkonen S, Mäkinen H, Lunqvist SO, Saranpää P (2014) Wood density of Norway spruce in uneven-aged stands. Can J For Res 44:136–144

Piispanen R, Heikkinen J, Valkonen S (2020) Deformations of boards from uneven-aged Norway spruce stands. European Journal of Wood and Wood Products 78:533–544

Pretzsch H, Rais A (2016) Wood quality in complex forests versus even-aged monocultures: review and perspectives. Wood Sci Technol 50:845–880

Pyörälä J (2013) Eri-ikäiskasvatuksen vaikutus kuusen puuaineen laatuun. Abstract: wood quality of Norway spruce in uneven-aged forests. M.Sc. thesis. University of Helsinki. p 66. (In Finnish with English abstract). https://helda.helsinki.fi/handle/10138/40716

Pyörälä J, Piispanen R, Valkonen S, Lundqvist S (2022) Tracheid dimensions of Norway spruce in uneven-aged stands. Can J For Res 52:346–356

Sarvas R (1949) Siemenpuuhakkuu männikön uudistushakkuuna Etelä-Suomessa. Summary: seed-tree cutting as a regeneration method in scots pine forests of Southern Finland. Communications Instituti Forestalis Fenniae 37(6):43

Schütz JP (2001) Der Plenterwald und weitere Formen strukturierter und gemischter Wälder (Selection silviculture and other forms of forest management with structurally complex and mixed-species stands). Parey Buchverlag, Berlin. (In German) ISBN 3-8263-3347-0

Seeling U (2001) Transformation of plantation forests—expected wood properties of Norway spruce (Picea abies (L.) karst.) within the period of stand stabilisation. For Ecol Manage 151(1–3):95–210. https://doi.org/10.1016/S0378-1127(00)00708-8

Vadla K (1992) Virkeskvalitet i bledningsskog og skog etablert etter skjermstillingshogst - en litteraturstudie. Summary: Quality of wood with selection cutting and shelterwood cutting- a literature survey. Rapport II fra forskningsprogrammet «Skogøkologi og flersidig skogbruk». Rapport fra Skogforsk 10/92:21. (In Norwegian with English summary)

Valkonen S (2000) Effect of retained scots pine trees on regeneration, growth, form, and yield of forest stands. Resumen: Efecto de la retención de pies de Pino silvestre sobre la regeneración, crecimiento, forma y producción Investigación agraria, Sistemas y recursos forestales, Fuera de serie no. 1–2000: 121–145

Valkonen S (ed) (2020) Metsän jatkuvasta kasvatuksesta (Continuous-Cover Forestry). Metsäkustannus and Luonnonvarakeskus, Helsinki, p 127. (In Finnish)

Valkonen S, Ruuska J, Siipilehto J (2002) Effect of retained trees on the development of young scots pine stands in southern Finland. For Ecol Manag 166:227–243

Varmola M (1996) Nuorten viljelymänniköiden tuotos ja laatu. Abstract: yield and quality of young scots pine cultivations. Dr. Sc. Thesis, vol 585. Metsäntutkimuslaitoksen tiedonantoja, p 70. + 6 appendix papers

Voegeli H (1961) Die Schattenerziehung der Föhre (Shade tolerance of Scots pine). Schweiz Z Forstwes 112(5/6):350–363. (In German)

Wiedemann E (1925) Die praktischen Erfolge des Kieferndauerwaldes (the practical implications of the *Dauerwald* management system with scots pine). Verlag von Friedr. Vieweg & Sohn, Braunschweig, p 184

Wimmer R, Downes GM (2003) Temporal variation of the ring width—wood density relationship in Norway spruce grown under two levels of anthropogenic disturbance. IAWA J 24(1):53–61

Chapter 10
Multiple Use of Forests

Seija Tuulentie, Therese Bjärstig, Inger Hansen, Unni Lande, Paul McLean,
Jani Pellikka, Rainer Peltola, and Jasmine Zhang

Abstract

- Sustainable forest management approaches, regardless of whether they involve continuous cover forestry (CCF) or rotation forestry (RF), require a holistic landscape perspective that acknowledges the multiple interests, values, and uses that depend on the locally relevant economic, ecological, and socio-cultural circumstances. These must be considered alongside the use of forests and forest landscapes as a resource for rural development.
- Forests provide a wide range of goods and services. Those addressed here (i.e. tourism, recreation, health, grazing, non-timber forest products, and societal protection from natural hazards) are a subset of all of those potential services that are already considered to be of special significance for the Nordic region.
- Most recreational users consider variation in the forest landscape and long-distance views as visually attractive but think that clearcuttings and soil tilling are harmful.
- In general, CCF favours bilberries, while lingonberries and some mushrooms benefit from even-aged forestry.

S. Tuulentie (✉) · R. Peltola
Natural Resources Institute Finland (Luke), Rovaniemi, Finland
e-mail: seija.tuulentie@luke.fi; rainer.peltola@luke.fi

T. Bjärstig
Department of Political Science, Umeå University, Umeå, Sweden
e-mail: therese.bjarstig@umu.se

I. Hansen · U. Lande · P. McLean
Norwegian Institute of Bioeconomy Research (NIBIO), Ås, Norway
e-mail: inger.hansen@nibio.no; unni.lande@nibio.no; paul.mclean@nibio.no

J. Pellikka
Natural Resources Institute Finland (Luke), Helsinki, Finland
e-mail: jani.pellikka@luke.fi

J. Zhang
Swedish University of Agricultural Sciences (SLU), Uppsala, Sweden
e-mail: jasmine.zhang@slu.se

P. Rautio et al. (eds.), *Continuous Cover Forestry in Boreal Nordic Countries*,
Managing Forest Ecosystems 45, https://doi.org/10.1007/978-3-031-70484-0_10

169

- Owing to the many and varied demands relating to forests and forest landscapes in Norway, Sweden, and Finland, CCF-supported multiple-use strategies and planning will need to consider stakeholder requirements more, now and in the future, than is currently the case.

Keywords Recreation · Nature-based tourism · Grazing · Non-timber forest products · Natural hazards

10.1   Introduction

In an international context, "multiple use" in terms of land use is often defined as the provision of a combination of different goods and services. This involves two general strategies. Multiple use can be based on either "integrated" or "differentiated" land use, defined by the spatial distribution of land use including forest management (Hoogstra-Klein et al. 2017). Integrated land use means that different benefits are produced simultaneously in the same area, normally as a compromise between the required uses, while differentiated land use means individual services are specifically targeted in individual areas, with each service optimised within the respective areas. CCF generally comes under the integrated strategy, while traditional clearcutting typically forms part of a differentiated strategy. Rather than a sharp division based on the extremes, there is often a gradation between integrated and differentiated land use.

   In the Nordic countries, forests are used for more than one purpose. Multiple-use forest management aims at simultaneously producing several goods and services (Huuskonen et al. 2021) as well as values and functions (Solbär et al. 2019). In Norway, the term multiple use has been used since the 1970s. Today's multiple use is regulated by the Forestry Act of 1989 (Lov om skogbruk 2005). In Sweden, the national forest programme defines multiple use as: "…that the forest is used for several different purposes (e.g. forestry, nature tourism, reindeer husbandry, nature conservation, wood processing, hunting, enjoyment, non-profit community networks, recreation, public health, arena for cultural expressions). The purposes may be commercial or non-commercial." (Andersson et al. 2016, p. 13). In the Finnish Forest Strategy by 2035, multiple use is not emphasised as such, but the multi-purpose nature of forests is assumed to increase and the need to coordinate different forms of use will be emphasised in the future. It is also noted that forests are increasingly subject to usage restrictions and various conservation pressures (Kansallinen metsästrategia 2035 2022).

   Multiple use is not a new phenomenon. In both Sweden and Norway, multiple use of forests in the form of, e.g., timber extraction for household needs and forest grazing for domestic animals was dominant until the rise of modern industrial forestry in the 1950s (Fritzbøger and Søndergaard 1995, Kardell and Bishop 2014). In Finland, extraction of wood for household construction and fuelwood, hunting, and the use of land for silvopasture, slash-and-burn agriculture, and tar production were dominant forms of forest use before the emergence of modern forest industry (Tasanen 2015).

In line with the international trend toward sustainable forestry, initiatives in Sweden have been aimed at developing multifunctional forestry. The political goals have mainly focused on strengthening the goods and services that provide biological values of forests, while the ones seen as holding social values have not been prioritised in the same way (Beland-Lindahl et al. 2017). One explanation for this imbalance is that the social value of forests is a concept that lacks national consensus and a consistent definition (Sténs et al. 2016). Within the framework of the Swedish national forest programme, aspects such as employment, social policy, and rural development are now also included as important parts, which indicates a wider scope of what can be considered as services provided by forests. Multiple use is clearly a concept with a broad definition which many different actors can relate to, yet these actors can have quite different interpretations of what multiple use means (Hoogstra-Klein et al. 2017).

Multiple use of forests has been promoted for various reasons. In the Swedish forest policy (SOU 1992, p. 76), for instance: "the fact that almost all forest land in Sweden is used economically and the great importance of forestry for the social economy makes it necessary to combine different land use interests on the forest land, i.e., to apply multiple use." Differentiated land use, on the other hand, means that the forest landscape becomes a mosaic where different benefits are produced in different areas, i.e., that the forest is zoned so that forest production is carried out in certain stands while other stands are set aside for nature conservation or other functions; this is often referred to as "triad" forest management (Himes et al. 2022). A spatial differentiation of different uses is a form of multiple use, as well as spatial integration of two or more uses (Hoogstra-Klein et al. 2017).

Integrated and differentiated land use can be seen as two extremes along a continuum, whereas in reality there is a range of different variants of mixed use between the two extremes (Ekvall and Bostedt 2009, Klemperer 1996). This sheds light on implications for forest management methods, as some stands can be managed with a focus on intensive timber production, while others are managed (or left for free development) to provide several other benefits. Such a zonal approach is easier to implement in large areas with a single owner. It can be challenging or even impossible to implement across multiple owners, as collaborations and possible sharing of revenues are required. This puts Norway, Sweden, and Finland in slightly different positions in terms of differentiated land use. In Norway, 12% of the forest area is publicly owned i.e., state, municipal or church (Ministry of Agriculture and Food 2011, p. 18), whereas in Finland and Sweden the corresponding figures are 32% and 28%, respectively (Finnish Forest Association 2019, Swedish Forest Industries 2023). Integrated approaches, which can more easily be considered within CCF systems, could have an advantage. Another variant is where there is a dominant benefit, e.g., timber production, while other benefits are produced in addition when possible. This, however, means that emphasis must be placed on one goal beforehand, and can thereby be understood as multiple use but with restrictions. A further variant is to diversify the production of different benefits in time rather than in space.

An important issue for forest use across the Nordic countries is allemannsretten (Norway), allemansrätten (Sweden), and jokaisenoikeus (Finland). These terms literally translate to "everyone's rights" in English but are more accurately described as "the public right of access" (Norwegian Environment Agency 2023). The public right of access permits mostly unrestricted access to the outdoors (i.e., countryside, woodlands, coast, waterways, and mountains), irrespective of who owns the land. The exclusions are military zones, cultivated farmland and private gardens. This is an important privilege for the public throughout the Nordic countries, who value the importance of outdoor recreation.

The Nordic public right of access encompasses more than just unrestricted access; it also involves the gathering of specific forest resources, such as wild berries, mushrooms, and herbs. Hunting and fishing are not part of these rights, with slight variations in different Nordic countries. The access extends to almost all forest land, with certain restrictions in nature and cultural conservation areas. Even though the term "everyone's right" became common only in the early twentieth century, the concept's roots can be traced back to centuries-old customary rights. Behind these rights lay sparse settlement, the absence of large, private land ownership, and practices like a widespread hunting culture and slash-and-burn agriculture. The public right of access is regulated by formal written law only in Norway, while in Finland or Sweden it is based on customary law and integrated in some specific legislation (Tuulentie and Rantala 2013). It should also be noted that, in Sweden, different aspects of the public right of access are supported legally in the national public interests within Chaps. 3 and 4 of the Swedish Environmental Code (Svensson et al. 2020).

There have, however, been concerns regarding the Nordic public right of access and whether commercial activity on private land should be allowed without the landowner's permission, or indeed without renumerating the landowners due to gains from that activity. This particular aspect of commercial use based on the public right of access is perceived to be a barrier for development of multiple use forestry. Many non-timber uses could generally have a more prominent role in the owners' management choices if associated alternative and non-timber incomes could be provided to those owners, so that they are less reliant on timber sales. This has been an active debate in Sweden for the past two decades, where forest owners' property rights can sometimes be considered as compromised by other commercial uses of forests in Sweden, such as berry picking by international businesses (Sténs and Sandström 2013), outdoor recreation, and tourism operations (Sténs et al. 2016). Despite this, the Swedish authorities have not made any attempt to limit or regulate the public right of access, indicating both the importance of this concept to the national psyche and the significant challenge of differentiating between commercial and recreational activity on the ground. Focusing on the Finnish case of whether free berry-picking was already an existing tradition alongside the public right of access, La Mela (2014) argues that the practice of berry picking only became open to everyone with the new economic expectations in the late nineteenth century, where having berries as an open resource would sustain the poor as well as be commercialised for the markets. In a survey of forest actors regarding their perceptions of multiple-use forestry, the public right of access is seen as an important foundation for enabling forest owners to understand multiple forms of forest use and to both encourage and facilitate communications and agreements regarding land use (Zhang et al. 2022).

The concept of forest "ecosystem services" is more comprehensive than the concept of multiple use (Huuskonen et al. 2021). Forest ecosystems provide multiple services for many stakeholder groups, often with conflicting interests and thereby with limited capacity for sustainability in land use strategies and planning. Tradeoff models that evaluate both monetary and non-monetary measures have proven to be a useful way of reducing conflicts over ecosystem services (e.g. Pukkala 2011). Here, multiple use is understood as a part of ecosystem services that are related to cultural ecosystem services and non-timber forest use, and the relationship between these and timber production methods is considered.

The importance of forests as an environment for collecting non-timber forest products and for recreation and domestic tourism cannot be overemphasised. Furthermore, forests in Nordic countries are increasingly an operating environment for international tourism. The importance of protected areas and especially national parks as an attractive factor for tourism and outdoor recreation has grown, but non-protected forest areas are also important environments for various activities. National parks and other protected areas are often located far from large population centres, where adjacent forests have proven to be especially valuable recreation areas. In a similar way, since many outdoor activities take place in the vicinity of the home or place of leisure, forests close to populations centres are especially valuable. About two-thirds of domestic outdoor activities are directed to nearby recreation areas (Sievänen and Tyrväinen 2015).

From a philosophical perspective, Haila (2015) states that multiple use symbolises the forest environment as a public good, accessible to all citizens. Alongside the traditional multiple use interpretation, he also emphasises multisensory and spiritually deep forest relationships. Paying attention to multisensory and a deeper forest relationship has become increasingly important, both in research and in public debate. Recently, special attention has focused on the effects of forests on wellbeing and health (Simkin et al. 2021), and these effects are closely linked to the recreational and tourism use of forests.

In this chapter, we examine forest landscapes as areas for recreational and nature-based tourism use, as producers of berries, mushrooms and huntable game, as a basis for reindeer and sheep husbandry, and as bearers of aesthetic and cultural values. The issue of forestry in relation to avalanches, landslides and rockfalls is also addressed, with a particular focus on Norway. Where possible, we address the implication of CCF management vs. RF management. All these aspects must be considered when managing the forest resource on all forest land and on all forest impediments in the Nordic countries considered here.

10.2   Nature-Based Tourism and Recreation

For nature-based tourism, such as hiking, protected areas and especially national parks are popular destinations. In addition, many nature tourism activities take place in managed and non-protected forest areas. In a Swedish study, forests were recognised as a vital attribute for all kinds of nature-based tourism settings (Margaryan 2018). Landscape and accessibility are recognised as the most important forest factors for tourists and recreational users. Studies from Norway, Sweden and Finland show that a common preference among forests users was a forest stand with relatively large trees at an advanced stage of stand development, but also with the sense of accessibility and the provision of a view (Gundersen and Frivold 2008, Karjalainen and Store 2015, Silvennoinen 2017). Clearcuts have been seen as negative for tourists' "wilderness" experience, and preferably there should be minimal traces of forest harvesting, including stumps and residues in their preferred destinations (Edwards et al. 2012, Karjalainen and Store 2015). In addition, the condition of the understorey vegetation, in terms of greenness and uniformity of coverage, is highly valued (Karjalainen and Store 2015).

The value of forest nature for tourism and recreation can be increased by modifying forest management practices, for instance by making clearcuts smaller, which is a feature of all types of CCF and some kinds of RF. A variety of forest structures is also popular: the view that opens after a stretch of dense forest pleases people, and water bodies, wide landscapes and distant views are appreciated (Miina et al. 2020). A study has shown that Finnish people generally prefer pine and birch forests, with spruce forests as the least popular and, again, that relatively open forests are preferred for recreational use (Silvennoinen 2017).

In the far north, the landscape changes dramatically with the seasons, most markedly with the presence or absence of snow. The winter landscape is especially important for tourism in Finnish Lapland, with the main season for tourism in the north occurring between November to March (Lapland Tourism Strategy 2019). Efforts have been made to develop summer tourism for decades, but summer seems to be a less attractive season in most parts of Finnish Lapland. In a study of the preferences of international tourists, it was found that the differences relating to the seasons are greatest in forest regeneration areas, where snow covers the traces of forest harvesting (Tyrväinen et al. 2017). Tourists considered even-aged forests, advanced thinning stands, and mature stands to be suitable environments in both summer and winter (Tyrväinen et al. 2017). Therefore, clearcutting has no detrimental effect on winter tourism, but does impact summer tourism.

The nature of forestry operations along the routes and landscapes are particularly important. Ahtikoski et al. (2011) tried to calculate whether the increase in tourists on the routes between the major tourist centres of Levi and Ylläs in Finnish Lapland could compensate for the possible losses in timber revenue if harvesting operations along the routes were restricted. The result was that large-scale tourism with its multiplier effects compensated for those possible losses in timber revenue. Konu and Tyrväinen (2020) focused on key forest sites from a tourism point of view when studying the possibilities of assigning values to landscapes for recreation. They examined the outdoor trails and the views from scenic spots in mature forest stands which, according to the management recommendations, should be regenerated by clearcuts. The goal has been that not only professional tourism providers, but also self-guided tourists, cabin owners and local residents, would benefit from the preservation of diverse nature areas. According to Konu and Tyrväinen (2020), extending tourism income to local forest owners could also help reduce conflicts between the tourism and forest sectors.

A survey of Finnish Lapland's nature-based tourism companies (Kosenius et al. 2013) showed that these companies set stricter environmental and recreational requirements for the management of the state's productive forests than for the management of private forests. More than 90% of respondents felt that the routes and structures located in the state's productive forests promoted tourism, and that the needs of nature-based tourism companies should be better provided for by the operations of Metsähallitus, the Finnish administrator of state-owned land and water. About 65% of respondents felt that wider protection zones should be left along watercourses than at present, and clearcutting should be completely stopped in the state's productive forests. Among forestry measures, the avoidance of soil treatment and clearcutting was considered particularly important for nature-based tourism.

10.3 Non-timber Forest Products

10.3.1 Berry and Mushroom Picking

Non-timber forest products (NTFPs; e.g., wild berries, mushrooms, herbs, plants, craft material, game meat) are an important and widely used ecosystem service provided by forests. The direct monetary value of NTFPs compared to timber varies by country, but even in the most timber production-oriented boreal forests, it is difficult to overestimate their cultural value. For example, more than 3 million Finns—64% of the population—pick berries, mushrooms, and wild herbs every year (LVVI 2020). It is estimated that approximately 20–30% of Sweden's population (age 18–74) pick berries, and close to 40% pick mushrooms at some point during the year (Andersson 2018). Estimates for the Norwegian population are not available, but these are undoubtedly very popular activities. According Nordström et al. (2020), the majority of forest owners in Swedish rural areas state that they are completely or partially self-sufficient in terms of mushrooms, berries, and game meat. Even though wild berries for commercial purposes are primarily picked by foreign labour, wild berries are also a source of additional income for domestic pickers, especially in the direct sales market (Vaara 2015). However, domestic berry picking is mostly for recreational and household use in the Nordic countries. The most important wild berries are bilberry (*Vaccinium myrtillus*) and lingonberry (*Vaccinium vitis-idaea*).

Bilberries are found throughout the Nordic countries in grove-like and dry forests, heaths and fell forests. The clearcuts and soil tilling associated with RF damage the bilberry root system. Since bilberry does not tolerate the dry and hot environment often prevailing in clearcut areas, it does not usually return to the forest until after the first thinning in RF. In Finland, bilberry coverage decreased between the 1950s and 1990s from 18% to 8%, due to the transition to RF and the accompanying change in forest structure (Salemaa 2000a). However, since 1995, bilberry coverage appears to have increased (Mäkipää 2022).

Lingonberry can also be found all over the region, mostly in bright and dry forests, but coverage has also decreased since the 1950s. Again, clearcuts damage lingonberry plants, albeit less than bilberries, and lingonberries recover from damage faster (Salemaa 2000b). Lingonberry tolerates drought better than bilberry and the best fruit production is in areas where tree canopies do not shade the growth, so the most productive lingonberry growth is often in clearings or under seed trees and in seedling stands (Turtiainen et al. 2013) (Fig. 10.1).

There is little empirical research regarding the effects of CCF methods on berry yields. However, bilberry probably benefits from CCF systems where the forest canopy remains intact and there is no extensive soil tilling, and machines are restricted to a permanent trail network. In CCF, spruce-dominant forests for example are kept to a basal area of 10–22 m²/ha (Tapio 2019). According to coverage models, the coverage of bilberry increases up to basal areas of 26 m²/ha (Turtiainen et al. 2016). However, the optimal basal area for berry yield is 14 m²/ha, because in denser stands bilberry has poor yields despite good coverage (Miina et al. 2009). Based on the models, bilberry would seem to thrive well in CCF areas of spruce forests in both peat and mineral soils (Miina et al. 2020), and produce five times higher bilberry yields compared to RF. In contrast, the effect of silvicultural system had only a minor effect in pine forests (Pukkala et al. 2011).

Unlike bilberry, lingonberry does not thrive in shady habitats, so the best berry yields occur after regeneration and final fellings under RF. Eventually however, under RF, there will be a reduction in surface area coverage and productivity as the forest grows and less light is available (Turtiainen et al. 2013). Nonetheless,

Fig. 10.1 Lingonberry benefits from the availability of light in young stands at the seedling stage and may produce good yields in such forests (Photos: Erkki Oksanen / Luke)

according to models, RF is better than the CCF in terms of lingonberry yields (Peura et al. 2018).

At the end of the nineteenth century, mushrooms were still only occasionally eaten in most part of the Nordic region. However, attitudes changed during the twentieth century, and mushroom picking is now a popular outdoor activity and can even provide supplementary income (Svanberg and Lindh 2019). The most popular mushrooms are chanterelles (*Cantharellus tubaeformis, C. cibarius*), false morels (*Gyromitra esculenta*), black trumpets (*Craterellus cornucopioides*), lactarius mushrooms (*Lactarius trivialis, L. rufus, L. torminosus*), and pinewood king boletes (*Boletus pinophilus*, Turtiainen and Salo 2015, Svanberg and Løvaas 2023). In clearcuts, yields of all mushrooms except false morels drop significantly for several years. However, yields rebound in young forests (Miina et al. 2020) and, for example, the best *Boletus edulis* yields are obtained from 20–40-year-old spruce forests before the first thinning (Tahvanainen et al. 2016). Models applied in Nordic conditions have shown that, in terms of mushroom yields, RF is a better option than CCF (Peura et al. 2018). It is also noteworthy that the optimal stand management schedules for commercially marketed mushrooms are very similar to that of timber (Tahvanainen et al. 2018).

10.3.2 Forestry and Game Management

How the needs of wild animals, in particular moose (*Alces alces*, or the European elk), should be considered in forestry is a topic that has been actively discussed for over half a century. In Fennoscandia, moose hunting in forest environments plays an important role, both financially and in terms of the number of participants. Moose hunting is practiced by approximately 60,000 Norwegian hunters (Statistics Norway 2023), 250,000 Swedish hunters (Jagareforbundet.se), and some 110,000 Finnish hunters (Artell et al. 2020). It has been estimated that the total value of moose hunting for the 2019–2020 hunting season is NOK 1.1 billion (Pedersen et al. 2020), and the consumer surplus from moose hunting in Finland is EUR 260 million (Artell et al. 2020).

Moose eat many plant or tree species that are found in early successional stages following clearcuts, so moose have benefited from intensified forestry management since the beginning of the twentieth century (Lavsund et al. 2003, Bjørneraas et al. 2011). The relationship between forestry and the moose is ambiguous—on one hand, higher densities of moose can increase the economic value of forest property for hunting (Engelman et al. 2018); on the other, moose foraging in forest stands (mainly Scots pine used as winter habitats) can cause severe economic damage for the forest owner. For decades, non-industrial private forest owners have been advised on how to prevent damaging browsing of moose in forest estates and, particularly in Finland, how to manage forest patches to promote grouse (i.e., *Tetrao urogallus* or capercaillie, *Tetrao tetrix* or black grouse, *Tetrastes bonasia* or hazel grouse) habitats (Herrero et al. 2020). Management is increasingly emphasised in research literature as game-oriented, grouse-oriented, or grouse-friendly forest management (Haakana et al. 2020, Haara et al. 2021, Ikonen et al. 2023), and emphasises forestry practices where forest patches vary in tree age and species and have an unmodified canopy layer. This is therefore likely to benefit from CCF and mixed-species forest management.

However, the relationship between forestry practices and hunting not only comes down to the fact that certain types of game animals are most likely to be found in specific forest environments, but equally that some types of forestry practices maintain forests and canopy layers as a more pleasant place to walk. For example, a notable segment of hunting tourists in Finland appreciates forest landscapes in their hunting grounds that they regard as representing "wilderness", peaceful and not modified by intensive human land use (e.g. Pellikka et al. 2018). An example of the cultural value of wilderness areas and compatible forestry practices for hunters is the recurrence of a wilderness context in fictional hunting literature (in Finland: Varis 2003).

Hunters also evaluate and accept different forestry activities from the perspective of other roles. Some are trained foresters and are more positive to harvesting than the general public (e.g., Gundersen and Frivold 2008). Nearly half of Finnish hunters are non-industrial private forest owners (Toivonen 2009), and have multiple ownership goals, including ones related to income and economic security provided by forestry (Kouhia and Pellikka 2021). For these owners, providing a means to balance income with continual hunting opportunities would likely be a win-win situation, and CCF management might be a suitable prospect for them.

10.4   Grazing

In Norway, Sweden and Finland, forests are essential for reindeer husbandry and, to some extent, for livestock grazing, which makes such forest a contemporary cultural landscape (Nordström et al. 2020). This long-term cultural use of the forests is central to the landscape we see today, as the cultural use of the land has a special and unique historical basis. It has been used for multiple purposes for several thousand years. It is therefore of great importance in terms of natural and cultural values, in addition to the biological cultural heritage. This is often forgotten, but these cultural values should be considered in forest management practices.

10.4.1   Reindeer Grazing

Lichen-rich forests are a crucial resource in reindeer (*Rangifer tarandus*) herding systems. During the winter, reindeer primarily feed on lichens (Kivinen et al. 2010), although wintertime fodder feeding in fences is becoming more and more common. Forests rich in pendulous lichen are important grazing areas, especially during periods of reduced availability of ground pasture (including ground lichen) during snowy and/or icy conditions. Clearcuttings remove the lichen-bearing, old-growth trees. Forest management considering reindeer herding should take measures aiming at, e.g., promoting uneven-aged forest structure, retaining old-growth trees, avoiding leaving logging residue on the most terricolous lichen-rich sites, minimal soil preparation, and the use of natural regeneration (Turunen et al. 2020). Many of these are components of CCF management. However, CCF selection harvesting may have the same effect on the availability of lichen-bearing trees as clearcutting unless there is a separate a specific harvesting instruction to retain these trees.

10.4.2 Livestock Grazing

In Norway, nearly 2 million sheep graze on open hillsides and in forests for at least 3 months during the summer (Maurtvedt 1989). About 260,000 cattle graze on cultivated land or rangeland pastures for 5 weeks or more. The number of beef cattle is increasing at the expense of dairy cattle, hence the greater use of woodland pastures for cattle (ssb.no). In Finland and Sweden, livestock silvopasture is less important.

Sheep and cattle prefer the grasses of forest pastures (Wam et al. 2022). In coniferous forests, the decisive factor for good pasture is availability of light in the understorey layer and nitrogen in the soil. Therefore, clearcutting areas and the subsequent young forests are the most suitable grazing forests (Rekdal and Angeloff 2021), so CCF may not be the most optimal forest management method for livestock grazing. On the other hand, forest with greater cover offers shelter from weather and possible predators (Herfindal et al. 2017). More studies are needed to establish the optimal spatial structure mix of open and closed canopies for livestock. Some of the group felling methods within CCF could facilitate this.

It is well known that livestock grazing reduces regrowth of deciduous trees and herbaceous plants after clearcuttings (e.g., Belsky and Blumenthal 1997, Östlund et al. 1997). With respect to commercial forestry this is considered positive because it reduces competition for nutrients, water, and light in favour of the desired conifer seedlings (e.g., Zimmerman and Neuenschwander 1984, Prolux and Mazumder 1998). However, in CCF forests, if the density of livestock becomes too high, particularly of heavy cattle breeds, trampling and bedding, in addition to grazing, may lead to tree damage, erosion and soil packing (e.g., Fleischner 1994, Hester et al. 2000), potentially limiting the natural regeneration and tree recruitment desired in CCF systems. This was demonstrated in a Norwegian study (Hjeljord et al. 2014). Within fenced, clearcut areas with young spruce regeneration, the proportion of damage to the replanted trees was found to be positively related to cattle use in the area, but not so for sheep. On the most intensively used areas 80% of the young trees were damaged. Furthermore, the density of deciduous trees was five times lower inside the livestock fence than outside. It was concluded that livestock grazing may reduce resource competition in favour of spruce, whereas the current animal density clearly reduced forest regeneration in the study area. Intensive grazing is therefore likely to pose problems for CCF if the aim is to increase the variation in tree species and to rely predominantly on natural regeneration.

Interest is growing in using plant resources in the forest for grazing livestock (Larsson and Rekdal 2000). With higher animal densities, the risk of conflicts will increase, both within the livestock husbandry and between grazing interests and other uses of forests. Knowledge of the grazing value of different vegetation types provides a better basis for managing grazing resources within a multifaceted exploitation of forests. Larsson and Rekdal (2000) summarised the grazing value of eight forest vegetation types in south-eastern Norway; lichen forests had the poorest grazing value and tall herb forest the best.

CCF managed in a way that secures a variety of tree species, different stand ages and stand densities—including some small, felled spots—will provide vegetation types of good grazing value, but also a diverse habitat securing shelter and good resting areas for the grazing livestock. However, intensive grazing is likely to pose problems for CCF if the aim is to increase the variation in tree species and to rely predominantly on natural regeneration.

10.5   Health and Wellbeing

The role of forests and nature for health and wellbeing is becoming increasingly important (Nordström et al. 2020). The connection between forests and health is mainly made in relation to recreation and outdoor life, but some of the forest owners (mainly well-educated female forest owners) also highlight more innovative areas of use regarding social values, such as developing and adapting their forest to be a place of recovery and rehabilitation for people with stress and fatigue symptoms (Bjärstig and Sténs 2018, see also Nordh et al. 2009). These types of forests used for "forest therapy" purposes have clear potential to be enhanced by CCF management, both in improving accessibility on maintained roads and trails, and in aesthetic terms by creating a more varied and naturalised forest structure. Most of these forest owners also emphasise the role of the forest in allowing experiences of solitude and silence, and that visitors often have a spiritual and/or religious experience when in the forest (Bjärstig and Kvastegård 2016). An increasingly common example of the use for health and wellbeing is "forest bathing", a relaxation and/or therapeutic practice that engages all the five senses with the forest environment (Furuyashiki et al. 2019).

The demography of managing forests for health and wellbeing is important to consider. In the Nordic region it is primarily female forest owners who are interested in developing and adapting their management methods to make their forest a place for recovery and treatment for people suffering from depression and/or stress-related symptoms (Nordström et al. 2020). This result can be linked to, and is in line with, previous research on female forest entrepreneurs (e.g., Appelstrand 2015, Lidestav 1998, Lidestav 2010) as well as research on rehabilitation forests and the forest as a place for wellbeing and health (Sonntag-Öström et al. 2011, 2015).

According to the report by Nordström et al. (2020) the main question with this use concerns how these services can be commercialised so that additional jobs can be created, and how it contributes to sustainable rural development. Examples of guided activities include forest bathing, mindfulness spas in the forest, and health retreats with forest walks. Further non-commercial values relating to health are also prized by female forest owners, such as using herbal plants for alternative medical treatment, or a natural diet.

Studies are needed to explicitly examine the links between female forest owners and their attitudes and knowledge regarding CCF management methods. The above examples support the hypothesis that female forest owners tend to think and act "greener" (Umaerus et al. 2013, 2019), and therefore they might be more willing to choose CCF over RF, to focus on "softer" values of forest such as recreational, social, and environmental values, which are often interconnected (Hertog et al. 2022). However, for some uses relating to health and wellbeing or a "greener experience", any forest harvesting might be less desirable than having a completely natural forest, where management is restricted to creating access. In this case, techniques relating to conversion to CCF could still be highly beneficial in creating suitable areas from existing even-aged forests.

10.6   Societal Protection from Gravitational Hazards

Forests provide the most naturally aesthetic and cost-effective way to mitigate against gravitational hazards in and near steep terrain. The precise nature of gravitational hazards varies by location and by season, but in general those in which forest management has a significant role are snow avalanches, shallow landslides and rockfalls. In the context of these natural hazards, the definition of steep is terrain that is more than 25 degrees, which is the minimum angle at which movement of material (i.e. mass) down slopes will occur. The areas affected, or risk zones, include the slopes themselves and the "runout" or "deposit" zones (Fig. 10.2). The events can vary massively in terms of speed, scale, and consequences. In the worst cases, there can be catastrophic loss of life and massive damage to societal infrastructure, but serious consequences can equally be a degradation of land and natural environments. The protection of society and societal assets from gravitational hazards in mountainous and fjordic landscapes is therefore a significant non-timber forest product. This use of forests for protection has been relatively less publicised in the Nordic countries, compared to the European Alps for example. Consequently, it is not well understood and is probably taken for granted by the public. However, this use as protection is increasingly recognised by national governments and actors outside the forest industries. Legislation about using forests for protection is being considered in Norway (Nordrum et al. 2022), although many uncertainties remain about how this will be achieved in practice.

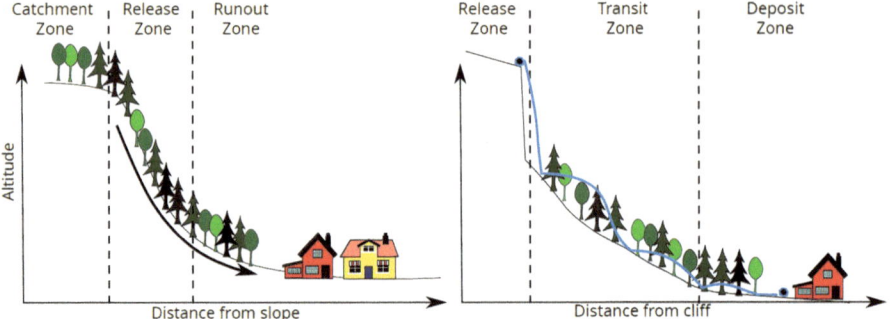

Fig. 10.2 Left – Forest structure in the release zone for snow avalanches and landslides affects the extent of the runout zone and, in many cases, the scale and velocity of the movement event. Sufficient forest cover can minimise the risk. For landslides forest management in the catchment area above a prone slope will also affect slope hydrology. **Right** (after Dorre et al. 2007): trees can catch large rocks and either prevent them travelling further downslope or substantially reduce their velocity, thereby reducing the risk to society. Removing all trees (e.g. through clearfelling) effectively removes most of the protection. Partial cutting to retain a sufficient forest structure to provide protection, and to create or maintain a stable forest over time, could bring many benefits to areas at risk

10.6.1 The Mechanisms of Protection and the Impacts of Forest Management

Forest management does not normally directly cause snow avalanches or rockfall, but it is one of the factors commonly associated with shallow landslides. This is because clearcutting forests contribute to a large and sudden change in slope hydrology (Rice 1977, Sidle 2005), and because of reduced mechanical stabilisation of the slope by tree roots (Rice 1977). Forest roads and wheel ruts resulting from mechanised harvesting can drastically alter drainage channels on a given slope (Fig. 10.3). It is important to consider that forests above slopes also contribute to the slope hydrology, even if they are seemingly not in steep terrain.

On slopes prone to avalanches, the protective function of trees is primarily achieved by intercepting snowfall in their branches, which later falls to the ground gradually. This prevents the formation and subsequent overloading of weak layers that are usually the cause of avalanches (Schneebeli and Bebi 2004). To a lesser extent trees also mechanically stabilise the slope (Teich et al. 2012, 2014). The forest can therefore be considered as providing a similar function to engineered structures such as avalanche fences and barriers. Given the prevalence of winter recreation in the Nordic countries, this protective function brings a significant societal benefit (Fig. 10.4). Trees provide a mechanical barrier during rockfalls, with the strength requirements of the trees depending on the sizes and weight of the rocks that are likely to fall (Dorre et al. 2007).

Fig. 10.3 **Left** excavator roads without drainage commonly used in Norwegian clearcuts on slopes contribute to water channelling and flow accumulation in new parts of the slopes. **Right** (Photo Paul McLean, NIBIO). Wheel tracks left by heavy forwarders used in clearcutting can channel water and accelerate erosion downslope. While these channels are frequently repaired by excavators following forest operations, such repairs primarily serve an aesthetic purpose as significant soil compaction has taken place. CCF does not promote temporary machine access for large volumes of wood, therefore the potential to prevent some of these problems is significant (Photo Stephan Hoffman, NIBIO)

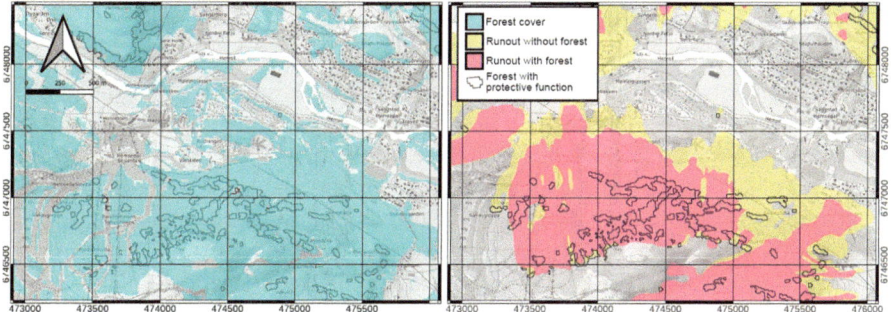

Fig. 10.4 An example of protection forest in Hemsedal, a popular alpine skiing resort in Norway. The forest provides some protection from avalanches for the many recreational activities in the area, including protection for both tourist and permanent accommodation. Without the forest, the risk zones would be larger, and the avalanches would be triggered more easily. However, a substantial portion of these forests are currently managed under RF, where eventual clearfelling will remove the protection for several decades until the next forest has regrown. Some CCF systems could help maintain or even enhance the protection. (Background map ©Kartverket, forest information produced from NIBIO's forest resource map SR16, avalanche risk zones are produced using NAKSIN 4, NVE 2023)

10.6.2   The Implications of CCF in Protection Forests

Certain forest structures are required to provide protection against these gravitational hazards (Table 10.1, derived from Nordrum et al. 2022). One of the main requirements is continually maintaining a minimum forest cover, which aligns clearly with the definition of continuous cover. The practice of clearcutting in protection forests results in a "protection gap" (Wohlgemuth et al. 2017) until the forest

Table 10.1 The effect of forest structure can be quantifiably linked to protection from gravitational hazards for snow avalanches (Issler et al. 2020) and rockfall (Berger and Dorren 2007) for Norwegian conditions, but this is not yet possible for landslides. While each slope is different some simple thresholds can be defined to guide forest management. CCF systems, and particularly selection systems, can promote and maintain these forest structures irrespective of whether timber is extracted

Natural hazard	Most important elements of forest structure	Continual canopy required (% of surface area)	Maximum gap size (diameter or side length) (m)	Comments
Avalanche	Species, canopy coverage	50	30 m	Spruce intercepts most of the snow
Landslides	Species, canopy coverage	60	Unknown	Deep rooting species are better, conifers intercept more water year-round.
Rockfall	Tree size and forest density	NA	40 m	A minimum 400 trees/ha is proposed, with a DBH of 20 cm

is regenerated and the protection restored, thereby temporarily increasing the risk to society. On the other hand, many of the systems considered under CCF (see Chap. 2) already promote creating and maintaining forest structures that are consistent with the requirements for protection (Table 10.1). This strongly favours using CCF in protection forests, particularly in place of clearcutting even-aged forests. However, steep terrain is challenging for forest operations if timber production is to remain an objective. The challenges relate to both cost and technical difficulty, and already apply to clearcutting. More complex silvicultural choices will increase both relative difficulty and cost (McLean and Hoffmann in press). This potential for extra costs poses problems in the Nordic region due to the fragmented and highly privatised forest ownership. The problems are the question about who pays and the requirement for cooperation between multiple owners around risk zones. To resolve this, other economic aspects, such as considering the value of protection, or the costs of alternative technical provision of that protection (Moos and Dorren 2022), need to be taken into consideration.

Currently there is no such framework within the Nordic region. However, it is highly likely on an economic basis alone that the value of protection will be greater than any short-term financial gain from clearcutting. Likewise, the value of protection will likely warrant more expensive timber extraction if this is desired. However, it should be noted that the extraction of wood is not essential for protection, and forest structures can be produced and maintained by felling the trees even if the felled timber is left in place. In fact, if felled timber is left perpendicular to slopes this can provide additional mechanical protection from mass movement in an eco-engineering solution (e.g. Berger et al. 2013) and could have strong synergies with objectives for biodiversity in terms of habitat provision (see Chap. 11). In summary, CCF could provide the means to both provide protection from gravitational hazards and produce timber, but the silvicultural options may not be so varied, and any economic benefits (see Chap. 8) will not be as easy to estimate compared to flat terrain with easy access.

10.7   Summary with Recommendations

All forestlands in Norway, Sweden, and Finland, through tradition and legal recognition, inherently provide many uses beyond wood biomass production. When managing Nordic forests for multiple use, all the functions of the forest should be considered. The important functions considered here are:

- Supply of renewable raw materials
- Habitat for plants and animals
- Venue for outdoor life
- Bearer of aesthetic and cultural values
- Protection from natural hazards
- The source of nutrition for reindeer and other grazing livestock
- Producer of berries, mushrooms and huntable game

The weighting given to the various functions in any area depends on the natural conditions in the landscape, the needs of local stakeholders and rightsholders (e.g., indigenous peoples), the local population density, and the forest site quality (SOU 1992, p. 76). Ultimately however, management decisions are driven by the individual owner's management objectives. Whether to consider RF or CCF will therefore come down to evaluating the relative pros and cons of each, which are considered within this chapter are summarised below (Fig. 10.5).

In terms of tourism and recreational use, it is somewhat indisputable that during the thaw and summer periods heavy tillage and clearcutting do not attract tourists. However, during the snowy season, these features are less of a hindrance. When summer tourism in the north is to be developed, it is essential to avoid clearcuts and logging residues in important tourist areas. Koivula et al. (2020) state that forest management with continuous cover and different age structure preserves the attractiveness of forests for recreational and tourist use better than clearcut logging.

So far, the preservation of landscapes or other intangible nature values for tourism has not received wide attention in forest policy or land use planning (Tyrväinen et al. 2020). In discussions about the use of forests, conflicts often arise between different forms of use, such as forestry, nature conservation and recreation (Sténs et al. 2016, Svensson et al. 2023, Stoessel et al. 2022). The challenges for future research are to improve understanding of the synergies between multiple use and forestry, and the issues of conflict and their location. Spatially-integrated multiple-use planning solutions are needed, as well as a development of forest management approaches that allow spatiotemporally simultaneous provisioning of other values than wood biomass production and a flexible management choice of priority value(s). Participatory GIS methods, for example, could be used more in

Fig. 10.5 The infograph shows how CCF favours most recreational uses while RF is suitable for at different stages. The situation also varies according to the season, as e.g. the cross-country skiing experience does not suffer from clearcuts whereas for false morels clearcuts are almost necessary

positioning. Wider use of participatory methods would also bring more understanding of the needs of different user groups.

Wild berries and mushrooms can produce good yields in both periodic and continuous cover silviculture, depending on the species. Based on the modelling, it seems that bilberry benefits more from CCF than RF, while lingonberries and most mushrooms produce better yields under RF. However, certain CCF systems can produce gaps with the right kind of conditions for lingonberries and mushrooms, so this should not be seen as an obstacle to implementation. Currently, there is little empirical information on the effects of CCF on berry and mushroom yields, so monitoring programmes should be set up in CCF areas. In addition to monitoring of berry and mushroom yield, pollinator monitoring should be set up in the same areas, as the success of pollination is the most important individual factor affecting wild berry yields. Further, as with hunting, picking berries and mushrooms is equally about exercise in nature and recreation. When promoting the picking of these natural products, the visual attractiveness of forests must also be taken into account, which is usually better in forests under CCF management than in clearcuts.

There can be no question that clearcutting systems are less effective in protecting from natural hazards than maintaining a continual forest cover, because clearcutting involves periodic, total removal of the protection. In areas where protection is critical, it is questionable whether timber harvesting is really sustainable (e.g. where significant erosion damage is caused in the process) and in the best interests of society. Nonetheless, there are large areas of even-aged, planted forests that are providing this protection, so converting these to multilayered forests with more self-sustaining structures using CCF techniques to avoid future protection gaps is a logical thing to do. Here, CCF techniques may also permit the steady production of timber if that remains an objective in these areas. RF techniques that exclude clearcutting can possibly also be used in areas requiring protection, as long as they can be designed to continually provide enough forest cover to prevent the relevant hazard or hazards. This would involve maximum gap sizes and sufficient overstorey retention until the next generation, or understorey, has grown sufficiently to replace the same protective function as that provided by the overstorey.

A transition towards increased multiple use forest management in Norway, Sweden and Finland requires full acknowledgement of the multiple values, services and goods associated with the forest ecosystem. CCF for multiple use is far from a simplified and normative forest management system that can be applied systematically across the forest landscape. However, it may facilitate the delivery of a better balance of the multiple uses when applied correctly in the correct places.

References

Ahtikoski A, Tuulentie S, Hallikainen V et al (2011) Potential trade-offs between nature-based tourism and forestry, a case study in northern Finland. Forests 2(4):894–912

Andersson J (2018) Möjligheter att med bär- och svampplockning skapa arbete för lågutbildade på Sveriges landsbygd (Opportunities to use berry and mushroom picking to create work for low-educated people in rural Sweden). Examensarbete 2018:24. Skogsmästarprogrammet, Skinnskatteberg. https://stud.epsilon.slu.se/14136/7/andersson_j_181220.pdf. Accessed 15 Feb 2024

Andersson G, Berg M, Bergkvist U et al (2016) Tillväxt, mångbruk, värdeskapande av skogen som resurs (Growth, multiple use, value creation of the forest as a resource). Underlagsrapport från arbetsgrupp 1 inom nationellt skogsprogram. Nationella Skogsprogrammet

Appelstrand M (2015) Osynliga entreprenörer i skogen? Förutsättningar för kvinnors företagande inom den skogliga sektorn (Invisible entrepreneurs in the forest? Conditions for women's entrepreneurship in the forestry sector). In: Andersson E (ed) Den öppna skogen. Kön, genus och jämställdhet i skogssektorn. Sveriges Lantbruksuniversitet, pp 109–120

Artell J, Lankia T, Pellikka J et al (2020) Hirvenmetsästys: tavat ja arvot 2019–2020 (Moose hunting: customs and values). Luonnonvara- ja biotalouden tutkimus 81/2020. Luonnonvarakeskus, Helsinki, p 32

Beland Lindahl K, Stens A, Sandström C et al (2017) The Swedish forestry model: more of everything? Forest Policy Econ 77:44–55

Belsky AJ, Blumenthal DM (1997) Effects of livestock grazing on stand dynamics and soils in upland forests of the interior west. Conserv Biol 11:315–327

Berger F, Dorren LKA (2007) Principles of the tool Rockfor.net for quantifying the rockfall hazard below a protection forest. Schweiz Z fur Forstwes 158(6):157–165. https://doi.org/10.3188/szf.2007.0157

Berger F, Dorren L, Kleemayr K et al (2013) Eco-engineering and protection forests against Rockfalls and snow avalanches. In: Cerbu G (ed) Management strategies to adapt alpine space forests to climate change risks. InTech. https://doi.org/10.5772/56275

Bjärstig T, Kvastegård E (2016) Forest social values in a Swedish rural context: the private forest owners' perspective. Forest Policy Econ 65:17–24

Bjärstig T, Sténs A (2018) Social values of forests and production of new goods and services: the views of Swedish family forest owners. Small-scale For 17(1):125–146

Bjørneraas K, Solberg EJ, Herfindal I et al (2011) Moose Alces alces habitat use at multiple temporal scales in a human-altered landscape. Wildl Biol 17(1):44–54

Dorre L, Berger F, Jonsson M et al (2007) State of the art in rockfall—forest interactions. Schweiz Z fur Forstwes 158(6):128–141. https://doi.org/10.3188/szf.2007.0128

Edwards D, Jay M, Jensen FS et al (2012) Public preferences for structural attributes of forests: towards a pan-European perspective. Forest Policy Econ 19:12–19

Ekvall H, Bostedt G (2009) Skogsskötselns ekonomi (The economics of forest management). Skogsskötselserien nr 18. Skogsstyrelsen, Jönköping

Engelman M, Lagerkvist CJ, Gren M (2018) Hunters' trade-off in valuation of different game animals in Sweden. Forest Policy Econ 92:73–81

Finnish Forest Association (2019) Forest ownership in Finland. https://forest.fi/article/forest-ownership-in-finland/#3114d049. Accessed 9 Jan 2024

Fleischner TL (1994) Ecological costs of livestock grazing in western North America. Conserv Biol 8:629–644

Fritzbøger B, Søndergaard P (1995) A short history of forest uses. In: Hytönen M (ed) Multiple-use forestry in the Nordic countries. METLA, Jyväskylä, Finland, pp 11–41

Furuyashiki A, Tabuchi K, Norikoshi K et al (2019) A comparative study of the physiological and psychological effects of forest bathing (Shinrin-yoku) on working age people with and without depressive tendencies. Environ Health Prev Med 24:1–11

Gundersen VS, Frivold LH (2008) Public preferences for forest structures: a review of quantitative surveys from Finland, Norway and Sweden. Urban For Urban Green 7(4):241–258

Haakana H, Huhta E, Hirvelä H et al (2020) Trade-offs between wood production and forest grouse habitats in two regions with distinctive landscapes. For Ecosyst 7(1):1–16

Haara A, Matala J, Melin M et al (2021) Economic effects of grouse-friendly forest management. Silva Fenn 55(3):14. https://doi.org/10.14214/sf.10468

Haila Y (2015) Sopeutuuko metsien monikäyttö biotalouteen? (will multiple use of forests adapt to the bioeconomy?). Metsätieteen aikakauskirja 4(2015):253–255

Herfindal I, Lande US, Solberg EJ et al (2017) Weather affects temporal niche partitioning between moose and livestock. Wildl Biol 2017(1):1–12

Herrero A, Pellikka J, Matala J (2020) Miten riistaa neuvotaan huomioimaan metsänhoidossa? (how is game advised to be taken into account in forest management?). Suomen Riista 66:81–96

Hertog IM, Brogaard S, Krause T (2022) Barriers to expanding continuous cover forestry in Sweden for delivering multiple ecosystem services. Ecosyst Serv 53:101392

Hester AJ, Edenius L, Buttenschøn RM et al (2000) Interactions between forests and herbivores: the role of controlled grazing experiments. Forestry 73:381–391

Himes A, Betts M, Messier C et al (2022) Perspectives: thirty years of triad forestry, a critical clarification of theory and recommendations for implementation and testing. For Ecol Manag 510:120103. https://doi.org/10.1016/j.foreco.2022.120103

Hjeljord O, Histøl T, Wam HK (2014) Forest pasturing of livestock in Norway: effects on spruce regeneration. J For Res 25(4):941–945

Hoogstra-Klein MA, Brukas V, Wallin I (2017) Multiple-use forestry as a boundary object: from a shared ideal to multiple realities. Land Use Policy 69:247–258

Huuskonen S, Domisch T, Finér L et al (2021) What is the potential for replacing monocultures with mixed-species stands to enhance ecosystem services in boreal forests in Fennoscandia? For Ecol Manag 479:118558

Ikonen P, Miettinen J, Luoma M et al (2023) Beliefs of Forest owners toward cooperative Capercaillie Lekking site management operations: a pilot study. Hum Dimens Wildl 28(6):620–634

Issler D, Gisnås K, Domaas U (2020) Approaches to including climate and forest effects in avalanche hazard indication maps in Norway. NGI Technical Note 20150457-10-TN. NGI, Norwegian Geotechnical Institute, Oslo, Norway

Kansallinen metsästrategia 2035 (2022) National Forest Strategy 2035. Ministry of Agriculture and Forestry, Finland. https://mmm.fi/hanke2?tunnus=MMM034:00/2022. Accessed 15 Feb 2024

Kardell Ö, Bishop K (2014) Mångbruk i skogen är ingen ny idé (multiple use in the forest is not a new idea). In: Mårald E, Nordlund C (eds) Idéer och värderingar. Future forests 20092012. Future forests rapportserie nr 3. Sveriges lantbruksuniversitet, Umeå

Karjalainen E, Store R (2015) Maisemanhoito metsätaloudessa. (Landscape management in forestry). In: Salo K (ed) Metsä. Monikäyttö ja ekosysteemipalvelut. Luke, Helsinki, pp 257–262

Kivinen S, Moen J, Berg A et al (2010) Effects of modern forest management on winter grazing resources for reindeer in Sweden. Ambio 39:269–278

Klemperer WD (1996) Forest resource economics and finance. McGraw-Hill, New York

Koivula M, Silvennoinen H, Koivula H et al (2020) Continuous-cover management and attractiveness of managed scots pine forests. Can J For Res 50(8):819–828

Konu H, Tyrväinen L (2020) Matkakohteen luontoympäristön vetovoimaisuuden ylläpitämi-nen maisema- ja virkistysarvokaupan avulla (Maintaining the attractiveness of the destination's natural environment through landscape and recreational value trading). Matkailututkimus 16(2):40–44

Kosenius AK, Juutinen A, Neuvonen M et al (2013) Virkistyskäyttöä edistävä metsänhoito valtion talousmetsissä: hyötyjen rahamääräinen arvo (Recreational forest management in state-owned forests: the monetary value of benefits). Metlan työraportteja 261

Kouhia A, Pellikka J (2021) Hirvitalous osana metsänomistamisen tavoitteita (The role of moose management among other ownership goals in Finnish non-industrial private forest owners). Suomen Riista 67:58–75

La Mela M (2014) Property rights in conflict: wild berry-picking and the Nordic tradition of alle-mansrätt. Scand Econ Hist 62(3):266–289

Lapland Tourism Strategy (2019) Regional Council of Lapland. https://arcticsmartness.eu/wp-content/uploads/Matkailu_tilannekuvaraportti_web.pdf. Accessed 15 Feb 2024

Larsson JY, Rekdal Y (2000) Husdyrbeite i barskog. Vegetasjonstyper og beiteverdi (Livestock grazing in coniferous forest. Vegetation types and grazing value). NIJOS-Rapport 9(2000):1–38

Lavsund S, Nygrén T, Solberg EJ (2003) Status of moose populations and challenges to moose management in Fennoscandia. Alces 39:109–130

Lidestav G (1998) Women as non-industrial private forest landowners in Sweden. Scand J For Res 13:66–73

Lidestav G (2010) In competition with a brother: Women's inheritance positions in contemporary Swedish family forestry. Scand J For Res 25:14–24

Lov om skogbruk (2005) https://lovdata.no/dokument/LTI/lov/2005-05-27-31. Accessed 15 Feb 2024

LVVI (2020) Luonnon virkistyskäytön valtakunnallinen inventointi 3 (National Inventory of Nature Recreation 3). https://www.luke.fi/fi/projektit/lvvi3p. Accessed 15 Feb 2024

Mäkipää R (2022) Metsä- ja suokasvillisuuden muutokset 1950-luvulta 2020-luvulle (Changes in forest and swamp vegetation from the 1950s to the 2020s). https://www.luke.fi/fi/projektit/operaatio-mustikka. Accessed 15 Feb 2024

Margaryan L (2018) Nature as a commercial setting: the case of nature-based tourism providers in Sweden. Curr Issue Tour 21(16):1893–1911

Maurtvedt A (ed) (1989) Saueboka. Landbruksforlaget

McLean JP, Hoffmann S (in press) Forest management & timber harvesting in steep terrain—risks and opportunities. Norwegian case studies NIBIO report

Miina J, Hotanen JP, Salo K (2009) Modelling the abundance and temporal variation in the production of bilberry (Vaccinium myrtillus L.) in Finnish mineral soil forests. Silva Fenn 43(4):577–593

Miina J, Tolvanen A, Kumpula J et al (2020) Metsien luonnontuotteet, virkistyskäyttö ja por-olaitumet jatkuvapeitteisessä ja jaksollisessa kasvatuksessa (Natural forest products, recreation and reindeer pastures under continuous and intermittent cover). Metsätieteen aikakauskirja:2020–10345

Ministry of Agriculture and Food (2011) Norwegian Forests. https://www.regjeringen.no/globalas-sets/upload/lmd/skogaaret/vedlegg/brosjyre_norsk_skogpolitikk_2011_engelsk.pdf. Accessed 16 Sep 2024

Moos C, Dorren L (2022) Cost-benefit analysis as a basis for risk-based Rockfall protection Forest management. In: Teich M et al (eds) Protective forests as ecosystem-based solution for disaster risk reduction (Eco-DRR). IntechOpen. https://doi.org/10.5772/intechopen.99513

Nordh H, Grahn P, Währborg P (2009) Meaningful activities in the forest, a way back from exhaustion and long-term sick leave. Urban For Urban Green 8(3):207–219

Nordrum R et al (2022) Forvaltningsmodeller for sikringsskog mot naturfarer (Management models for protection forests against natural hazards). 22/3–27. Landbruksdirektorate

Nordström EM, Bjärstig T, Zhang J (2020) Mångbruk av skog - om att utveckla skogens mervärden (Multiple use of forests - about developing the forest's added values). Sveriges Lantbruksuniversitet https://pub.epsilon.slu.se/19116/1/ffrapport_mangbruk-av-skog_2020-07-02.pdf. Accessed 15 Feb 2024

Norwegian Environment Agency (2023) https://www.miljodirektoratet.no/publikasjoner/2020/juni-2020/the-norwegian-right-to-roam-the-countryside/. Accessed 15 Feb 2024

Östlkulnd L, Zackrisson O, Axelsson AL (1997) The history and transformation of a Scandinavian boreal forest landscape since the 19th century. Can J For Res 27:1198–1206

Pedersen AS, Kjelsaas I, Guldvik MK et al (2020) Samfunnsøkonomisk verdi av elgjakt i Norge (Socio-economic value of moose hunting in Norway). Menon-Publikasjon 28

Pellikka J, Artell J, Rautiainen M et al (2018) Valtion maiden kanalintulupametsästäjät (Grouse hunters on state-owned land.). Metsähallituksen luonnonsuojelujulkaisuja. Series B 241

Peura M, Burgas D, Eyvindson K et al (2018) Continuous cover forestry is a cost-efficient tool to increase multifunctionality of boreal production forests in Fennoscandia. Biol Conserv 217:104–112

Prolux M, Mazumder A (1998) Reversal of grazing impact on plant species richness in nutrient-poor vs. nutrient-rich ecosystems. Ecology 79:2581–2592

Pukkala T, Lähde E, Laiho O et al (2011) A multifunctional comparison of even-aged and uneven-aged forest management in a boreal region. Can J For Res 41(4):851–862

Rekdal Y, Angeloff M (2021) Arealrekneskap i utmark. Utmarksbeite—ressursgrunnlag og beitebruk (land accounting in the outfields. Outdoor grazing - resource base and grazing use). NIBIO Rapport 7(208):1–108

Rice RM (1977) Forest management to minimize landslide risk. FAO, p 17

Salemaa M (2000a) Vaccinium myrtillus. Mustikka. (Vaccinium myrtillus. Bilberry). In: Reinikainen A, Mäkipää R, Vanha-Majamaa I et al (eds) Kasvit muuttuvassa metsäluonnossa. Tammi, Helsinki, pp 128–130

Salemaa M (2000b) Vaccinium vitis-idaea. Puolukka. (Vaccinium vitis-idaea. Lingonberry). In: Reinikainen A, Mäkipää R, Vanha-Majamaa I et al (eds) Kasvit muuttuvassa metsäluonnossa. Tammi, Helsinki, pp 136–138

Schneebeli M, Bebi P (2004) Hydrology | snow and avalanche control. In: Encyclopedia of forest sciences. Elsevier, pp 397–402. https://doi.org/10.1016/B0-12-145160-7/00271-4

Sidle RC (2005) Influence of forest harvesting activities on debris avalanches and flows. In: Debris-flow hazards and related phenomena. Springer Berlin Heidelberg (Springer Praxis Books), Berlin, Heidelberg, pp 387–409. https://doi.org/10.1007/3-540-27129-5_16

Sievänen T, Tyrväinen L (2015) Virkistyskäyttö ja luontomatkailu (Recreation and nature-based tourism). In: Salo K (ed) Metsä. Monikäyttö ja ekosysteemipalvelut, Luke, Helsinki, pp 262–266

Silvennoinen H (2017) Metsämaiseman kauneus ja metsänhoidon vaikutus koettuun maise-maan metsikkötasolla (Forest landscape beauty and the impact of forest management on perceived landscape at the stand level). Dissertationes Forestales 242. Suomen metsätieteellinen seura

Simkin J, Ojala A, Tyrväinen L (2021) The perceived restorativeness of differently managed forests and its association with Forest qualities and individual variables: a field experiment. Int J Environ Res Public Health 18:422

Solbär L, Marcianó P, Pettersson M (2019) Land-use planning and designated national interests in Sweden: arctic perspectives on landscape multifunctionality. J Environ Plan Manag 62(12):2145–2165

Sonntag-Öström E, Nordin M, Slunga Järvholm L et al (2011) Can the boreal forest be used for rehabilitation and recovery from stress-related exhaustion? A pilot study. Scand J For Res 26:245–256

Sonntag-Öström E, Nordin M, Dolling A et al (2015) Can rehabilitation in boreal forests help recovery from exhaustion disorder? The randomised clinical trial ForRest Scand. J For Res 30:732–748

SOU (1992) Skogspolitiken inför 2000-talet. Huvudbetänkande 1990 års skogspolitiska kommitté (Forest policy for the 21st century. Main report 1990 Forest policy committee). Ministry of Agriculture, Stockholm

Statistics Norway (2023) Active hunters. https://www.ssb.no/en/jord-skog-jakt-og-fiskeri/jakt/statistikk/aktive-jegere. Accessed 9 Jan 2024

Sténs A, Sandström C (2013) Divergent interests and ideas around property rights: the case of berry harvesting in Sweden. Forest Policy Econ 33:56–62

Sténs A, Bjärstig T, Sandström C et al (2016) In the eye of the stakeholder: the challenges of governing social forest values. Ambio 45(2):87–99

Stoessel M, Moen J, Lindborg R (2022) Mapping cumulative pressures on the grazing lands of northern Fennoscandia. Sci Rep 12:16044. https://doi.org/10.1038/s41598-022-20095-w

Svanberg I, Lindh H (2019) Mushroom hunting and consumption in twenty-first century post-industrial Sweden. J Ethnobiol Ethnomed 15(1):42

Svanberg I, Løvaas M (2023) Previously neglected—now increasingly popular the recent acceptance of funnel chanterelle, Craterellus tubaeformis (Fr.) Quél., as food in contemporary Scandinavia (Norway, Sweden). https://doi.org/10.21203/rs.3.rs-3173005/v1

Svensson J, Neumann W, Bjärstig T et al (2020) Landscape approaches to sustainability—aspects of conflict, integration and synergy in national public land-use interests. Sustain For 12:5113. https://doi.org/10.3390/su12125113

Svensson J, Neumann W, Bjärstig T et al (2023) Wind power distribution across subalpine, boreal, and temperate landscapes. Ecol Soc 28(4):18. https://doi.org/10.5751/ES-14452-280418

Swedish Forest Industries (2023) Insights about Swedish forests and forestry. https://www.forestindustries.se/forest-industry/forest-management/insights-swedish-forests-and-forestry. Accessed 9 Jan 2024

Tahvanainen V, Miina J, Kurttila M et al (2016) Modelling the yields of marketed mush-rooms in Picea abies stands in eastern Finland. For Ecol Manag 362:79–88

Tahvanainen V, Miina J, Pukkala T et al (2018) Optimizing the joint production of timber and marketed mushrooms in Picea abies stands in eastern Finland. J For Econ 32:34–41

Tapio (2019) Jatkuva kasvatus (Continuous cover forestry). In: Äijälä O, Koistinen A, Sved J, Vanhatalo K, Väi-sänen P et al (eds) Metsänhoidon suositukset, p 113

Tasanen T (2015) Metsät ennen, nyt ja tulevaisuudessa (Forests in the past, present, and future). In: Salo K (ed) Metsä. Monikäyttö ja ekosysteemipalvelut (Forest. Multiple use and ecosystem services). Luonnonvarakeskus (Luke), Helsinki, pp 25, 328 p–32

Teich M et al (2012) Snow avalanches in forested terrain: influence of Forest parameters, topography, and avalanche characteristics on runout distance. Arct Antarct Alp Res 44(4):509–519. https://doi.org/10.1657/1938-4246-44.4.509

Teich M et al (2014) Computational snow avalanche simulation in forested terrain. Nat Hazards Earth Syst Sci 14(8):2233–2248. https://doi.org/10.5194/nhess-14-2233-2014

Toivonen AL (2009) Suomalainen metsästäjä 2008 (Finnish hunter 2008). Riista- ja kalatalous—Selvityksiä 19/2009. 22 p

Turtiainen M, Salo K (2015) Ruokasienten käyttö ja kauppa (Use and trade of edible mushrooms). In: Salo K (ed) Metsä. Monikäyttö ja ekosysteemipalvelut. Luonnonvarakeskus, Helsinki, pp 167–171

Turtiainen M, Miina J, Salo K et al (2013) Empirical prediction models for the coverage and yields of cowberry in Finland. Silva Fenn 47(3):1005

Turtiainen M, Miina J, Salo K et al (2016) Modelling the coverage and annual variation in bilberry yield in Finland. Silva Fenn 50(4):1573

Turunen MT, Rasmus S, Järvenpää J et al (2020) Relations between forestry and reindeer husbandry in northern Finland–Perspectives of science and practice. For Ecol Manag 457:117677

Tuulentie S, Rantala O (2013) Will "free entry into the forest" remain? In: Müller DK, Lundmark L, Lemelin RH (eds) New issues in polar tourism. Springer, Dordrecht, pp 177–188

Tyrväinen L, Silvennoinen H, Hallikainen V (2017) Effect of the season and forest management on the visual quality of the nature-based tourism environment: a case from Finnish Lapland. Scand J For Res 32(4):349–359

Tyrväinen L, Mäntymaa E, Juutinen A et al (2020) Private landowners' preferences for trading forest landscape and recreational values: a choice experiment application in Kuusamo, Finland. Land Use Policy 107:104478

Umaerus P, Lidestav G, Eriksson LO et al (2013) Gendered business activities in family farm forestry: from round wood delivery to health service. Scand J For Res 28(6):596–607

Umaerus P, Högvall Nordin M, Lidestav G (2019) Do female forest owners think and act 'greener'? Forest Policy Econ 99:52–58

Vaara M (2015) Luonnonmarjojen käyttö kotitalouksissa ja teollisuudessa (Use of wild berries in households and industry). In: Salo K (ed) Metsä. Monikäyttö ja ekosysteemipalvelut. Luonnonvarakeskus, Helsinki, pp 139–142

Varis M (2003) Ikävä erätön ilta: suomalainen eräkirjallisuus (An unpleasant night without wild: Finnish wilderness literature). Suomalaisen Kirjallisuuden Seura, Helsinki

Wam HK, Herfindal I, Hjeljord O et al (2022) Skogsbeite—plass til både elg og husdyr? NIBIO POP 8(10):1–4

Wohlgemuth T et al (2017) Post-windthrow management in protection forests of the Swiss Alps. Eur J For Res 136(5–6):1029–1040. https://doi.org/10.1007/s10342-017-1031-x

Zhang JJ, Mårald E, Bjärstig T (2022) The recent resurgence of multiple-use in the Swedish Forestry discourse. Soc Nat Resour 35(4):430–446

Zimmerman GT, Neuenschwander LF (1984) Livestock grazing influences on community structure, fire intensity, and fire frequency within the Douglas-fir/ninebark habitat type. J Range Manag 37:104–110

Chapter 11
Biodiversity

Matti Koivula, Adam Felton, Mari Jönsson, Therese Löfroth, Fride Høistad Schei, Juha Siitonen, and Jörgen Sjögren

Abstract

- This chapter summarises biodiversity responses to continuous cover forestry (CCF). The comparator throughout this chapter is rotation forestry (RF) and its main harvesting method—clearcutting—unless otherwise stated.
- Research on the biodiversity effects of logging methods applied in CCF (mostly selection or gap cutting) mainly concerns the short-term effects of measures taken in mature, originally fairly even-aged forests, at best 10–15 years after cutting. Thus far, no surveys or chronosequences cover the whole rotation period (60–100 years).
- Continuous cover forestry is likely to benefit species that suffer when the tree cover is removed, such as bilberry and its associated species. Species requiring spatial continuity in host trees or canopy cover may also benefit.

M. Koivula (✉) · J. Siitonen
Natural Resources Institute Finland (Luke), Helsinki, Finland
e-mail: matti.koivula@luke.fi; juha.siitonen@luke.fi

A. Felton
Southern Swedish Forest Research Centre, Swedish University of Agricultural Sciences (SLU), Lomma, Sweden
e-mail: adam.felton@slu.se

M. Jönsson
Swedish Species Information Centre, Swedish University of Agricultural Sciences (SLU), Uppsala, Sweden
e-mail: mari.jonsson@slu.se

T. Löfroth · J. Sjögren
Department of Wildlife, Fish and Environmental Studies, Swedish University of Agricultural Sciences (SLU), Umeå, Sweden
e-mail: therese.lofroth@slu.se; jorgen.sjogren@slu.se

F. H. Schei
Norwegian Institute of Bioeconomy Research (NIBIO), Ås, Norway
e-mail: fride.schei@nibio.no

© The Author(s) 2025
P. Rautio et al. (eds.), *Continuous Cover Forestry in Boreal Nordic Countries*,
Managing Forest Ecosystems 45, https://doi.org/10.1007/978-3-031-70484-0_11

- Selection cutting may preserve the majority of species in the mature forest, but the most sensitive species may decline or even disappear. Gap cutting (diameter 20–50 m) affects forest-interior species relatively little, but species' abundances in gaps change with increasing gap size. Shelterwood cutting seems to closely resemble selection cutting in terms of species responses. In the long term, however, shelterwood cutting results in an even-aged and sparse overstorey, which does not produce the biodiversity benefits of CCF.
- Species that have declined due to forestry mostly require large living and dead trees. The preservation of these species is not ensured by CCF alone, but requires deliberately maintaining these structural features.
- A mosaic of different forest-management practices within landscapes may provide complementary ways to maintain rich biodiversity.

Keywords Forest-interior species · Red list of species · Resource continuity · Retention trees · Structural features

11.1 Recent Changes in Forest Structure and Biodiversity in Fennoscandia

In Fennoscandia, a rapid transition from continuous cover forestry (CCF), mainly in the form of diameter-limit cutting, to even-aged rotation forestry (RF) began in the 1950s (Lundmark et al. 2013). The currently dominant forest-management regime is based on clearcutting, soil preparation, and regeneration favouring conifers by planting or sowing, followed by thinning from below (Fig. 11.1). As a result, most productive forest land in Finland and Sweden is structurally simplified, even-aged and even-structured (Gustafsson et al. 2010; Kuuluvainen et al. 2012). Areas of forest with natural dynamics are greatly reduced and fragmented (Timonen et al. 2011). The conifers Norway spruce (*Picea abies*) and Scots pine (*Pinus sylvestris*) dominate planted forests, although Fennoscandian temperate and hemi-boreal climatic zones (southern Sweden) host more tree species than boreal Fennoscandia, including European beech (*Fagus sylvatica*), pedunculate oak (*Quercus robur*), and sessile oak (*Quercus petraea*). Palaeoecological studies indicate that both beech and the two oak species have declined greatly over recent centuries in the southern region (Lindbladh and Foster 2010). Along with the increasingly conifer-dominated timber stocks, these declines emphasise the decreasing share of broadleaf trees in the southern region (Lindbladh et al. 2014).

Following all these changes, species diversity has been severely affected, and about 10% of Fennoscandian forest species are nationally red listed and often largely confined to remnant structures and habitats where forestry has been less intense (Gustafsson 2002; Puumalainen et al. 2003; Mikusiński et al. 2007; Timonen et al. 2011). The main threats to forest-dwelling species are the scarcity of old forests, large old trees, and large decaying wood, and the reduction of wildfire areas

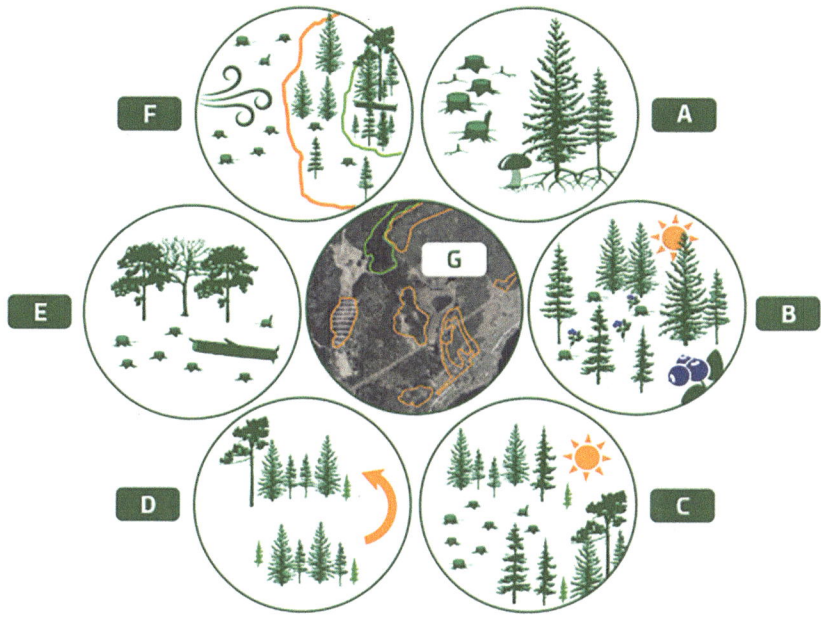

A ⊕ Soil mycorrhizal fungi.
   ⊖ Unknown interactions with damage agents and biodiversity.

B ⊕ Selection cutting supports semi-shade plants and their associates.
   ⊖ Long-term effects unknown.

C ⊕ Gap cutting supports both shade- and sun-favoring species.
   ⊖ Long-term effects unknown.

D ⊕ Connectivity for species dependent on large trees.
   ⊖ Responses of many species groups unknown.

E ± Red-listed species do not benefit without separately
     considering deciduous and old trees, and deadwood.

F ⊕ Buffer to maintain conditions in key habitats or reserves.
   ⊖ Responses of many species groups unknown.

G ⊕ Landscape mosaics of different management practices support biodiversity.
   ⊖ Landscape applications unknown.

Fig. 11.1 Infographic summarizing biodiversity benefits of CCF. Plus (+) symbols indicate a positive response to CCF when contrasting with RF, whereas minus (−) symbols indicate a negative response or a knowledge gap

and young forests originated from natural disturbances (Hyvärinen et al. 2019). The dominance of clearcutting practices has also disadvantaged species that thrive in closed-canopy or semi-open old forests with small-scale disturbance dynamics (Kuuluvainen 2009; Kuuluvainen and Aakala 2011; Kuuluvainen et al. 2012; Brunet 2023). In present-day Nordic forests, the preservation of species requiring the afore-mentioned resources requires conservation actions and modifications to forest man-agement, whereas species occupying many types of habitat, or species associated with open areas such as recent clearcuts, thrive in current landscapes of man-aged forest.

11.2 Opportunities for Safeguarding Forest Biodiversity

Nature-conservation efforts primarily follow two different approaches: "land spar-ing" and "land sharing" (also referred to as wildlife-friendly farming; Green et al. 2005). Land sparing is a segregation approach that aims for spatial separation of areas for production and conservation, where conservation efforts are restricted to protected areas. Land sharing is an integrative approach to nature conservation, rel-evant to multifunctional productive forests where integration of productive and con-servation goals is achievable and sought within the same area.

The most effective single means of protecting species and habitats is probably land sparing, i.e., the establishment of protected areas. Nordic countries have increasingly adopted voluntary measures to conserve forest biodiversity, but these actions can be considered slow. Finland serves as a representative example here. In northern Finland, there is about 1.9 million ha of strictly-protected forest and shrub-lands (17% of the forest area), while in southern Finland the corresponding figure is 0.4 million ha (4%; Niinistö et al. 2021). Of this area, 0.9 million ha (8%) is forest land in the north and 0.3 million ha (3%) in the south. It has been estimated that 10–30% of the forest area needs to be protected to preserve threatened forest species (Hanski 2011). In southern Finland, where the greatest abundance of threatened species and habitats is found, a 10% protection rate could be achieved, for example, by protecting 404,000 ha of forest over 120 years old and 471,000 ha of other mature forest (Kotiaho et al. 2021). This would take more than 100 years to achieve using only the resources of the state-driven programme for voluntary protection of southern-Finnish forests (METSO; Anttila et al. 2021).

Due to the slow designation of protected areas, integrative conservation actions in managed forests—land sharing—have become important for the fate of forest species in Fennoscandia. In addition to conservation measures, expectations have been placed on CCF harvesting methods—notably selection cutting and gap cut-ting. Developing the management practices of production forests affects a much larger area than protection and restoration measures allow.

11.3 Theoretical Justifications for Continuous Cover Forestry

Within Fennoscandia, CCF is often advocated based on the expectation that its increased use would mitigate the negative effects of rotation forestry (RF) on forest biodiversity. Three primary justifications underlie the expected biodiversity benefits. First, over the last 30 years there has been a paradigm shift in the scientific understanding of natural disturbance dynamics in European boreal forests. It was previously thought that stand-replacing disturbances were the norm (Angelstam and Kuuluvainen 2004). This understanding, prevalent since the 1950s, favoured clearcutting and even-aged forests, as this form of silviculture was thought to most closely approximate large-scale stand-replacing disturbance events and subsequent forest regrowth (Kuuluvainen 2009; Kuuluvainen and Aakala 2011). In contrast, the most recent assessments have found that non-stand-replacing disturbances resulting in gap and cohort dynamics are more typical of natural disturbance regimes in northern-European forests (Berglund and Kuuluvainen 2021). As CCF would better approximate the finer spatio-temporal scales of natural disturbance regimes, namely via selection or gap cutting, forest species adapted to such dynamics may be expected to benefit (Felton et al. 2016).

Second, conservation of Fennoscandian forest lands is largely done by leaving uncut patches at stand felling, which are relatively small and often isolated from formally or voluntarily protected and set-aside areas (Gustafsson and Perhans 2010). Whereas such small-scale conservation actions are crucial to meet biodiversity goals, their net contributions can be dampened by the intensity of the production forest matrix surrounding them (Felton et al. 2020; Kärvemo et al. 2021). For example, if production stands are clearcut, this can reduce the diversity of signal species (i.e., species that reflect conditions favourable for red-listed species at a given site; Skogsstyrelsen 2020), such as bryophytes and lichens, in adjacent buffer zones (Hylander and Weibull 2012; Johansson et al. 2018), as well as fungi in adjacent reserve edges (Ruete et al. 2016). For this reason, applying CCF instead of clearcutting in production forests neighbouring set-aside forest patches may support the biological communities of protected forests by reducing the severity of edge effects (cf. Koelemeijer et al. 2022, 2023).

Third, CCF should enhance stand- and landscape-scale habitat availability by providing distinct resources and environments from those provided by RF. These include increased within-stand horizontal and vertical structural heterogeneity (Joelsson et al. 2017), improved landscape-scale forest connectivity (Lindenmayer and Franklin 2002), and reduced distances among mature trees and large deadwood (Atlegrim and Sjöberg 2004). Furthermore, by providing understorey microclimates more commonly associated with mature tree cover, CCF may favour species associated with mature forests that are otherwise lost and require re-establishment after clearcutting (Kuuluvainen et al. 2012; Joelsson et al. 2017). CCF can potentially buffer forest understories from the ambient climate, with future regional temperature increases likely making the environment for many species after clearcutting even more hostile (De Frenne et al. 2021; Hylander et al. 2021). Thus, CCF may support biodiversity in production forests because of its (1) consistency with natural disturbance regimes, (2) potential capacity to buffer small-scale conservation actions surrounded by production forest, and (3) provision of specific forest habitats and resources rare in even-aged forestry.

The conclusions of studies comparing the effects of silvicultural practices on species diversity, and the generalisability of the results depend on several factors. These include the structure and species composition of the study stands before treatment, cutting methods and how they are implemented, the species group studied, the length of the study period, and presumably also the wider landscape context. What matters is how well the forest-management methods are able to maintain species over the entire rotation period, both at the stand and landscape level. Unfortunately, stands or entire forest areas that have been treated with CCF for a long time are rarely available to study. Old (i.e., more than 100 years) RF stands are similarly rare. However, the long-term effects of treatments aiming at continuous tree-canopy cover can be evaluated based on how the recurrent cuttings will affect the structures of direct relevance to habitat availability and environmental conditions within the stand (Fig. 11.2).

When evaluating the effects of CCF on species diversity, the common question "Would CCF be better for biodiversity than RF?" may result in misleading answers. Instead, one should ask under which conditions, for which species, using which method, and over what timeframe, could CCF better conserve biodiversity than RF based on clearcutting—or vice versa. Here, we examine the effects of selection, gap and shelterwood cutting on stand-level forest biodiversity. We evaluate them separately because the methods differ in their spatial distribution and size of the openings and intermediate areas created during harvesting. In addition, soil preparation can be used in gap cutting, and its biodiversity effects are also discussed below. The widespread use of RF is a baseline for the potential benefits of CCF in Fennoscandia.

11.4   Impacts of CCF Logging Methods on Species Communities

11.4.1   CCF Research, and the Impact of Logging Intensity

A recent literature review found several experiments comparing the short-term (single cutting) biodiversity effects of continuous cover and even-aged cutting methods in the Nordic boreal region (Koivula and Vanha-Majamaa 2020). In most experimental designs, the treatments compared have been selection cutting and clearcutting compared with untreated reference forest. Three experiments also assessed gap cutting, whereas shelterwood cutting has rarely been examined. Monitoring periods have so far been short, usually only the first 1–5 years after harvesting and at most 10 years (Koivula and Vanha-Majamaa 2020; Savilaakso et al. 2021).

Fig. 11.2 Bilberry (*Vaccinium myrtillus*) and associated species may benefit from continuous cover forestry. (**a**) Hazel grouse (*Bonasa bonasia*) chicks feed on insect larvae that feed on bilberry. Finland, July 2021 © Mika Nieminen. (**b**) Moose (*Alces alces*) forage on tree leaves and forest-floor grasses and shrubs. Ungulate browsing affects tree regeneration (see Chap. 12 on forest damage) and, in turn, tree-species composition and associated biodiversity. © Erkki Oksanen / Luke

The studied taxa have included herbivorous insect larvae living on bilberry (Atlegrim and Sjöberg 1995, 1996a; Kvasnes and Storaas 2007), lichens and mosses (Jalonen and Vanha-Majamaa 2001; Vanha-Majamaa et al. 2017), soil organisms (Siira-Pietikäinen et al. 2001, 2003; Siira-Pietikäinen and Haimi 2009; Kim et al. 2021), spiders (Matveinen-Huju and Koivula 2008), beetles (Atlegrim et al. 1997, Koivula 2002a, 2002b, Joelsson et al. 2017, 2018; Jokela et al. 2019; Koivula et al.

2019), birds (Versluijs et al. 2020) and vascular plants (Atlegrim and Sjöberg 1996b, Jalonen and Vanha-Majamaa 2001; Vanha-Majamaa et al. 2017). Based on these studies, the species composition of mature spruce-dominated boreal forests remains almost unchanged in the short term in treatments where the proportion of removed trees is 33–50% by volume (Ekholm et al. 2023). However, the most sensitive moss and liverwort species disappear even under the most conservative logging treatments (Vanha-Majamaa et al. 2017). We discuss nuances of logging methods below.

11.4.2   Selection Cutting

Studies in boreal Sweden compared beetle faunas in spruce-dominated (1) 120–130-year-old stands that had been selectively cut 2–15 years earlier, (2) forests of similar age that had not been treated during the last 50 years, and 50–60 year-old stands that were either (3) clearcut or (4) thinned 6–7 years earlier (Joelsson et al. 2017, 2018). Thinned previously-clearcut forests in these study forests were at most 50–60 years old, so the species present in these forests reflected the longest post-clearcutting period. The results showed that fresh clearcuts clearly differed from the other forest classes in terms of species composition. However, species composition did not differ significantly between thinned stands and selectively-cut stands. The recovery of the beetle fauna therefore appears to be rapid in thinned and selectively-cut stands, whereas in clearcuts, species composition had not recovered fully 50 years after logging.

Vegetation and fungi are inconsistently affected by selection cutting. A before-after control-impact study contrasting unharvested mature forest and selection cutting 3–4 years after harvesting detected no significant differences in the community compositions of vascular plants, bryophytes or wood-inhabiting fungi (Ekholm et al. 2022). In contrast to clearcutting, recent partial harvest of 30% in uneven-aged Norway-spruce forests in mid-boreal Sweden maintained soil chemical properties and fungal communities similar to unmanaged forests (Kim et al. 2021). The composition of both fruiting-body fungal and soil-fungal communities were similar in continuous cover and unmanaged forest, but substantially different in clearcut areas (Kim et al. 2021). Similarly, Sterkenburg et al. (2019) showed that tree retention enabled the maintenance of the most frequent ectomycorrhizal fungi, whereas less abundant species were progressively lost at random with increasing harvest intensity (100%, 60%, 30% and 0% retained trees) in Scots pine forests in northern Sweden.

The long-term effects of old selection cutting (diameter limit cutting) on species composition have also been investigated in the Nordic countries. These studies have compared the stand structure and species composition of spruce-dominated forests selectively cut 50–100 years ago with natural old-growth forests. The results show that selection cutting, in which about 50% of the stand volume had been removed, still has negative effects on dead wood and polypore fungi many decades later (Bader et al. 1995; Lindblad 1998; Sippola et al. 2001; Josefsson et al. 2010). In old selectively-cut forests, there were fewer large-diameter, medium-decayed trees, and both the total number of polypore species and the number of threatened species were significantly lower than in old natural forests. Similarly, selection cutting had reduced the biomass of epiphytic lichens growing on old trees with thick branches (Esseen et al. 1996), and consequently the number of invertebrates living in the lichen microhabitat (Pettersson et al. 1995; Pettersson 1996). It should be noted, however, that these studies did not include a comparison with 50–100-year-old even-aged forests after clearcutting. What these results mainly show is that even selection cutting of mature trees has negative long-term impacts on structural features that are important for species, and hence on species composition.

11.4.3   Gap Cutting

The main responses of species assemblages to gap cutting are increased abundance of open-habitat and generalist species, and decreased numbers of closed-forest species, especially in the cleared gaps. The species composition in retained parts of a gap-cut stand remains almost unchanged. Larger gaps show a greater change in species composition (e.g., Jokela et al. 2019). Light-demanding tree, pioneer or early-successional plant species would benefit from such gaps (Brunet 2023). Due to the relative novelty of the method, it is not yet possible to assess its long-term effects, not even by comparing the diversity of species present in different-aged gaps.

Most Nordic publications on the species effects of gap cutting are from the Finnish project "Monimuotoisuus talousmetsien uudistamisessa" (MONTA; Diversity in the regeneration of managed forests). In this project, three small gaps of about 0.16 ha (32 × 50 m or 40 × 40 m) were made in a 1-ha core of mature Norway-spruce stands, leaving the rest of the stand untreated. About 50% of the stand volume was removed, which is more than in the selection-cutting treatment in the same project (about 33%). During the first three summers following harvesting, the composition of common closed-forest species remained largely unchanged (Jalonen and Vanha-Majamaa 2001; Koivula 2002a; Koivula and Niemelä 2003; Matveinen-Huju and Koivula 2008). However, individuals of open-habitat species quickly appeared in the gaps. The soil decomposer communities in gaps and untreated parts of the stands also differed slightly (Siira-Pietikäinen et al. 2001). After 10 years, species-community structure in gaps and retained areas showed clearer differences in soil-decomposing organisms (Siira-Pietikäinen and Haimi 2009), vascular plants, mosses and lichens (Vanha-Majamaa et al. 2017) and ground beetles (Koivula et al. 2019). In gaps, common species of open, sun-exposed environments became more abundant, while species of shaded environments often decreased compared to untreated forest (ibid.).

Muurinen et al. (2019) studied understorey vegetation in a northern Finnish spruce-dominated forest area that had burned in 1919 and subsequently regenerated naturally. In 1953 (about 34 years post-fire) an experiment was set up in this area, with four 0.1-ha unlogged reference plots and four 0.1-ha gap-cut plots (about 30%

of tree volume was removed; the gaps were long, narrow strips). Vegetation was inventoried 8 years later. Compared to the reference plots, the plant and lichen community of gap-cut plots remained mostly unchanged. The contrast with the 10-year effects in the MONTA project may be due to different gap shapes, different gap origins (fire or harvesting) and/or different dominant tree ages (and associated plant communities) at the time of harvesting. Regrettably, the gap cuts reported in Muurinen et al. (2019) were thinned in 1987, so their long-term resampling is not possible.

At the beginning of the "Management models based on natural forest disturbance dynamics" (DISTDYN) experiment, established on Finnish state lands in 2009–2010, openings about 25–35 m and 40–60 m in diameter were cut in Norway spruce and Scots pine stands, with logging intensity varying between 20% and 35% of volume. During the first 3 years following logging, the amount of coarse (diameter ≥ 15 cm) deadwood in spruce stands had a greater effect on the number of saproxylic beetle species than the felling method (Jokela et al. 2019). Polypore species showed similar patterns 7 years after logging (Pasanen et al. 2019). Similar observations have also been made in a Swedish gap-cutting experiment in Norway-spruce forests (gap diameter about 20 m, logging intensity about 20%). During the first three summers after felling, gaps differed little in beetle-community composition, though some groups, including phloem-feeding beetles (e.g., bark beetles), became more abundant in the stands managed using gap cutting (Hjältén et al. 2017; Kärvemo et al. 2017). Their increase in abundance is linked to the fresh dead wood produced by logging and is therefore temporary (Jokela et al. 2019). In the Swedish study mentioned above, the stand-level bird community composition did not differ significantly between mature reference forests and gap cuts 5 years after felling (Versluijs et al. 2017). Similarly, vascular plants and mosses showed no detectable difference 8 years after felling (del Alba et al. 2021) (Fig. 11.3).

11.4.4 Shelterwood Cutting

Ecological research on CCF logging methods in the Nordic countries has so far largely focused on selection and gap cutting, but shelterwood cutting has also occasionally been studied. Based on these, the short-term effects of shelterwood cutting resemble selection cutting. Forest species communities will be additionally impacted when the shelter trees are removed, whereas selection cutting likely better maintains shade and micro-climatic conditions.

Ectomycorrhizal fungi of old (157–174 years), and shelterwood and clearcut (both logged 50 years earlier) Scots pine forests were studied in central Sweden (Varenius et al. 2016). This study showed that the fungal communities differed between old and harvested forests, but sites harvested using the two methods hosted similar communities, with only occasional species-level differences. In hemiboreal forests of mainland Estonia, shelterwood logging was examined in dry pine- and mesic spruce-dominated forests (Tullus et al. 2018). These authors inventoried

Fig. 11.3 Some specialised species may benefit from continuous cover forestry. (**a**) Pale-footed horsehair (*Bryoria fuscescens*), an epiphytic lichen. Finland, September 2020 © Taru Rikkonen. (**b**) Chanterelle (*Cantharellus cibarius*), a mycorrhizal fungus. Norway, September 2023 © Fride Høistad Schei

liverworts, mosses and vascular plants in 2-ha plots in each of 30 stands: ten each of mature managed stands, stands where shelterwood logging had been done 5–9 years earlier (about 42% of tree volume was harvested), and stands where shelterwood logging had been done and the shelter trees removed 4–14 years later (in total about 95% was harvested). Generally, the richness of mosses and liverworts dropped following logging. Spruce and pine forests showed similar drops, although richness was consistently higher in spruce forests (Tullus et al. 2018). Harvesting reduced the richness of species of conservation concern by half in spruce-dominated forests, but less so in pine-dominated forests. Differences between the two shelterwood treatments were small, but liverworts showed a significant additional decline after shelter tree removal (Tullus et al. 2018). These shelterwood studies suggest that (1) before shelter trees are removed, shelterwood cutting better maintains the forest-species community than clearcutting, whereas (2) species of conservation concern may decline, and (3) the removal of shelter trees causes an additional negative impact on sensitive forest species.

Shelterwood management has been employed in southern-Swedish beech forests since the 1840s (Brunet and Berlin 2005), largely replacing semi-open pasture beechwoods with dense even-aged beech stands ready for harvest after approximately 100–140 years (Brunet et al. 2012). This involves canopy thinning after a mast year to favour dense natural regeneration, after which the remaining seed trees are cut (Brunet et al. 2010). In their comparative review of management of European beech forests, Brunet et al. (2010) contrasted shelterwood with selection or gap cutting and concluded that selection cutting resulting in an uneven-aged stand structure is likely to benefit forest biodiversity. This is due to its resultant higher contribution of (1) multiple canopy layers, (2) variable tree sizes, (3) spatial heterogeneity, and (4) advance regeneration (Brunet et al. 2010). The general sensitivity of different species groups to shelterwood forestry in broadleaf forests roughly increases in the following order: herbaceous plants < soil macrofungi < ground-dwelling arthropods < land snails < saproxylic fungi < cavity-nesting birds and saproxylic insects < epiphytic lichens and bryophytes < epixylic bryophytes (Brunet et al. 2010).

11.4.5   Comparison of CCF Logging Methods

One important question concerning biodiversity is whether closed-forest species are equally well preserved in forests harvested with selection, gap or shelterwood cutting. We are not aware of any experiments where these treatments differ only in the spatial distribution of unharvested crop trees but not in the number of trees removed, but a cautious comparison can be made using data from the Finnish DISTDYN and MONTA projects. In the DISTDYN project, gap size appeared to have a larger (positive) effect on beetle-species abundance than the proportion of removed trees (Jokela et al. 2019). Beetle-species composition was affected in the same direction by both variables. Koivula (2012) compared logging methods using published MONTA-project data (Koivula 2002a, 2002b) and a model with stand volume (m^3/ha) as a covariate and logging method and year (1995–1998) as factors. In the analysis, the logging method did not affect the ground-beetle species, but the variation in stand volume—which changes in concert with logging method—had a strong effect, mainly by increasing the abundance of open-habitat species with decreasing stand volume. If the spatial distribution of the stand has a large effect, this should have been reflected in the significance of the logging method. It is important to note that this was a short-term result, covering only the first three post-harvest summers; in gap-cut stands seven summers later (2006), the gaps hosted beetle assemblages distinctive from unharvested areas (Koivula et al. 2019). On the other hand, in the same stands, moss assemblages 10 years after harvesting had lost more species in selection than in gap cutting, even though more trees had been left in the selection-cut stands (Vanha-Majamaa et al. 2017). This result suggests that, when applied once in a mature forest, gap cutting is better at conserving moss species than selection cutting, but it is not yet possible to assess the effect of gap cutting on the whole logging cycle.

Unlike selection cutting, gap cutting allows the soil to be mechanically prepared to aid seedling establishment. Therefore, we briefly discuss experimental results on the biodiversity effects of soil preparation. The impact of soil preparation on common plant and ground-beetle species in a forest is primarily reflected in the variation in abundance within the stand, with pioneer species found mostly in prepared microsites and shade-demanding species in unprepared areas (Koivula 2002a, Pihlaja et al. 2006; Vanha-Majamaa et al. 2017). On the other hand, soil preparation negatively affects several plant and insect groups associated with closed forests (Hautala et al. 2011; Vanha-Majamaa et al. 2017; Tullus et al. 2018). All soil-preparation methods are particularly destructive to downed dead wood (Hautala et al. 2004; Rabinowitsch-Jokinen and Vanha-Majamaa 2010). Even patch scalping, which is lighter than ploughing, may reduce the species richness of epiphytic lichens and epixylic mosses by more than 50% (Rabinowitsch-Jokinen et al. 2012). It can also reduce the amount of advanced-decay-state deadwood—a necessary substrate for these species—by more than 50% (Rabinowitsch-Jokinen and Vanha-Majamaa 2010; Hautala et al. 2011).

11.5 The Importance of Stand Structural Components in Continuous Cover Forestry

CCF poses some of the same problems for biodiversity as clearcutting-driven RF. In both cases, the trees are harvested before they have had time to develop characteristics associated with senescence. This in turn means that the amount of deadwood and old trees will be considerably smaller than in an old-growth forest, yet such features are particularly important for many threatened forest species (e.g., Hyvärinen et al. 2019). In Finland, for example, the average deadwood volume in all forests is about 6 m^3/ha, whereas in natural conditions the average volume would be about 95 m^3/ha (Mönkkönen et al. 2022), and the respective average densities for large (> 40 cm) trees are about 7 and 42 trees/ha (Henttonen et al. 2019). However, the average density of trees that are both large and old (> 150 years) is currently about 1 tree/ha in Finland, of which about half are in production forests (Henttonen et al. 2019). Based on the natural density of large trees (Mönkkönen et al. 2022), and assuming that the share of old trees among all large trees in the 1910s (Henttonen et al. 2019) was similar to pristine forests, the density of large and old trees in natural conditions would be about 5–55 trees/ha, depending on geographical region and site type.

According to Nordic national legislation and certification standards (Forest Stewardship Council [FSC] and Programme for the Endorsement of Forest Certification [PEFC]), general considerations in CCF should be similar to those of clearcutting-based RF. They should thus consider rare and declining biotopes, retention of large-sized deadwood and living trees of certain minimum size and number, retaining trees of particular biodiversity value (nature value trees, very old or large trees), and creating artificial snags ("high stumps") and buffer zones along watercourses. The effect of such conservation actions can be expected to differ between CCF stands (logged using selection, gap or shelterwood cutting) and clearcuts, for example because the retained and created substrates will be largely shaded in the former. However, while practices such as deadwood enrichment have been studied in stands harvested using CCF methods, no studies have so far compared stand structure more broadly in CCF and RF.

11.5.1 *Deadwood Enrichment*

The Finnish DISTDYN experiment has assessed the effects of deadwood abundance (including both deadwood of natural origin and artificial snags) in spruce-dominated, selectively- and gap-cut stands and clearcut stands, up to seven summers after harvest. Two years after logging, 10-ha clearcuts had 1.5 times higher species richness of saproxylic beetles and polypore fungi than gaps of 0.01 ha (selection cutting); selection and gap cutting did not differ significantly from each other in this respect (Jokela et al. 2019; Pasanen et al. 2019). Simultaneously, an increase in deadwood from 3.0 m^3/ha to 15.0 m^3/ha resulted in about a 1.3-fold increase in beetle richness (Jokela et al. 2019). Seven years after logging, a regression model for polypore richness found only the number of deadwood pieces to be a significant predictor, while cutting intensity, and pre-harvest species richness had no impact as explanatory variables (Pasanen et al. 2019). The increase in species richness in bigger gaps probably resulted from the logging-caused increased sunlight reaching the ground and the input of nutrients and organic matter from logging residue and cut stumps. The gap-size effect on richness appears ephemeral, as a Swedish study, also in Norway spruce forests, showed that 7 years after logging, the beetle richness appeared similar in selectively-cut and clearcut stands (Joelsson et al. 2017).

A Swedish deadwood-enrichment experiment assessed biodiversity responses to the addition of artificial snags and logs in clearcuts, mature managed forests and old-growth reserves (Gibb et al. 2006). The mature managed stands in this experiment had presumably not previously been clearcut, but selection cutting had occurred several decades earlier. For this reason, the results from this experiment can be considered to indicate the long-term effects of selection cutting. Results from this experiment showed that artificial snags and logs on clearcuts hosted different beetle assemblages and species of conservation concern than artificial snags and logs in mature managed stands and reserves (Gibb et al. 2006; Johansson et al. 2006; Hjältén et al. 2010; see Pasanen et al. [2019] for a similar DISTDYN result). Moreover, deadwood in reserves and selectively-cut mature managed stands hosted similar parasitoid wasp, beetle and fungus assemblages but the reserves tended to host larger populations (Hilszczański et al. 2005; Hjältén et al. 2010; Olsson et al. 2012), apparently due to higher volumes of deadwood (Gibb et al. 2006; Stenbacka et al. 2010).

Gap cutting combined with deadwood enrichment was evaluated in a Swedish spruce-forest experiment (Hägglund et al. 2015). The results showed that the combination supports a beetle fauna similar to unharvested mature closed forest (Hjältén et al. 2017; Hägglund and Hjältén 2018). Additionally, in this experiment, artificial snags and girdled trees hosted relatively similar beetle assemblages. The two most important determinants of the beetle assemblage composition were the tree species the wood came from and whether it was standing or lying (Hägglund and Hjältén 2018). Artificial snags are also an important source of deadwood in the managed forest landscape, as they provide habitats for early-successional generalists and several red-listed beetles (Gibb et al. 2006; Johansson et al. 2006; Lindbladh et al. 2007; Lindbladh and Abrahamsson 2008). However, they rapidly become too dry for many polypore fungus species (Pasanen et al. 2019).

Deadwood enrichment efficiently promotes common polypore species, but generally fails to provide the deadwood diversity and substrates for more specialised species associated with late decay stages, at least in the short term (Olsson et al. 2012; Pasanen et al. 2014, 2018; Baber et al. 2016). Logs generally host more species than artificial snags, but the species richness increases with wood diameter (Lindhe et al. 2004; Juutilainen et al. 2014). The red-listed polypore *Fomitopsis rosea* is an example of a wood-decaying fungus that prefers larger logs (Edman et al. 2006). Some—mostly common—species also occupy fine woody debris (Kruys and Jonsson 1999; Berglund et al. 2011; Juutilainen et al. 2014). These results imply that, for wood-decaying fungi, CCF mostly only supports common generalist species because there will be a limited supply of large-diameter logs, despite the likely more continuous availability of medium-diameter logs compared to RF.

Regarding coarse woody debris, selection cutting and even-aged shelterwood management of beech forests result in similar levels of stem, branch and crown structures in living trees, similar amounts of fallen coarse woody debris and similar densities of snags (Brunet et al. 2010). A key determinant of achieved deadwood volumes is the extent of debris-creating conservation actions, including green-tree retention, and creation of artificial snags, among others (Fig. 11.4).

Fig. 11.4 Continuous cover forestry does not directly support most red-listed forest species, but these species can be supported in forest management through various actions. (**a**) Cut-and-downed logs to increase deadwood in a selectively-cut stand—in this case, to support the green shield-moss *Buxbaumia viridis*. Sweden, June 2023 © Mari Jönsson. (**b**) Manipulation of tree-species composition through selection cutting to increase the proportion of pine and birch in spruce-dominated forest. Finland, June 2012 © Matti Koivula

11.5.2   Other Considerations

Minimising forest-floor damage (Sect. 11.4.5), protecting riparian zones, retaining permanently trees with high habitat value, manipulating the tree-species composition, and using prescribed burning may be more challenging in CCF than in RF, yet are equally important.

Selection cutting has been suggested in forests along watercourses where completely-uncut 25–35 m wide buffer zones are not possible for some reason (e.g., Jyväsjärvi et al. 2020; Kuglerová et al. 2020, 2022). Wise application of selection cutting adjacent to such sites could help expand and improve buffer zones along waterways, for instance by promoting deciduous trees.

In selection, gap and shelterwood cutting in CCF, the largest trees will be the most attractive for harvesting in each round. This requires specific considerations to spare very large (> 40 cm) and very old (> 100 years old deciduous, > 150 years old coniferous) trees through repeated fellings. These very large and old trees have had sufficient time to develop deeply-furrowed bark upon which many rare epiphytic lichens depend (e.g., Kuusinen and Siitonen 2009; Lie et al. 2009; Fritz et al. 2009; Nascimbene et al. 2013). New techniques with individual-tree selection and clear marking of such trees might circumvent this problem. Another method is permanent retention of the most valuable within-stand patches. In beech forests, selection cutting combined with permanent retention may provide the highest habitat quality relative to either selection cutting without retention, or shelterwood cutting with or without retention (Brunet et al. 2010). These projected habitat benefits help a broad range of taxa including herbaceous plants, ground arthropods, land snails, cavity-nesting birds, epixylic bryophytes, epiphytic bryophytes and lichens, saproxylic fungi, beetles, and flies (Brunet et al. 2010).

A recently considered alternative to single-tree selection cutting in temperate deciduous forests aims to create an oak-dominated, mixed-species, uneven-aged production forest. This can involve only 13–35 production oak trees/ha, occupying 20–70% of the overstorey at harvest, around which regular thinning is used to promote the timber trees (Löf et al. 2016). In the remaining areas of the stand, regenerating trees are left for relatively free development, with understorey trees allowed to persist as long as they do not interfere with the crowns of the production stems. The denser areas of the stand could promote competition and self-thinning, resulting in more dead wood, whereas the associated free development would protect canopy gaps from wind and tree disease (Löf et al. 2016). This option would improve tree-species diversity, stand structural heterogeneity, deadwood, and trees having cavities and large dead branches (Löf et al. 2016). These features would in turn benefit a wide range of taxonomic groups, including vascular plants, birds, epiphytic lichens and bryophytes, as well as saproxylic fungi and beetles (Löf et al. 2016). However, such features can be equally easily supported in CCF and even-aged RF.

11.6 Ecological Effects of Systemic Conversion from RF to CCF

Most of the above-described studies do not examine full CCF cycles, but rather assess early phases of conversion from even-aged, mature forests toward uneven-aged and often mixed-species forests. They also usually assess stand-level, not landscape-scale, conversion. The results generally suggest rather small ecological effects in the early phases of conversion, although some sensitive species decline. Better understanding of ecological effects of conversion would require monitoring over multiple decades, possibly a period similar to a full rotation of RF. Assessments of larger landscapes, dominated by RF or CCF (such as the set-up of DISTDYN), are crucial for understanding the ability of species to survive in forest landscapes treated in various ways.

11.7 Which Species Can Benefit from Continuous Cover Forestry?

11.7.1 Species Requiring Shade or Spatial Continuity of Trees, and Below-Ground Biota

Compared to current dominant, clearcut-driven forestry practices, several groups of species can benefit from CCF. Those are species requiring (1) canopy cover and shading or (2) forest stand or single-tree continuity (roots, stems and canopies), and (3) less-disturbed soils. The latter two groups include species with poor dispersal capacity and/or low establishment probability. Logging-caused changes in the abundance of common species can also have significant ecological consequences. In particular, the abundance of bilberry affects many other forest biota. Bilberry regenerates mainly vegetatively through their underground stems, and both above- and below-ground parts suffer from clearcutting and soil preparation, as well as from exposure to direct sunlight (Atlegrim and Sjöberg 1996b, Tonteri et al. 2016). The decline of bilberry after logging is often directly proportional to the intensity of logging (Bergstedt and Milberg 2001).

Bilberry is the most important plant species for herbivores in the field layer of northern coniferous forests. The quantity and quality of bilberry as a food source collapse following clearcutting. The leaves of sun-exposed bilberry shoots have less water and more phenol content than those growing in shade (Atlegrim and Sjöberg 1996b). Due to lower bilberry abundance and possibly also its lower nutritional value, the number of moth and sawfly larvae in clearcut stands is about five times lower than in old-growth stands, whereas after selection cutting the abundance of these larvae remains almost unchanged (Atlegrim and Sjöberg 1995, 1996a; Kvasnes and Storaas 2007). Herbivorous insect larvae, in turn, are an important food source for several insectivorous bird species (Atlegrim and Sjöberg 1995), so bilberry abundance may also indirectly affect the reproductive success of forest birds. During their first weeks of life, chicks of forest grouse species, notably the capercaillie (*Tetrao urogallus*), grow rapidly due to feeding mainly on protein-rich sawfly and moth larvae (Kvasnes and Storaas 2007). In addition, bilberry thickets provide the shelter needed by the chicks. Bilberry and other ericaceous shrubs are important food resources for large herbivores hunted in boreal forests, directly contributing to cultural and provisioning services.

An ecologically important group of species that would probably benefit from switching from CCF to RF are epiphytic lichens growing on tree branches and trunks. Abundant lichens growing on large trees in turn support a diverse and abundant invertebrate fauna, which is an essential food source for many bird species such as the willow tit (*Poecile montanus*), crested tit (*Lophophanes cristatus*), Eurasian treecreeper (*Certhia familiaris*), and Siberian jay (*Perisoreus infaustus*) that forage in the canopy, especially in winter (Pettersson et al. 1995).

Several species groups can use continuous cover, uneven-aged forests as dispersal routes or habitat. Furthermore, in such forests, source populations might be frequently close by, so that distances between mature stands may not prevent the colonisation of new stands by these species. The presence and abundance of several epiphytic lichen species are limited by the ability to disperse and colonise (e.g., Dettki et al. 2000; Öcklinger et al. 2005; Fritz et al. 2008). Similarly, the movements and dispersal of the Siberian flying squirrel (*Pteromys volans*) are hampered by open clearcuts but not by tree-covered areas of sub-optimal quality for foraging or breeding (Selonen and Hanski 2003, 2004).

CCF also has implications for below-ground root-associated biodiversity and ecosystem function, notably promoting more abundant and diverse mycorrhizal communities compared to clearcutting (Sterkenburg et al. 2019; Kim et al. 2021). Mycorrhizal fungi represent a large fraction of biodiversity in boreal forests. Their mycelia form a foundation for soil food webs and play a critical symbiosis-driven role in forest production and nutrient cycling, as mediators of nutrient and water uptake by trees. The largely hidden life of mycorrhizal fungi in soils has hampered understanding of their diversity, biology and ecosystem function, as well as the consequences of forest-management practices, including CCF.

11.7.2 Threatened Species

Species that are preserved in CCF are mainly common, non-threatened forest species which also survive in even-aged managed forests. Given current conditions, no changes in forestry are required to secure the existence or population viability of common forest species and generalists, or most species requiring shade. However, if the aim is to halt and reverse the decline of European (Muys et al. 2022) or national forest biodiversity, the focus should be on declining (mostly red-listed) species. According to the Finnish Red Data, 436 (53%) vulnerable, endangered and critically endangered, primarily forest species are declining and rare due mostly to a scarcity of certain forest structural features, notably old-growth forests, very large and old trees or coarse deadwood, and natural tree-species composition (Hyvärinen et al. 2019). The numbers are quite similar in Sweden and Norway (SLU Artdatabanken 2020; Artsdatabanken 2021). These species are evolutionarily adapted to particular types, natural densities and amounts of the listed features.

Key questions are (1) whether and how quickly the species can recover from logging disturbance, and (2) whether CCF can maintain mature and old-forest species that are significantly reduced or unable to live in a commercial forest landscape based on RF. For most species that are declining and threatened by forestry, suitable structural features can be maintained and enhanced in either CCF or RF. In the case of RF, in phases of thinning and regeneration it is possible to retain individual or groups of trees that develop into old and large trees within a younger generation of trees and, when these trees die, they produce coarse deadwood. Old trees and coarse deadwood may also develop in retained forest patches outside the logging area. It is clear that CCF alone will not solve the challenges of biodiversity conservation in commercial forests. Productive and economically efficient CCF can often require repeated and intensive treatments. Management of structural features is important for biodiversity, but is neither an inherent part of RF nor CCF, so it needs to be addressed in addition to the choice of production system (cf. Gustafsson et al. 2020).

11.8 Conclusions and Research Needs

Compared to clearcutting-driven RF, CCF likely benefits species requiring shade and continuous availability of relatively big trees, such as some epiphytic lichens, including a few red-listed species. CCF also likely supports the long-distance dispersal of certain species, such as some epiphytic lichens and the Siberian flying squirrel, and provides shelter for the soil and its tree-root associated species community. Furthermore, if applied next to a protected area, it may help maintain the environmental conditions and species communities in that area by dampening edge effects. All these benefits, of course, depend on method specifics, such as gap sizes, tree structures and logging intensity. However, CCF alone does not maintain or increase structural features crucial for most red-listed forest species, particularly large and old trees and large-diameter deadwood. It is therefore unlikely to reverse negative trends of forest biodiversity, unless these features are intentionally retained and produced.

The main scientific gap in CCF concerns the cumulative biodiversity effects of decades of cutting. In CCF, the retained trees are small, so the ability of the regeneration method to maintain the microclimatic conditions of a closed forest and the species living on large trees, for example, is uncertain. On the other hand, it is not known to what extent CCF affects regional-level biodiversity, and little is known about species' responses to logging methods on soils other than heathlands; very little research has so far been carried out on peatlands. In the general forestry context, certain aspects would warrant more research, such as landscape topography, unwanted side effects of logging (e.g., wind or drought), continuities of logging intensity, and possible harvesting-associated changes in interspecific interactions (e.g., food webs) and ecosystem services such as pollination. In addition to these, several taxonomic groups (e.g., birds, mammals, slugs and snails, and soil organisms) are under-studied and need more attention to more completely describe the effects of CCF on biodiversity.

References

Angelstam P, Kuuluvainen T (2004) Boreal forest disturbance regimes, successional dynamics and landscape structures—a European perspective. Ecol Bull 51:117–136

Anttila S Koskela T Simkin J et al. (eds) (2021) METSO-tilannekatsaus 2020: Etelä-Suomen metsien monimuotoisuuden toimintaohjelma 2008–2025 (METSO situation report 2020: the Forest Biodiversity Action Plan for Southern Finland 2008–2025). Luonnonvara- ja biotalouden tutkimus 36/2021. (In Finnish)

Artsdatabanken (2021) Norwegian Red List for Species. https://www.biodiversity.no/Pages/135380/Norwegian_Red_List_for_Species. Accessed 20 Sep 2023

Atlegrim O, Sjöberg K (1995) Effects of clear-cutting and selective felling in Swedish boreal coniferous forest: response of invertebrate taxa eaten by birds. Entomol Fenn 6:79–90

Atlegrim O, Sjöberg K (1996a) Effects of clear-cutting and single-tree selection harvests on herbivorous insect larvae feeding on bilberry (*Vaccinium myrtillus*) in uneven-aged boreal *Picea abies* forests. For Ecol Manag 87:139–148

Atlegrim O, Sjöberg K (1996b) Response of bilberry (*Vaccinium myrtillus*) to clear-cutting and single-tree selection harvests in uneven-aged boreal *Picea abies* forests. For Ecol Manag 86:39–50

Atlegrim O, Sjöberg K (2004) Selective felling as a potential tool for maintaining biodiversity in managed forests. Biodivers Conserv 13:1123–1133

Atlegrim O, Sjöberg K, Ball J (1997) Forestry effects on a boreal ground beetle community in spring: selective logging and clear-cutting compared. Silva Fenn 8:19–26

Baber K, Otto P, Kahl T et al (2016) Disentangling the effects of forest-stand type and dead-wood origin of the early successional stage on the diversity of wood-inhabiting fungi. For Ecol Manag 377:161–169

Bader P, Jansson S, Jonsson BG (1995) Wood-inhabiting fungi and substratum decline in selectively logged boreal spruce forests. Biol Conserv 72:355–362

Berglund H, Kuuluvainen T (2021) Representative boreal forest habitats in northern Europe, and a revised model for ecosystem management and biodiversity conservation. Ambio 50:1003–1017

Berglund H, Jönsson MT, Penttilä R et al (2011) The effects of burning and dead-wood creation on the diversity of pioneer wood-inhabiting fungi in managed boreal spruce forests. For Ecol Manag 261:1293–1305

Bergstedt J, Milberg P (2001) The impact of logging intensity on field-layer vegetation in Swedish boreal forests. For Ecol Manag 154:105–115

Brunet J (2023) Generell naturvårdshänsyn i hyggesfritt skogsbruk i Mellaneuropa (General nature conservation considerations in clear-cut free forestry in Central Europe). SLU, Alnarp. (In Swedish)

Brunet J Berlin G (2005) Skånes skogar—historia, mångfald och skydd (Forests of Scania—history, biodiversity and conservation). Länsstyrelsen i Skåne län, rapport 12. (In Swedish)

Brunet J, Fritz Ö, Richnau G (2010) Biodiversity in European beech forests—a review with recommendations for sustainable forest management. Ecol Bull 53:77–94

Brunet J, Felton A, Lindbladh M (2012) From wooded pasture to timber production—changes in a European beech (*Fagus sylvatica*) forest landscape between 1840 and 2010. Scand J For Res 27:245–254

De Frenne P, Lenoir J, Luoto M et al (2021) Forest microclimates and climate change: importance, drivers and future research agenda. Glob Chang Biol 27:2279–2297 del Alba CE, Hjältén J, Sjögren J (2021) Restoration strategies in boreal forests: differing field and ground layer response to ecological restoration by burning and gap cutting. For Ecol Manag 494:119357

Dettki H, Klintberg P, Esseen P-A (2000) Are epiphytic lichens in young forests limited by local dispersal? Ecoscience 7:317–325

Edman M, Möller R, Ericson L (2006) Effects of enhanced tree growth rate on the decay capacities of three saprotrophic wood-fungi. For Ecol Manag 232:12–18

Ekholm A, Axelsson P, Hjältén J et al (2022) Short-term effects of continuous cover forestry on forest biomass production and biodiversity—applying single-tree selection in forests dominated by *Picea abies*. Ambio 51:2478–2495

Ekholm A, Lundqvist L, Axelsson EP et al (2023) Long-term yield and biodiversity in stands managed with the selection system and the rotation forestry system: a qualitative review. For Ecol Manag 537:120920

Esseen P-A, Renhorn K-A, Pettersson RB (1996) Epiphytic lichen biomass in managed and old-growth boreal forests: effect of branch quality. Ecol Appl 6:228–238

Felton A, Gustafsson L, Roberge J-M et al (2016) How climate change adaptation and mitigation strategies can threaten or enhance the biodiversity of production forests: insights from Sweden. Biol Conserv 194:11–20

Felton A, Löfroth T, Angelstam P et al (2020) Keeping pace with forestry: multi-scale conservation in a changing production forest matrix. Ambio 49:1050–1064

Fritz Ö, Gustafsson L, Larsson K (2008) Does forest continuity matter in conservation? A study of epiphytic lichens and bryophytes in beech forests of southern Sweden. Biol Conserv 141:655–668

Fritz Ö, Niklasson M, Churski M (2009) Tree age is a key factor for the conservation of epiphytic lichens and bryophytes in beech forests. Appl Veg Sci 12:93–106

Gibb H, Pettersson RB, Hjältén J et al (2006) Conservation-oriented forestry and early successional saproxylic beetles: responses of functional groups to manipulated dead wood substrates. Biol Conserv 129:437–450

Green RE, Cornell SJ, Scharlemann JP et al (2005) Farming and the fate of wild nature. Science 307:550–555

Gustafsson L (2002) Presence and abundance of red-listed plant species in Swedish forests. Conserv Biol 16:377–388

Gustafsson L, Perhans K (2010) Biodiversity conservation in Swedish forests: ways forward for a 30-year-old multi-scaled approach. Ambio 39:546–554

Gustafsson L, Kouki J, Sverdrup-Thygeson A (2010) Tree retention as a conservation measure in clear-cut forests of northern Europe: a review of ecological consequences. Scand J For Res 25:295–308

Gustafsson L, Bauhus J, Asbeck T et al (2020) Retention as an integrated biodiversity conservation approach for continuous-cover forestry in Europe. Ambio 49:85–97

Hägglund R, Hjältén J (2018) Substrate specific restoration promotes saproxylic beetle diversity in boreal forest set-asides. For Ecol Manag 425:45–58

Hägglund R, Hekkala A-M, Hjältén J et al (2015) Positive effects of ecological restoration on rare and threatened flat bugs (Heteroptera: Aradidae). J Insect Conserv 19:1089–1099

Hanski I (2011) Habitat loss, the dynamics of biodiversity, and a perspective on conservation. Ambio 40:248–255

Hautala H, Jalonen J, Laaka-Lindberg S et al (2004) Impacts of retention felling on coarse woody debris (CWD) in mature boreal spruce forests in Finland. Biodivers Conserv 13:1541–1554

Hautala H, Laaka-Lindberg S, Vanha-Majamaa I (2011) Effects of retention felling on epixylic species in boreal spruce forests in southern Finland. Restor Ecol 19:418–429

Henttonen HM, Nöjd P, Suvanto S et al (2019) Large trees have increased greatly in Finland during 1921–2013, but recent observations on old trees tell a different story. Ecol Indic 99:118–129

Hilszczański J, Gibb H, Hjältén J et al (2005) Parasitoids (hymenoptera, Ichneunionoidea) of sap-roxylic beetles are affected by forest successional stage and dead wood characteristics in boreal spruce forest. Biol Conserv 126:456–464

Hjältén J, Stenbacka F, Andersson J (2010) Saproxylic beetle assemblages on low stumps, high stumps and logs: implications for environmental effects of stump harvesting. For Ecol Manag 260:1149–1155

Hjältén J, Hägglund R, Löfroth T et al (2017) Forest restoration by burning and gap cutting of voluntary set-asides yield distinct immediate effects on saproxylic beetles. Biodivers Conserv 26:1623–1640

Hylander K, Weibull H (2012) Do time-lagged extinctions and colonizations change the interpre-tation of buffer strip effectiveness?—a study of riparian bryophytes in the first decade after logging. J Appl Ecol 49:1316–1324

Hylander K, Greiser C, Christiansen DM et al (2021) Climate adaptation of biodiversity conserva-tion in managed forest landscapes. Conserv Biol 36:e13847

Hyvärinen E, Juslén A, Kemppainen E et al (eds) (2019) The 2019 red list of Finnish species. Ministry of Environment and Finnish Environment Institute

Jalonen J, Vanha-Majamaa I (2001) Immediate effects of four different felling methods on mature boreal spruce forest understorey vegetation in southern Finland. For Ecol Manag 146:25–34

Joelsson K, Hjältén J, Work T et al (2017) Uneven-aged silviculture can reduce negative effects of forest management on beetles. For Ecol Manag 391:436–445

Joelsson K, Hjältén J, Gibb H (2018) Forest management strategy affects saproxylic beetle assem-blages: a comparison of even and uneven-aged silviculture using direct and indirect sampling. PLoS One 13:e0194905

Johansson T, Gibb H, Hilszczanski J et al (2006) Conservation-oriented manipulations of coarse woody debris affect its value as habitat for spruce-infesting bark and ambrosia beetles (Coleoptera: Scolytinae) in northern Sweden. Can J For Res 36:174–185

Johansson V, Wikström CJ, Hylander K (2018) Time-lagged lichen extinction in retained buffer strips 16.5 years after clear-cutting. Biol Conserv 225:53–65

Jokela J, Siitonen J, Koivula M (2019) Short-term effects of selection, gap, patch and clear cutting on the beetle fauna in boreal spruce-dominated forests. For Ecol Manag 446:29–37

Josefsson T, Olsson J, Östlund L (2010) Linking forest history and conservation efforts: long-term impact of low-intensity timber harvest on forest structure and wood-inhabiting fungi in north-ern Sweden. Biol Conserv 143:1803–1811

Juutilainen K, Mönkkönen M, Kotiranta H et al (2014) The effects of forest management on wood-inhabiting fungi occupying dead wood of different diameter fractions. For Ecol Manag 313:283–291

Jyväsjärvi J, Koivunen I, Muotka T (2020) Does the buffer width matter: testing the effective-ness of forest certificates in the protection of headwater stream ecosystems. For Ecol Manag 478:118532

Kärvemo S, Björkman C, Johansson T et al (2017) Forest restoration as a double-edged sword: the conflict between biodiversity conservation and pest control. J Appl Ecol 54:1658–1668

Kärvemo S, Jönsson M, Hekkala A-M et al (2021) Multi-taxon conservation in northern forest hot-spots: the role of forest characteristics and spatial scales. Landsc Ecol 36:989–1002

Kim S, Axelsson EP, Girona MM et al (2021) Continuous-cover forestry maintains soil fungal communities in Norway spruce dominated boreal forests. For Ecol Manag 480:118659

Koelemeijer IA, Ehrlén J, Jönsson M et al (2022) Interactive effects of drought and edge exposure on old-growth forest understory biodiversity. Landsc Ecol 37:1839–1853

Koelemeijer IA, Ehrlén J, De Frenne P et al (2023) Forest edge effects on moss growth are ampli-fied by drought. Ecol Appl 33:e2851

Koivula M (2002a) Boreal carabid-beetle (Coleoptera, Carabidae) assemblages in thinned uneven-aged and clear-cut spruce stands. Ann Zool Fenn 39:131–149

Koivula M (2002b) Alternative harvesting methods and boreal carabid beetles (Coleoptera, Carabidae). For Ecol Manag 167:103–121

Koivula MJ (2012) Under which conditions does retention harvesting support ground beetles of boreal forests? Baltic J Coleopterol 12:7–26

Koivula M, Niemelä J (2003) Gap felling as a forest harvesting method in boreal forests: responses of carabid beetles (Coleoptera, Carabidae). Ecography 26:179–187

Koivula M, Vanha-Majamaa I (2020) Experimental evidence on biodiversity impacts of variable retention forestry, prescribed burning, and deadwood manipulation in Fennoscandia. Ecol Process 9:11

Koivula MJ, Venn S, Hakola P et al (2019) Responses of boreal ground beetles (Coleoptera, Carabidae) to different logging regimes ten years post-harvest. For Ecol Manag 436:27–38

Kotiaho JS Ahlvik L Bäck J et al. (2021) Metsäluonnon turvaava suojelun kohdentaminen Suomessa. (Conservation targeting to secure forest biodiversity). Suomen Luontopaneelin julkaisuja 4/2021. (In Finnish with English and Swedish summaries)

Kruys N, Jonsson BG (1999) Fine woody debris is important for species richness on logs in managed boreal spruce forests of northern Sweden. Can J For Res 29:1295–1299

Kuglerová L, Jyväsjärvi J, Ruffing C et al (2020) Cutting edge: a comparison of contemporary practices of riparian buffer retention around small streams in Canada, Finland, and Sweden. Water Resour Res 56:e2019WR026381

Kuglerová L, Nilsson G, Maher Hasselquist E (2022) Too much, too soon? Two Swedish case studies of short-term deadwood recruitment in riparian buffers. Ambio 52:440–452

Kuuluvainen T (2009) Forest management and biodiversity conservation based on natural ecosystem dynamics in northern Europe: the complexity challenge. Ambio 38:309–315

Kuuluvainen T, Aakala T (2011) Natural forest dynamics in boreal Fennoscandia: a review and classification. Silva Fenn 45:823–841

Kuuluvainen T, Tahvonen O, Aakala T (2012) Even-aged and uneven-aged forest management in boreal Fennoscandia: a review. Ambio 41:720–737

Kuusinen M, Siitonen J (2009) Epiphytic lichen diversity in old-growth and managed *Picea abies* stands in southern Finland. J Veg Sci 9:283–292

Kvasnes MAJ, Storaas T (2007) Effects of harvesting regime on food availability and cover from predators in capercaillie (*Tetrao urogallus*) brood habitats. Scand J For Res 22:241–247

Lie MH, Arup U, Grytnes J-A et al (2009) The importance of host tree age, size and growth rate as determinants of epiphytic lichen diversity in boreal spruce forests. Biodivers Conserv 18:3579–3596

Lindblad I (1998) Wood-inhabiting fungi on fallen logs of Norway spruce: relations to forest management and substrate quality. Nord J Bot 18:243–256

Lindbladh M, Abrahamsson M (2008) Beetle diversity in high-stumps from Norway spruce thinnings. Scand J For Res 23:339–347

Lindbladh M, Foster DR (2010) Dynamics of long-lived foundation species: the history of *Quercus* in southern Scandinavia. J Ecol 98:1330–1345

Lindbladh M, Abrahamsson M, Seedre M et al (2007) Saproxylic beetles in artificially created high-stumps of spruce and birch within and outside hotspot areas. Biodivers Conserv 16:3213–3226

Lindbladh M, Axelsson A-L, Hultberg T et al (2014) From broadleaves to spruce—the borealization of southern Sweden. Scand J For Res 29:686–696

Lindenmayer DB, Franklin JF (eds) (2002) Conserving Forest biodiversity: a comprehensive multiscaled approach. Island Press

Lindhe A, Åsenblad N, Toresson H-G (2004) Cut logs and high stumps of spruce, birch, aspen and oak—nine years of saproxylic fungi succession. Biol Conserv 119:443–454

Löf M, Brunet J, Filyushkina A et al (2016) Management of oak forests: striking a balance between timber production, biodiversity and cultural services. Int J Biodivers Sci Ecosyst Serv Manag 12:59–73

Lundmark H, Josefsson T, Östlund L (2013) The history of clear-cutting in northern Sweden—driving forces and myths in boreal silviculture. For Ecol Manag 307:112–122

Matveinen-Huju K, Koivula M (2008) Effects of alternative harvesting methods on boreal forest spider assemblages. Can J For Res 38:782–794

Mikusiński G, Pressey RL, Edenius L et al (2007) Conservation planning in forest landscapes of Fennoscandia and an approach to the challenge of countdown 2010. Conserv Biol 21:1445–1454

Mönkkönen M, Aakala T, Blattert C et al (2022) More wood but less biodiversity in forests in Finland: a historical evaluation. Memoranda Soc Fauna Fl Fenn 98(Suppl. 2):1–11

Muurinen L, Oksanen J, Vanha-Majamaa I et al (2019) Legacy effects of logging on boreal forest understorey vegetation communities in decadal time scales in northern Finland. For Ecol Manag 436:11–20

Muys B, Angelstam P, Bauhus J et al (2022) Forest biodiversity in Europe. From science to policy 13. European Forest Institute

Nascimbene J, Thor G, Nimis PL (2013) Effects of forest management on epiphytic lichens in temperate deciduous forests of Europe—a review. For Ecol Manag 298:27–38

Niinistö T, Peltola A, Räty M et al (eds) (2021) Forest statistical yearbook 2021. Natural Resources Institute Finland (Luke), Helsinki, p 202

Öcklinger E, Niklasson M, Nilsson SG (2005) Is local distribution of the epiphytic lichen *Lobaria pulmonaria* limited by dispersal capacity or habitat quality? Biodivers Conserv 14:759–773

Olsson J, Johansson T, Jonsson BG et al (2012) Landscape and substrate properties affect species richness and community composition of saproxylic beetles. For Ecol Manag 286:108–120

Pasanen H, Junninen K, Kouki J (2014) Restoring dead wood in forests diversifies wood-decaying fungal assemblages but does not quickly benefit red-listed species. For Ecol Manag 312:92–100

Pasanen H, Junninen K, Boberg J et al (2018) Life after tree death: does restored dead wood host different fungal communities to natural woody substrates? For Ecol Manag 409:863–871

Pasanen H, Juutilainen K, Siitonen J (2019) Responses of polypore fungi following disturbance-emulating harvesting treatments and deadwood creation in boreal Norway spruce dominated forests. Scand J For Res 34:557–568

Pettersson RB (1996) Effect of forestry on the abundance and diversity of arboreal spiders in the boreal spruce forest. Ecography 19:221–228

Pettersson RB, Ball JP, Renhorn K-E et al (1995) Invertebrate communities in boreal forest canopies as influenced by forestry and lichens with implications for passerine birds. Biol Conserv 74:57–63

Pihlaja M, Koivula M, Niemelä J (2006) Responses of boreal carabid beetle assemblages (Coleoptera, Carabidae) to clear-cutting and top-soil preparation. For Ecol Manag 222:182–190

Puumalainen J, Kennedy P, Folving S (2003) Monitoring forest biodiversity: a European perspective with reference to temperate and boreal forest zone. J Environ Manag 67:5–14

Rabinowitsch-Jokinen R, Vanha-Majamaa I (2010) Immediate effects of logging, mounding and removal of logging residues and stumps on coarse woody debris in managed boreal Norway spruce stands. Silva Fenn 44:51–61

Rabinowitsch-Jokinen R, Laaka-Lindberg S, Vanha-Majamaa I (2012) Immediate effects of logging, mounding and removal of logging residues on epixylic species in managed boreal Norway spruce stands in southern Finland. J Sustain For 31:205–229

Ruete A, Snäll T, Jönsson M (2016) Dynamic anthropogenic edge effects on the distribution and diversity of fungi in fragmented old-growth forests. Ecol Appl 26:1475–1485

Savilaakso S, Johansson A, Häkkilä M et al (2021) What are the effects of even-aged and uneven-aged forest management on boreal forest biodiversity in Fennoscandia and European Russia? A systematic review. Environ Evid 10:1

Selonen V, Hanski IK (2003) Movements of the flying squirrel *Pteromys volans* in corridors and in matrix habitat. Ecography 26:641–651

Selonen V, Hanski IK (2004) Young flying squirrels (*Pteromys volans*) dispersing in fragmented forests. Behav Ecol 15:564–571

Siira-Pietikäinen A, Haimi J (2009) Changes in soil fauna 10 years after forest harvestings: comparison between clear felling and green-tree retention methods. For Ecol Manag 258:332–328

Siira-Pietikäinen A, Pietikäinen J, Fritze H et al (2001) Short-term responses of soil decomposer communities to forest management: clear felling versus alternative forest harvesting methods. Can J For Res 31:88–99

Siira-Pietikäinen A, Haimi J, Siitonen J (2003) Short-term responses of soil macroarthropod community to clear felling and alternative forest regeneration methods. For Ecol Manag 172:339–353

Sippola A-L, Lehesvirta T, Renvall P (2001) Effects of selective logging on coarse woody debris and diversity of wood-decaying polypores in eastern Finland. Ecol Bull 49:243–254

Skogsstyrelsen (2020) Ett urval av naturvårdsarter och andra indikatorarter (Selection of protected and other indicator species). Skogsstyrelsen. (In Swedish)

SLU Artdatabanken (2020) The Swedish Red List 2020. https://doi.org/10.15468/jhwkpq. Accessed via GBIF.org 20 Sep 2023

Stenbacka F, Hjältén J, Hilszczański J et al (2010) Saproxylic and non-saproxylic beetle assemblages in boreal spruce forests of different age and forestry intensity. Ecol Appl 20:2310–2321

Sterkenburg E, Clemmensen KE, Lindahl BD et al (2019) The significance of retention trees for survival of ectomycorrhizal fungi in clear-cut scots pine forests. J Appl Ecol 56:1367–1378

Timonen J, Gustafsson L, Kotiaho JS et al (2011) Hotspots in cold climate: conservation value of woodland key habitats in boreal forests. Biol Conserv 144:2061–2067

Tonteri T, Salemaa M, Rautio P et al (2016) Forest management regulates temporal change in the cover of boreal plant species. For Ecol Manag 381:115–124

Tullus T, Rosenvald R, Leis M et al (2018) Impacts of shelterwood logging on forest bryoflora: distinct assemblages with richness comparable to mature forests. For Ecol Manag 411:67–74

Vanha-Majamaa I, Shorohova E, Kushnevskaya H et al (2017) Resilience of understory vegetation after variable retention felling in boreal Norway spruce forests—a ten-year perspective. For Ecol Manag 393:12–28

Varenius K, Kåren O, Lindahl B et al (2016) Long-term effects of tree harvesting on ectomycorrhizal fungal communities in boreal scots pine forests. For Ecol Manag 380:41–49

Versluijs M, Eggers S, Hjältén J et al (2017) Ecological restoration in boreal forest modifies the structure of bird assemblages. For Ecol Manag 401:75–88

Versluijs M, Hekkala A-M, Lindberg E et al (2020) Comparing the effects of even-aged thinning and selective felling on boreal forest birds. For Ecol Manag 475:118404

Chapter 12
Forest Damage

Jarkko Hantula, Malin Elfstrand, Anne-Maarit Hekkala, Ari M. Hietala, Juha Honkaniemi, Maartje Klapwijk, Matti Koivula, Juho Matala, Jonas Rönnberg, Juha Siitonen, and Fredrik Widemo

Abstract

- Heterobasidion root and butt rot pose a greater risk in continuous cover forestry (CCF) than in rotation forestry (RF) in conifer-dominated forests, regardless of whether selective, gap or shelterwood cutting is used.

J. Hantula (✉) · J. Honkaniemi · M. Koivula · J. Siitonen
Natural Resources Institute Finland (Luke), Helsinki, Finland
e-mail: ext.jarkko.hantula@luke.fi; juha.honkaniemi@luke.fi; matti.koivula@luke.fi; juha.
siitonen@luke.fi

M. Elfstrand
Department of Forest Mycology and Plant Pathology, Swedish University of Agricultural
Sciences (SLU), Uppsala, Sweden
e-mail: malin.elfstrand@slu.se

A.-M. Hekkala · F. Widemo
Department of Wildlife, Fish and Environmental Studies, Swedish University of Agricultural
Sciences (SLU), Umeå, Sweden
e-mail: anne.maarit.hekkala@slu.se; fredrik.widemo@slu.se

A. M. Hietala
Division of Biotechnology and Plant Health, Norwegian Institute of Bioeconomy Research,
Ås, Norway
e-mail: ari.hietala@nibio.no

M. Klapwijk
Department of Ecology, Swedish University of Agricultural Sciences (SLU),
Uppsala, Sweden
e-mail: maartje.klapwijk@slu.se

J. Matala
Natural Resources Institute Finland (Luke), Joensuu, Finland
e-mail: juho.matala@luke.fi

J. Rönnberg
Southern Swedish Forest Research Centre, Swedish University of Agricultural Sciences
(SLU), Lomma, Sweden
e-mail: jonas.ronnberg@slu.se

P. Rautio et al. (eds.), *Continuous Cover Forestry in Boreal Nordic Countries*,
Managing Forest Ecosystems 45, https://doi.org/10.1007/978-3-031-70484-0_12

- Damage from wind, snow, spruce bark beetle, and large pine weevil are likely to be less severe in CCF than in RF. However, the conversion of RF to CCF may briefly expose stands to windthrow.
- Browsing by large herbivores on saplings may limit regeneration of tree species other than spruce in continuous cover forestry and reduce tree species diversity, but alternative silvicultural practices may also increase forage availability in the field and shrub layer. Browsing damage outcomes for saplings in CCF are difficult to predict.
- For many types of damage in CCF, substantial knowledge gaps complicate the assessment of damage risk.

Keywords Abiotic damage · Heterobasidion root rot · Forest disease · European spruce bark beetle · Cervid damage

Forest damage can be roughly divided into abiotic and biotic damage. Abiotic damage refers to harmful effects that arise from non-living, environmental factors. Abiotic stressors that can lead to tree mortality and habitat destruction include wind, snow, frost, flood, and drought. Biotic damage that weakens or kills trees, and in some cases triggers large-scale forest die-offs, can be caused by insects, fungi, oomycetes and mammals. Multiple stressors can coincide and interact to some degree at tree, forest, or landscape scales.

Tree species, populations of trees or even individual trees vary in their genetic resistance to specific abiotic and biotic stressors, depending on past selection pressures (Mageroy et al. 2023). Tree resistance can also change with age: older trees may be able to fight off a specific attacker through certain combinations of constitutive and induced defence, whereas young seedlings may lack crucial defence mechanisms and succumb, or the stressor in question may only concern trees of a specific age or size. Biotic damage arises through complex interactions between the tree, the biotic damaging agent, and the environment. Conditions that predispose trees and favour the propagation and spread of biotic damaging agents are normally the precursor to biotic outbreaks.

At the individual-tree level, resistance to abiotic or biotic stressors can be defined as the ability of a tree to limit damage or to prevent the damage or attack altogether. At the stand or landscape level, forest resilience refers to the ability to absorb perturbations and maintain desired forest functioning, structure, and composition. As forest resilience to specific abiotic and biotic stressors depends on the tree-species composition and age structure, forest operations strongly influence resilience in managed forests. While CCF is not a novel concept, there is generally very little documented information about tree resistance and forest resilience toward abiotic and biotic stressors in CCF conditions. The current understanding of factors that lead to abiotic and biotic outbreaks is almost exclusively based on studies in even-aged conifer monocultures subjected to rotation forestry (RF), the backbone of forestry in the northern hemisphere.

In this chapter we summarise the current knowledge and make some predictions about how structural features of CCF forests and associated forest operations (outlined in other chapters of this book), can either strengthen or weaken forest resilience to specific abiotic or biotic stressors.

12.1 Windthrow, Snow and Drought

Wind and snow are, by area affected, the most significant abiotic causes of forest damage in Fennoscandia (Díaz-Yáñez et al. 2016; Korhonen et al. 2021). In addition, drought is increasingly damaging pine and spruce. Like all abiotic damage sources, wind, snow and drought are strongly associated with exceptional weather conditions, forest and landscape structure, and the physical properties of trees. There are regional differences in the prevalence of these damage types.

The importance of wind damage increases as wind speed reaches storm levels. Snow damage occurs when the snow load suddenly becomes exceptionally large. Forest structure, particularly basal area and stem density, affects both types of damage significantly (Peltola 2006; Suvanto et al. 2021; 2019). Wind blows faster through less dense forests. Recently thinned forests, where trees have not had time to adapt to the new spatial pattern and the altered living conditions, are particularly susceptible. Also snow damage is usually most severe in recently thinned stands, especially if thinning has been delayed. There may also be indirect interactions between biotic and abiotic damage agents. For example, pre-commercial thinning is sometimes employed to mitigate browsing damage, but may lead to snow damage (e.g., Päätalo et al. 1999; Wallentin and Nilsson 2014).

Windthrow, in particular, increases in mosaic-like landscapes, where clearcutting creates sharp forest edges where the wind can blow strongly (Zeng et al. 2010, 2007). In terms of the physical properties of trees, stem diameter and height, their ratio, and root depth are the most significant factors affecting both wind and snow damage (Peltola 2006). Thicker trees, and trees with higher height-diameter ratios, increase the risk, while a deeper root system anchors the tree and reduces windthrow risk.

Abiotic damage has received little attention in the continuous cover forestry context. The risk of wind damage in CCF is likely to be smaller than in RF stands (Hahn et al. 2021; Hanewinkel et al. 2014; Nevalainen 2017; Pukkala et al. 2016), with the exceptions of shelterwood cutting (Hånell and Ottosson-Löfvenius 1994) and the conversion phase (Potterf et al. 2022). This is due, among other things, to differences in silvicultural methods, which affect the mean diameter and basal area of the growing stock (Pukkala et al. 2016). A diverse forest structure reduces wind speed inside the forest, reducing the risk of wind damage (Dobbertin 2002; Hanewinkel et al. 2014; Pukkala et al. 2016).

CCF-managed forests could be expected to recover more quickly from drought than even-aged stands, as they generally have a lower density of mature trees (Hlásny et al. 2014). Young naturally regenerated seedlings on a site managed using

CCF would be at lower drought risk due to the shade and shelter provided by the overstorey.

The importance of landscape structure has not been studied, but it is likely to impact damage risks significantly. If all forest management were to shift directly to CCF, the results of stand-level studies could be generalised. However, it is likely that diversification of forest management will lead to a mosaic of management practices. In this case, for example, clearcuts may still abut selectively cut stands, increasing wind damage risk to the latter. On the other hand, mature stands benefit from CCF in neighbouring stands.

12.2   Root Rot

12.2.1   The Biology of Root Rot Fungi

There are two *Heterobasidion* species in the Nordic countries. *Heterobasidion parviporum* causes root and butt rot in spruce and larch. *Heterobasidion annosum* has a wider host range. It causes root rot in pine but can also kill birch and juniper trees, and cause root and butt rot in spruce similar to *H. parviporum*. Despite its narrower host range, *H. parviporum* can also kill pine seedlings, compromising tree species diversification via natural regeneration in CCF (Korhonen 1978).

12.2.2   Root Rot in Rotation Forestry

Heterobasidion species spread to new sites by spores. In unmanaged forests, this usually occurs via bare-wood surfaces unprotected by bark. However, in managed forests, mechanised operations have opened a new and more efficient route for the spread of *Heterobasidion* root rot. Logging creates unprotected stump surfaces which are at risk of infection during the sporulation season. As a result, the prevalence of root rot in Nordic forests has significantly increased compared to what it would have been without human influence. Root rot has also become one of the most economically significant tree diseases in forestry, causing direct economic losses (reduced growth, lost timber value due to rot, mortality), and making trees more vulnerable to other damage agents, particularly wind (Fig. 12.1).

Heterobasidion root rot fungi arrive at a new site when their windborne basidiospores land on freshly cut tree stumps. Following spore germination, mycelia penetrate the stump and its root system, and can continue into other trees of the current and next generation through root contacts. These events result in disease hotspots up to tens of metres in diameter that may persist for hundreds of years. In individual stumps, these root rot fungi can survive for more than 50 years (Greig and Pratt 1976; Piri 1996). In Finland, prevention of root rot infection is mandatory in

Fig. 12.1 Root rot caused by *Heterobasidion* fungi weakens the root system, making trees more susceptible to windthrow. Photo: Jarkko Hantula

cuttings during snow-free periods (Piri et al. 2019). In Sweden, stump treatment is also practised in snow-free thinning operations, but not yet in final harvests. In Norway, stump treatment is currently voluntary and rarely used (Hietala et al. 2016).

There are both chemical (urea) and biological (RotStop®, *Phlebiopsis gigantea*) stump-treatment methods. Basinox is a new treatment that can be applied in Sweden. It is based on one strain of an unidentified *Pseudomonas* bacterium. All these agents provide good but incomplete protection against spore-mediated infection when conducted properly. However, these methods do not eradicate established root rot fungus. Existing mycelia in root systems continue to spread despite stump treatment.

Cultivating a *Heterobasidion*-resistant tree species for one rotation effectively eliminates the disease from a site. For a site infected with *H. parviporum*, both Scots pine and all deciduous trees are suitable alternatives, but not larch or lodgepole pine. The situation is worse in sites with *H. annosum*, as this species infects all conifers, causing similar root rot as *H. parviporum* in spruce. *Heterobasidion annosum* also infects birch but does not spread from a harvested pure birch stand to subsequently planted conifers due to the quick decay of hardwood stumps.

12.2.3   Root Rot in Continuous Cover Forestry

In CCF, selective cutting and gap cutting during snow-free periods can facilitate spore infection by *Heterobasidion*, either through stump surfaces or root damage caused by logging machinery. The total stump surface area exposed by logging in CCF, whether selective cutting, gap cutting or shelterwood preparation, tends to be similar or slightly smaller over time than that in RF. For this reason, the risk of stump-surface infection by spores is generally only slightly lower in CCF. On the other hand, the risk of stem and root damage by forest machinery to the remaining trees is significantly greater in CCF than in RF. This probably largely evens out the difference in root rot risk between CCF and RF (Dwyer et al. 2004). Therefore, further information is needed on possible differences in the spore-mediated spread of *Heterobasidion* species between continuous cover and even-aged forests.

Forest regeneration in CCF relies on existing undergrowth in selective cuttings, increasing its *Heterobasidion* risk compared to RF. Understorey vegetation favours the effective vegetative spread of root rot in to growing trees, as well as the persistence of *Heterobasidion* infection in infected root systems (Piri 1996; Piri and Valkonen 2013; Fig. 12.2). In CCF, dense sapling stands form where the soil has been exposed during logging or conditions are otherwise favourable for undergrowth. The next generation of trees in RF can become infected via root contacts with stumps of infected prior-generation trees. In CCF, the several-years-long co-presence of susceptible overstorey and understorey trees facilitates the root-to-root spread of infection between tree generations. In addition, understorey trees can be quite old, allowing their root systems much more time to become infected with root rot compared to trees in even-aged forestry.

In gap cuts or strip clearcuts, the difference in risk of root rot compared to even-aged forestry is smaller when broadleaf trees are left to grow in the openings as mixed stands (Piri et al. 1990) or even if new spruce seedlings are planted. However,

Fig. 12.2 The dense understorey in continuous cover forestry provides ideal conditions for the vegetative spread of root rot through the formation of a network of interconnected roots. Photo: Erkki Oksanen / Luke

using only undergrowth saplings for regeneration bears the same risk as selection cutting. The consequences of variation in management are also affected by which species of *Heterobasidion* is present. *Heterobasidion annosum* may also attack birch and can potentially spread both to and from birch and spruce or pine.

Controlling root rot that has already entered a site is practically impossible in a selection-cut managed forest because changing the tree species is not possible. However, control can be tried at small root rot centres including only a few trees by creating small openings around them and removing all surrounding trees susceptible to root rot, and regenerating with resistant tree species. Also, future attenuation of existing disease by new control measures like *Heterobasidion*-debilitating viruses (Vainio et al. 2018; Piri et al. 2023) could change the situation.

Controlling widespread *Heterobasidion* root rot, on the other hand, requires a clearcut and subsequent switch to a resistant tree species. It is never advisable to establish continuous cover stands on a site already infected with *Heterobasidion* if only susceptible tree species can efficiently reproduce. Examples of this situation include spruce on nutrient-rich mesic soil or pine on a dry nutrient-poor site. Even on a healthy site, it is safest to conduct forest management activities only during the winter, when spore dispersal is negligible. However, a warming climate or growing popularity of CCF poses logistical challenges for timber harvesting, making it impossible to avoid summer logging. In such a case, the quality of forest management becomes more important, especially regarding stump treatment and quick action to contain emerging root rot centres.

12.3 Other Significant Forest Diseases

Compared to *Heterobasidion* root rot, knowledge about other forest diseases in continuous cover forestry is limited. General knowledge can still inform understanding of factors that favour the spread of the specific causative agents and predispose the trees to infection. Before treating these predictions as credible scientific information, it is essential to subject them to rigorous testing in controlled experiments.

12.3.1 *Scleroderris Canker*

Shoot mortality due to the fungus *Gremmeniella abietina* causing scleroderris canker on pine trees varies annually from insignificant to extremely severe epidemics in boreal forests. The last serious outbreaks occurred in Finland in the 1980s and in Sweden at the turn of the 2000s. The risk of shoot mortality is mainly influenced by the location of the forest stand and the weather of preceding summers (Witzell and Karlman 2000; Thomsen 2009). In addition, the quality of forest management, such as timely thinning, and the provenance of the trees significantly affect damage risk.

Pine provenances from much further south than the growth site are particularly susceptible (Uotila 1985). Continuous cover forestry is usually based on natural regeneration, so it carries a low risk in terms of tree origin. In the Nordic countries, even-aged forests based on seeding or planting are typically regenerated with site-suitable seed origins, so scleroderris canker risk should not differ between continuous cover or even-aged forests. However, the risk of shoot mortality increases in even-aged forests if they are regenerated with too-southern seed origins, for example for climate change adaptation purposes.

The risk of shoot mortality by *G. abietina*, like many other shoot and needle diseases, increases in moist microclimates and under the shade of other trees (Read 1968; Niemelä et al. 1992). In continuous cover forests, air humidity does not differ significantly from that of even-aged seedling stands or young forests. However, shading by overstorey trees may slightly increase shoot-mortality risk for suppressed pine seedlings owing to a microclimate that may favour the pathogen and predispose the tree. Careful forest management can override these factors, as seedling stands grown on dry soils with continuous cover do not differ significantly from even-aged forest cultivation stands. Therefore, thinning and selection cuts in continuous cover stands must be carefully timed, just as in even-aged stands, to ensure that the microclimate does not become favourable for scleroderris canker.

Nevalainen (2017) compared CCF and RF, mentioning the risk of *G. abietina* infection originating from nurseries. This undoubtedly occurs despite the health requirements of seedlings, but the actual damage caused by *G. abietina* depends crucially on the prevailing conditions at the site, as noted above. Therefore, the contribution of nursery infections to *G. abietina* damage is marginal, and in this respect, there should be no significant difference in scleroderris canker risk between continuous cover and even-aged forest management.

The information in this section comes from data on even-aged forests and logical reasoning derived from it, not on experimental research on different forest structures, which is therefore clearly needed.

12.3.2 Scots Pine Blister Rust

Scots pine blister rust (also called pine stem rust, cronartium rust or resin top disease) is caused by *Cronartium pini*. This fungus has two forms (Hantula et al. 2002; Samils et al. 2021). The heteroecious form has a complete life cycle that includes the sexual sporulation stage on a herbaceous intermediate host, usually white swallow-wort (*Vincetoxicum hirundinaria*) or small cow-wheat (*Melampyrum sylvaticum*, Kaitera et al. 2005). The autoecious variant with an incomplete life cycle spreads directly from pine to pine. The heteroecious and autoecious forms are clearly differentiated and usually occur in separate populations locally, but with no clear separation at a larger geographic scale in Finland and Sweden (Samils et al. 2021).

The worst pine stem rust damage in Finland and Sweden has been linked to the heteroecious form; the presence of intermediate host plants on a site is a risk factor. Thus, it is not advisable to establish pine forests on overly nutrient-rich soils (Kaitera et al. 2005; Samils and Stenlid 2022). If the forest management method increases or decreases the occurrence of intermediate host plants, it may also impact the susceptibility of trees to cronartium rust. Small cow-wheat has become increasingly rare in Lapland due to forest densification (Jalkanen 2014), but so far there is no research-based evidence on whether the increased light in continuous cover forests could make this intermediate host more common.

Knowledge about the genetic basis of resistance to Scots pine blister rust is poor, but experience from North America indicates that there is genetic resistance against the closely related white pine blister rust (Sniezko and Liu 2022). Therefore, it is not advisable to use blister-affected trees as seed trees, either in RF or in CCF because of the likely susceptibility of their offspring.

Taken together, the need for scientific information on pine stem rust and the role of management strategies in the damage patterns is urgent.

12.4 Insect Damage

12.4.1 *European Spruce Bark Beetle (*Ips typographus*)*

Insect damage in continuous cover forestry has so far been studied relatively little. However, the risk of damage can be assessed through the ecology of the pests and the forest structures produced by CCF. Based on this knowledge, the general perception among Finnish forestry professionals is that the risk of insect damage is lower in CCF than in RF (Nevalainen 2017). This difference may result from CCF having relatively lower effects of bottom-up forces (resource quality and quantity) than RF compared to top-down forces (natural enemy pressure, Klapwijk et al. 2016). We will examine below how well founded this view is, and present evidence for bark beetles, defoliators and regeneration pests (*sensu* Björkman et al. 2015).

The spruce bark beetle is the most economically damaging insect in the Nordic countries (e.g., Uotila 1994; Nuorteva et al. 2022). Other insects usually cause relatively economically insignificant growth decline, needle loss, and individual tree deaths. The economic losses caused by the spruce bark beetle vary annually but have been estimated to be EUR 5–21 million in recent years in Finland (Hantula et al. 2023). Because healthy trees may also be harvested to compensate for the cost of harvesting small-scale damage, this estimate—based on cutblock sizes rather than volumes or shares of damaged trees—is likely an overestimate. The real impact is difficult to assess due to limited inventory data. However, with climate change, damage events caused by insects are expected to become more common (Jönsson et al. 2007, 2009, 2011; Seidl et al. 2008, 2011, Öhrn et al. 2014; Økland et al. 2015; Ruosteenoja et al. 2016; Venäläinen et al. 2020; Hlásny et al. 2021). In Finland, problems with the spruce bark beetle have so far remained relatively small and local (Viiri et al. 2019), although in the 2010s windthrow and drought appear to have increased secondary beetle damage (Nuorteva et al. 2022). In Sweden, spruce bark beetle damage has increased remarkably during the last decade. The total volume of spruce-bark-beetle-killed trees has been estimated at 32 Mm^3 between 2018–2022 (Wulff and Roberge 2021), while in the 1990s the estimate was 1.5 Mm^3 (Kärvemo and Schroeder 2010).

Large-scale bark beetle outbreaks are usually induced by windstorms (Ravn 1985; Furuta 1989; Krehan et al. 2010), but can also be caused by exceptionally warm summers and drought (Matthews et al. 2018). Normally the spruce bark beetle produces one generation per summer (Annila 1969) in northern Europe, but in some summers of the 2010s it has been able to produce two generations due to warm and dry weather (e.g., Pouttu and Annila 2011; Neuvonen et al. 2015; Neuvonen and Viiri 2017; Nuorteva et al. 2022). Mitigation of climate change would help limit forest damage, but forest structure can greatly determine how susceptible forests are to windthrows and bark beetles. From a forest management perspective, multi-species forests have been suggested to be less susceptible than monocultures for both the spruce bark beetle and storm damage (Seidl et al. 2011; Dobor et al. 2020; Müller et al. 2022). In extreme drought conditions, however, these generalisations may not hold (de Groot et al. 2023).

The spruce bark beetle prefers relatively large and already-weakened spruce trees (at least 15 cm in diameter at breast height) for various reasons (Netherer and Hammerbacher 2022; Nuorteva et al. 2022) (Fig 12.3). Such weakened trees can be found, for example, on the tops of hills, on sunny southwestern slopes, on the edges of clearcuts, in mature spruce stands, and on sites that are too dry for spruce (Wermelinger 2004). Advanced age and fungal infections can also weaken spruce trees. If there are significant numbers of nearby trees (e.g., tens of cubic metres) that have recently been killed by the spruce bark beetle, live trees are at risk because they may share the conditions favourable for the bark beetle. On the other hand, recurring hot and dry summers can expose large spruce trees to extensive damage at any site (Wermelinger 2004).

12.4.2 Spruce Bark Beetle Impacts on Uneven- and Even-Aged Spruce Stands

Even-aged, structurally homogeneous monocultures are generally more susceptible to bark beetle outbreaks than mixed-species or uneven-aged stands, due to the host specialisation of the beetle species described above (Raffa et al. 2015). For the spruce bark beetle, a mature, pure spruce stand on a dry site would be particularly favourable, but even a mixed-aged spruce stand on dry sandy soils would not be immune to attack. However, continuous cover spruce stands differ from even-aged stands in several key respects. First, in continuous cover stands most trees are always too small for the spruce bark beetle, and second, potential host trees—harvest-ready spruce trees—are considerably less frequent than in mature even-aged stands. Considering the whole rotation cycle, CCF never produces large dense clusters of mature trees, unlike clearcut-based RF. Third, the harvesting in continuous cover stands is focused on the largest and thus most potentially suitable tree species for the spruce bark beetle. According to the resource availability hypothesis (Endara and Coley 2011; Begon and Townsend 2021), low host-tree density limits the growth potential of the spruce bark beetle population; where there are fewer host trees, fewer new beetles will hatch. When beetle density is sufficiently low, the population is further limited by the Allee effect. According to this theory, mate selection and reproductive success are weaker in a sparse population than in a dense one (Stephens et al. 1999).

For these reasons, CCF seems to be better suited than RF for avoiding bark beetle damage in spruce stands. However, CCF may begin with a relatively even-aged forest, as is the case in a significant part of Nordic commercial forests. These forests may be treated by selective cutting, which aims to retain trees of as wide a size range as possible, or by gap cutting. If the retained trees are already quite large, root and stem damage to the residual stand caused by harvesting may increase the risk of bark beetles. Harvest residues release volatile compounds, attracting pioneer insects that seek suitable nearby trees. Strip roads open the canopy, increasing sun exposure, which stresses south-facing trees and exposes them to wind damage. Stressed trees release terpene alcohols and ketones that attract insects like the spruce bark beetle (Schiebe et al. 2019). The combined effect of creating strip roads and harvesting in uneven-aged forests may be similar to the impact of thinning in even-aged stands, but this topic warrants more study.

12.4.3 Other Locally Economically Important Insects

The risk of insects other than the spruce bark beetle under selective cutting has not been assessed, but a few experiments have assessed gap cutting from this perspective in the Nordic countries. Other bark beetle species potentially causing economic damage in spruce-dominated stands are the six-toothed bark beetle (*Pityogenes chalcographus*) and the small spruce bark beetle (*Polygraphus poligraphus*). The large pine weevil (*Hylobius abietis*) is also fairly common throughout the region, causing remarkable economic losses by killing conifer saplings (e.g., Björklund et al. 2003; Långström and Day 2004). A few other insect species cause growth loss or occasionally death of various tree species, notably sawflies (particularly the common pine sawfly *Diprion pini* and the European pine sawfly *Neopridion sertifer*) and the nun moth (*Lymantria monacha*, Hantula et al. 2023).

The six-toothed bark beetle and the small spruce bark beetle are common in spruce-dominated forests throughout the Nordic countries (Rassi et al. 2015; Artdatabanken 2023; Artsdatabanken 2023). The six-toothed bark beetle attacks and can kill smaller trees than the spruce bark beetle (Schebeck et al. 2022). Also

Fig. 12.3 Norway spruce trees killed by the spruce bark beetle (Ips typographus) in central Sweden. Photo: Anne-Maarit Hekkala

the small spruce bark beetle can kill weakened trees, particularly in warm and dry summers (Lekander 1959). Both species benefit from prolonged drought, higher-than-average summer temperatures, sun scorch at mature-forest edges (Schebeck et al. 2022), and from attacks to host trees by the spruce bark beetle (Hedgren 2004).

In a restoration experiment in Sweden, coniferous stands with small gaps (radius 20 m, 19% of the stand area) were examined for bark beetles. In the second post-harvest summer, the abundance and richness of primary bark beetles (notably those discussed above) had increased, but compared to reserves with a closed tree canopy, gap cutting did not affect the number of trees occupied by these beetles (Kärvemo et al. 2017). Three years later, however, the number of occupied trees was higher in gap-gut stands than in reserves or in burned stands (Hekkala et al. 2021). These findings, combined with the above discussion of spruce bark beetles in CCF and RF, suggest that clearings of varying sizes—from small gaps (as in CCF) to large clearcuts (as in RF)—benefit these beetles. One explanation could be that trees at gap edges abruptly become exposed to sun scorch, wind gusts and drought. This is potentially specific to this particular experiment, however, as it involved intentional killing of substantial amounts of trees to increase deadwood for biodiversity.

The large pine weevil feeds on the bark and phloem of trees, and is the most important pest of newly planted conifer saplings in Europe (Lalik et al. 2021). It is common in regeneration areas throughout Fennoscandia, with most damage occurring in the south, and benefits from warmer-than-average summers (Långström 1982; Rautio et al. 2014; Luoranen et al. 2023). In Nordic boreal forests, the consumption of seedlings and saplings peaks in the first growing season following logging but continues for 2–3 years (Långström 1982; Örlander and Nilsson 1999; Luoranen et al. 2017, 2022). Sapling consumption by the beetle in clearcuts increases from edges to centres and decreases under shelter trees (Nordlander et al. 2003a, 2003b). The large pine weevil uses conifer stumps up to 5 years old as a breeding substrate. In RF, seedlings are commonly planted 1 year after clearcutting, so regeneration may suffer if the pine weevil population is high. In CCF, however, seedlings of natural origin emerge more slowly in canopy openings (see Chap. 3), often after the peak of the pine weevil population. These findings suggest that CCF is less prone than RF to the damage caused by the large pine weevil.

Sawflies and the nun moth have not been examined in a CCF context. Their outbreaks are determined by similar factors to other forest insects: host trees of suitable species, age and density, abiotic factors or disturbances (drought, wind, fire), and biotic factors, such as predators, entomopathogens, viruses, bacteria and resource competitors (e.g., Biedermann et al. 2019). Generally, the host trees are less favourable for outbreaks of these species in CCF than in RF (see above).

The common and European pine sawflies lay their eggs on pine trees, and larvae feed on their needles. This slows tree growth, but if needles are extensively consumed in several consecutive summers, trees may die (Nevalainen et al. 2015). The growth decrease may be 4–40% or more in warm conditions (Perot et al. 2013; Blomqvist et al. 2022). As these sawflies occupy different-sized pine trees, CCF and RF may not differ much in terms of sawfly-outbreak likelihood, but this issue merits empirical research.

The nun moth is spreading northward in the Nordic countries and is common in the south. In Finland, for example, the species has expanded its distribution about 200 km northward in the past two decades (Fält-Nardmann et al. 2018). Thus far the nun moth has only caused occasional deaths of host trees in the Nordic countries (e.g., Melin et al. 2020), but in central Europe the species has for hundreds of years had peak years with vast areas of tree deaths (e.g., Nakládal and Brinkeová 2015). It feeds on many woody tree species, including the genera *Abies*, *Picea*, *Pinus*, *Betula*, *Prunus* and *Quercus*, as well as bilberry (*Vaccinium myrtillus*, Keena 2003). Features that predispose forested landscapes to nun moth outbreaks include strong dominance of one or a few tree species, and young forests (Hentschel et al. 2018). General preparedness for the nun moth by the Nordic forestry sector would benefit from empirical nun moth research into different logging methods, manipulations of tree-species composition, and CCF.

12.5 Damage by Deer

The effects of deer on forests have been extensively studied in connection with RF (Markgren 1974; Heikkilä 2000; Nikula et al. 2008), but there is very little research so far on deer damage in CCF (Nevalainen 2017; Komonen et al. 2020). Hence, the following evaluation of deer damage in CCF is based on studies done in even-aged forests. Comparing RF with CCF from the point of view of deer damage is difficult and not always meaningful. In RF, the biggest concern is usually damage to pine seedling stands, and in CCF, the areas that are usually naturally regenerated by spruce to have hardwoods or pine among the regeneration are of concern. There is an urgent need for data on how different forest management methods create forest structural features that predispose them to deer damage and how deer affect the regeneration potential of different tree species.

Nevalainen's (2017) expert survey combined with a literature-based analysis suggested that some even-aged forest management activities, such as tillage, can increase the risk of deer damage. This can happen by increasing the abundance of deciduous seedlings that impede the development of pine seedlings, increasing damage risk (Jalkanen et al. 2005; Nikula et al. 2008; Nevalainen et al. 2016). With increasing intensity of tree removal in spruce-dominated CCF stands, deciduous trees regenerated after felling experienced increasingly severe browsing pressure (Komonen et al. 2020). In this case, the problem is balancing a sufficiently high thinning intensity for regeneration and the increased risk of deer damage. There is a risk that if deer browsing prevents deciduous-tree regeneration, the biodiversity benefits sought from CCF will not be achieved.

At a larger scale, the impact of extensively practised continuous cover forestry on deer food resources and thus the risk of damage needs to be examined. This is particularly important when spruce, which is poorly suited to deer, is the main species regenerated in CCF, and pine regeneration areas are small scale. The current moose-damage situation depends on the balance between moose density and the area of pine-dominated seedling stands as pine is their preferred food resource (Nikula et al. 2021). Landscapes with more suitable food resources per moose is expected to have less moose damage. Similarly, if the area of pine-dominated seedling stands decreases, more damage is to be expected in the remaining pine stands.

Regeneration using Scots pine can be expected to increase with climate change, as pine is less sensitive than spruce to extreme heat and drought (Dyderski et al. 2018). This probably applies to both RF and CCF. For pine, the risk of deer damage in gap cuts applied in CCF can be assessed based on studies in different-sized clearcuts in even-aged forests. For example, smaller regeneration areas (below 0.5 ha) correlate with increased deer damage (Díaz-Yáñez et al. 2017). The explanation may be that small open areas create a favourable environment for moose, where seedling stands (providing food) and mature forests (providing shelter) are close to each other (Edenius et al. 2002). On the other hand, when moose are particularly abundant, no association between the risk of damage and the size of the clearcuts has been found (Andren and Angelstam 1993).

Furthermore, mixed stands are likely to gain wider use for improving ecosystem service delivery and increasing resilience of forests against environmental changes and damage agents (e.g., Jonsson et al. 2019). In both CCF and RF, this will increase important forage plant species in the field layer, such as dwarf shrubs (Atlegrim and Sjöberg 1996). Competition over dwarf shrubs from smaller deer increases the proportion of pine in the diet of moose (Spitzer et al. 2021) and results in more damage (Pfeffer et al. 2021). Thus, any silvicultural practice improving access to forage in the field layer is likely to decrease damage. The degree to which improved forage availability resulting from CCF compensates for decreased availability of browsable recruiting trees under even-aged forestry is difficult to predict (Roberge et al. 2016). These dynamics may also vary depending on local deer species composition and environmental conditions such as snow cover.

12.6   Conclusions and Future Perspectives

The risks of damage in continuous cover and even-aged forests differ significantly, but there is not much research comparing them. Therefore, reliably assessing many damage agents requires new research that focuses on their associated risks in CCF, as well as ways to manage them in both mineral soil and peatland forests.

As discussed above, root rot fungi are a particularly challenging problem for CCF. A site that is already affected by root rot should not be transferred to CCF under any circumstances. Additionally, it is possible that deer browsing on tree seedlings limits the regeneration of tree species other than spruce in CCF, but increasing forage availability in the field may reduce this effect. It is unclear whether the negative effects of browsing would be larger in CCF than in even-aged forestry. More research on landscape-level forage availability under different silvicultural practices is needed to understand how browsing may limit CCF.

Many damage-causing factors, such as *Cronartium pini* or *Gremmeniella abietina*, appear to be little impacted by the choice of CCF or RF; site quality and forest management are much more important. In addition, wind and snow damage, as well as the risk of damage caused by *Ips typographus*, are likely to be less significant in uneven-aged than in even-aged forests. It is important to understand that future forest landscapes will consist of a mosaic of different forest management practices and their stages, and therefore the risks of damage will vary considerably, depending on both stand- and landscape-level factors.

References

Andren H, Angelstam P (1993) Moose browsing on scots pine in relation to stand size and distance to forest edge. J Appl Ecol 30:133–142

Annila E (1969) Influence of temperature upon the development and voltinism of *Ips typographus* L. (Coleoptera, Scolytidae). Ann Zool Fenn 6:161–208

Artdatabanken (2023) SLU Artdatabanken – ett kunskapscentrum för arter och naturtyper (SLU Species databank – a knowledge centre for species and habitats). https://artfakta.se/artinformation/taxa/106558/detaljer?lang=sv. Accessed 28 Sep 2023

Artsdatabanken (2023) Artsdatabanken. Kunnskapsbank for naturmangfold (Species databank Knowledge bank for natural diversity) https://www.artsdatabanken.no/. Accessed 28 Sep 2023

Atlegrim O, Sjöberg K (1996) Response of bilberry (*Vaccinium myrtillus*) to clear-cutting and single-tree selection harvests in uneven-aged boreal *Picea abies* forests. For Ecol Manag 86:39–50. https://doi.org/10.1016/S0378-1127(96)03794-2

Begon M, Townsend CR (2021) Ecology: from individuals to ecosystems, 5th edn. John Wiley & Sons

Biedermann PHW, Müller J, Grégoire J-C, Gruppe A, Hagge J, Hammerbacher A, Hofstetter RW, Kandasamy D, Kolarik M, Kostovcik M, Krokene P, Sallé A, Six DL, Turrini T, Vanderpool D, Wingfield MJ, Bässler C (2019) Bark beetle population dynamics in the Anthropocene: challenges and solutions. Trends Ecol Evol 34:914–924

Björklund N, Nordlander G, Bylund H (2003) Host-plant acceptance on mineral soil and humus by the pine weevil *Hylobius abietis* (L.). Agric For Entomol 5:61–66

Björkman C, Bylund H, Nilsson U, Nordlander G, Schroeder M (2015) Effects of new forest management on insect damage risk in a changing climate. Climate change and insect pests. CABI, Wallingford UK, pp 248–266

Blomqvist M, Lyytikäinen-Saarenmaa P, Kosunen M, Kantola T, Holopainen M (2022) Defoliation-induced growth reduction of *Pinus sylvestris* L. after a prolonged outbreak of *Diprion pini* L. — a case study from Eastern Finland. Forests 13:839

de Groot M, Ogris N, Diaci J, Castagneyrol B (2023) When tree diversity does not work: the interacting effects of tree diversity, altitude and amount of spruce on European spruce bark beetle outbreaks. For Ecol Manag 537:120952

Díaz-Yáñez O, Mola-Yudego B, Eriksen R, González-Olabarria JR (2016) Assessment of the main natural disturbances on Norwegian Forest based on 20 years of national inventory. PLoS One 11:e0161361

Díaz-Yáñez O, Mola-Yudego B, González-Olabarria JR (2017) What variables make a forest stand vulnerable to browsing damage occurrence? Silva Fenn 51:1693

Dobbertin M (2002) Influence of stand structure and site factors on wind damage comparing the storms Vivian and Lothar. For Snow Landsc Res 77:187–205

Dobor L, Hlásny T, Zimová S (2020) Contrasting vulnerability of monospecific and species-diverse forests to wind and bark beetle disturbance: the role of management. Ecol Evol 10:12233–12245

Dwyer JP, Dey DC, Walter W-D, Jensen RG (2004) Harvest impacts in uneven-aged and even-aged Missouri Ozark forests. North J Appl For 21:187–193

Dyderski MK, Paź S, Frelich LE, Jagodziński AM (2018) How much does climate change threaten European forest tree species distributions? Glob Chang Biol 24:1150–1163

Edenius L, Bergman M, Ericsson G, Danell K (2002) The role of moose as a disturbance factor in managed boreal forests. Silva Fenn 36:57–67

Endara MJ, Coley PD (2011) The resource availability hypothesis revisited: a meta-analysis. Funct Ecol 25:389–398

Fält-Nardmann JJJ, Tikkanen O-P, Ruohomäki K, Otto LF, Leinonen R, Pöyry J, Saikkonen K, Neuvonen S (2018) The recent northward expansion of *Lymantria monacha* in relation to realised changes in temperatures of different seasons. For Ecol Manag 427:96–105

Furuta K (1989) A comparison of endemic and epidemic populations of the spruce beetle (*Ips typographus japonicus* Niijima) in Hokkaido. J Appl Entomol 107:289–295. https://doi.org/10.1111/j.1439-0418.1989.tb00258.x

Greig BJW, Pratt JE (1976) Some observations on the longevity of *Fomes annosus* in conifer stumps. Eur J For Pathol 6:250–253

Hahn T, Eggers J, Subramanian N, Toraño Caicoya A, Uhl E, Snäll T (2021) Specified resilience value of alternative forest management adaptations to storms. Scand J For Res 36:585–597

Hånell B, Ottosson-Löfvenius M (1994) Windthrow after shelterwood cutting in *Picea abies* peatland forests. Scand J For Res 9:1–4

Hanewinkel M, Kuhn T, Bugmann H, Lanz A, Brang P (2014) Vulnerability of uneven-aged forests to storm damage. Forestry 87:525–534

Hantula J, Kasanen R, Kaitera J, Moricca S (2002) Analyses of genetic variation suggest that pine rusts *Cronartium flaccidum* and *Peridermium pini* belong to the same species. Mycol Res 106:203–209

Hantula J, Ahtikoski A, Honkaniemi J, Huitu O, Härkönen M, Kaitera J, Koivula M, Korhonen KT, Lindén A, Lintunen J, Luoranen J, Matala J, Melin M, Nikula A, Peltoniemi M, Piri T, Räsänen T, Sorsa J-A, Strandström M, Uusivuori J, Ylioja T (2023) Metsätuhojen kokonaisvaltainen arviointi: METKOKA-hankkeen loppuraportti (Comprehensive assessment of forest damages: final report of the METKOKA project). Luonnonvaraja biotalouden tutkimus 46/2023

Hedgren PO (2004) The bark beetle *Pityogenes chalcographus* (L.) (Scolytidae) in living trees: reproductive success, tree mortality and interaction with *Ips typographus*. J Appl Entomol 128:161–166

Heikkilä R (2000) Männyn istutustaimikoiden metsänhoidollinen tila hirvivahingon jälkeen Etelä-Suomessa (The silvicultural condition of scots pine (*Pinus sylvestris*) planting stocks after moose damage in Southern Finland). Metsätieteen aikakauskirja 2(2000):259–267

Hekkala A-M, Kärvemo S, Versluijs M, Weslien J, Björkman C, Löfroth T, Hjältén J (2021) Ecological restoration for biodiversity conservation triggers response of bark beetle pests and their natural predators. Forestry 94:115–126

Hentschel R, Möller K, Wenning A, Degenhardt A, Schröder J (2018) Importance of ecological variables in explaining population dynamics of three important pine pest insects. Front Plant Sci 9:1667

Hietala AM, Solheim H, Talbot B (2016) Råte i granskog: Det er store forskjeller i kjennskap til forekomst og kontrolltiltak innen norsk skogbruk (Decay in spruce forest: there are significant differences in knowledge regarding occurrence and control measures within Norwegian forestry). NIBIO POP 2(28)

Hlásny T, Mátyás C, Seidl R, Kulla L, Merganicova K, Trombik J, Dobor L, Barcza Z, Konôpka B (2014) Climate change increases the drought risk in central European forests: what are the options for adaptation. Lesnicky Casopis For J 60:5–18

Hlásny T, Zimová S, Merganicová K, Stepánek P, Modlinger R, Turcáni M (2021) Devastating outbreak of bark beetles in The Czech Republic: drivers, impacts, and management implications. For Ecol Manag 490:119075

Jalkanen R (2014) Ensiharvennus vai uudistaminen—aggressiivinen tervasroso mäntytaimikoiden ja nuorten metsien kimpussa (Thinning or regeneration—tackling the aggressive Cronartium rust in young scots pine stands and forests). Metlan työraportteja 321:31–37

Jalkanen R, Aalto T, Hallikainen V, Hyppönen M, Mäkitalo K (2005) Viljelytaimikoiden hirvituhot Lapissa ja Kuusamossa (Moose damage in cultivated seedling stands in Lapland and Kuusamo). Metsätieteen aikakauskirja 4(2005):399–411

Jönsson AM, Harding S, Bärring L, Ravn HP (2007) Impact of climate change on the population dynamics of *Ips typographus* in southern Sweden. Agric For Meteorol 146:70–81

Jönsson AM, Appelberg G, Harding S, Bärring L (2009) Spatio-temporal impact of climate change on the activity and voltinism of the spruce bark beetle, *Ips typographus*. Glob Chang Biol 15:486–499

Jönsson AM, Harding S, Krokene P, Lange H, Lindelöw A (2011) Modelling the potential impact of global warming on *Ips typographus* voltinism and reproductive diapause. Clim Chang 109:695–718

Jonsson M, Bengtsson J, Gamfeldt L, Moen J, Snäll T (2019) Levels of forest ecosystem services depend on specific mixtures of commercial tree species. Nat Plants 5:141–147. https://doi.org/10.1038/s41477-018-0346-z

Kaitera J, Nuorteva H, Hantula J (2005) Distribution and frequency of *Cronartium flaccidum* on *Melampyrum* spp. in Finland. Can J For Res 35:229–234

Kärvemo S, Björkman C, Johansson T, Weslien J, Hjältén J (2017) Forest restoration as a double-edged sword: the conflict between biodiversity conservation and pest control. J Appl Ecol 54:1658–1668

Kärvemo S, Schroeder LM (2010) A comparison of outbreak dynamics of the spruce bark beetle in Sweden and the mountain pine beetle in Canada (Curculionidae: Scolytinae). Entomologisk tidskrift 131:215–224

Keena MA (2003) Survival and development of *Lymantria monacha* (Lepidoptera: Lymantriidae) on north American and introduced Eurasian tree species. J Econ Entomol 96:43–52

Klapwijk MJ, Bylund H, Schroeder M, Björkman C (2016) Forest management and natural bio-control of insect pests. Forestry 89:253–262

Komonen A, Paananen E, Elo M, Valkonen S (2020) Browsing hinders the regeneration of broadleaved trees in uneven-aged forest management in southern Finland. Scand J For Res 35:134–138

Korhonen K (1978) Intersterility groups of *Heterobasidion annosum*. Metsäntutkimuslaitoksen julkaisuja 94:6

Korhonen KT, Ahola A, Heikkinen J, Henttonen HM, Hotanen J-P, Ihalainen A, Melin M, Pitkänen J, Räty M, Sirviö M, Strandström M (2021) Forests of Finland 2014–2018 and their development 1921–2018. Silva Fenn 55:10662

Krehan H, Steyrer G, Tomiczek C (2010) Borkenkäfer-Kalamität 2009: Ursachen für unterschiedliche regionale Befallsentwicklungen (Causes for varying regional infestation developments). Forstschutz Aktuell 49:9

Lalík M, Galko J, Kunca A, Nikolov C, Rell S, Zúbrik M, Dubec M, Vakula J, Gubka A, Leontovyč R, Longauerová V, Konôpka B, Holuša J (2021) Ecology, management and damage by the large pine weevil (*Hylobius abietis*) (Coleoptera: Curculionidae) in coniferous forests within Europe. Cent Eur For J 67:91–107

Långström B (1982) Abundance and seasonal activity of adult *Hylobius*-weevils in reforestation areas during first years following final felling. Communicationes Instituti Forestalis Fenniae 106:1–22

Långström B, Day KR (2004) Damage, control and management of weevil pests, especially Hylobius abietis. In: Lieutier F, Day KR, Battisti A, Grégoire JC, Evans HF (eds) Bark and wood boring insects in living trees in Europe, a synthesis. Springer, Dordrecht, pp 415–444

Lekander B (1959) Der doppeläugige Fichtenbastkäfer *Polygraphus poligraphus* L. Ein Beitrag zur Kenntnis seiner Morphologie, Anatomie, Biologie und Bekämpfung (The small spruce bark beetle *Polygraphus poligraphus* L.: a contribution to the understanding of its morphology, anatomy, biology, and control). Meddelanden från Statens Skogsforskningsinstitut 48(9)

Luoranen J, Viiri H, Sianoja M, Poteri M, Lappi J (2017) Predicting pine weevil risk: effects of site, planting spot and seedling level factors on weevil feeding and mortality of Norway spruce seedlings. For Ecol Manag 389:260–271

Luoranen J, Laine T, Saksa T (2022) Field performance of sand-coated (Conniflex®) Norway spruce seedlings planted in mounds made by continuously advancing mounder and in undisturbed soil. For Ecol Manag 517:120259

Luoranen J, Riikonen J, Saksa T (2023) Damage caused by an exceptionally warm and dry early summer on newly planted Norway spruce container seedlings in Nordic boreal forests. For Ecol Manag 528:120649

Magerøy HM, Nagy NE, Steffenrem A, Krokene P, Hietala AM (2023) Conifer defences against pathogens and pests — mechanisms, breeding, and management. Curr For Rep 9:429–443

Markgren G (1974) The moose in Fennoscandia. Le Naturaliste Canadien 101:185–194

Matthews B, Netherer S, Katzensteiner K, Pennerstorfer J, Blackwell E, Henschke P, Hietz P, Rosner S, Jansson P-E, Schume H, Schopf A (2018) Transpiration deficits increase host susceptibility to bark beetle attack: experimental observations and practical outcomes for *Ips typographus* hazard assessment. Agric For Meteorol 263:69–89

Melin M, Viiri H, Tikkanen O-P, Elfving R, Neuvonen S (2020) From a rare inhabitant into a potential pest – status of the nun moth in Finland based on pheromone trapping. Silva Fenn 54:10262

Müller M, Olsson P-O, Eklundh L, Jamali S, Ardö J (2022) Features predisposing forest to bark beetle outbreaks and their dynamics during drought. For Ecol Manag 523:120480

Nakládal O, Brinkeová H (2015) Review of historical outbreaks of the nun moth (*Lymantria monacha*) with respect to host tree species. J For Sci 61:18–26

Netherer S, Hammerbacher A (2022) The Eurasian spruce bark beetle in a warming climate: phenology, behavior, and biotic interactions. In: Gandhi KJK, Hofstetter RW (eds) Bark beetle management, ecology, and climate change, 1st edn. Elsevier, Cambridge, pp 89–131

Neuvonen S, Viiri H (2017) Changing climate and outbreaks of forest pest insects in a cold northern country, Finland. In: Latola K, Savela H (eds) The interconnected Arctic — Arctic congress 2016. Springer Polar Sciences, pp 49–59

Neuvonen S, Tikkanen O-P, Pouttu A, Silver T (2015) Kirjanpainajatilanne 2014 ja vertailua aiempiin vuosiin (The bark beetle situation in 2014 and a comparison to previous years). In: Heino E, Pouttu A (eds) Metsätuhot vuonna 2014. Luonnonvara- ja biotalouden tutkimus 39/2015, pp 16–22

Nevalainen S (2017) Comparison of damage risks in even- and uneven-aged forestry in Finland. Silva Fenn 51:1741

Nevalainen S, Sirkiä S, Peltoniemi M, Neuvonen S (2015) Vulnerability to pine sawfly damage decreases with site fertility but the opposite is true with *Scleroderris* canker damage; results from Finnish ICP forests and NFI data. Ann For Sci 72:909–917

Nevalainen S, Matala J, Korhonen KT, Ihalainen A, Nikula A (2016) Moose damage in National Forest Inventories (1986–2008) in Finland. Silva Fenn 50:1410

Niemelä P, Lindgren M, Uotila A (1992) The effect of stand density on the susceptibility of *Pinus sylvestris* to *Gremmeniella abietina*. Scand J For Res 7:129–133

Nikula A, Hallikainen V, Jalkanen R, Hyppönen M, Mäkitalo K (2008) Modelling the factors predisposing scots pine to moose damage in artificially regenerated sapling stands in Finnish Lapland. Silva Fenn 42:587–603

Nikula A, Matala J, Hallikainen V, Ihalainen A, Pusenius J, Kukko T, Korhonen KT (2021) Modelling the effect of moose *Alces alces* population density and regional forest structure on the amount of damage in forest seedling stands. Pest Manag Sci 77:620–627

Nordlander G, Örlander G, Langvall O (2003a) Feeding by the pine weevil *Hylobius abietis* in relation to sun exposure and distance to forest edges. Agric For Entomol 5:191–198

Nordlander G, Bylund H, Örlander G, Wallertz K (2003b) Pine weevil population density and damage to coniferous seedlings in a regeneration area with and without shelterwood. Scand J For Res 18:438–448

Nuorteva H, Kytö M, Aarnio L, Hamberg L, Hantula J, Henttonen H, Huitu O, Kaitera J, Koivula M, Korhonen KT, Kuitunen P, Lehti P, Luoranen J, Melin M, Niemimaa J, Piri T, Poimala A, Poteri M, Strandström M, Terhonen E, Tikkanen O-P, Uimari A, Vainio E, Valtonen A, Velmala S, Vuorinen M, Ylioja T (2022) Metsätuhot vuonna 2020 (Forest damages in the year 2020). Luonnonvara- ja biotalouden tutkimus 2/2022

Öhrn P, Långström B, Lindelöw Å, Björklund N (2014) Seasonal flight patterns of *Ips typographus* in southern Sweden and thermal sums required for emergence. Agric For Entomol 16:147–157

Økland B, Netherer S, Marini L (2015) The Eurasian spruce bark beetle – role of climate. In: Björkman C, Niemelä P (eds) Climate change and insect pests. CABI, Wallingford, pp 202–218

Örlander G, Nilsson U (1999) Effect of reforestation methods on pine weevil (*Hylobius abietis*) damage and seedling survival. Scand J For Res 14:341–354

Päätalo M-L, Peltola H, Kellomäki S (1999) Modelling the risk of snow damage to forests under short-term snow loading. For Ecol Manag 116:51–70

Peltola H (2006) Mechanical stability of trees under static loads. Am J Bot 93:1501–1511

Perot T, Vallet P, Archaux F (2013) Growth compensation in an oak-pine mixed forest following an outbreak of pine sawfly (*Diprion pini*). For Ecol Manag 295:155–161

Pfeffer SE, Singh NJ, Cromsigt JPGM, Kalén C, Widemo F (2021) Predictors of deer damage on commercial forestry – a study linking management data. For Ecol Manag 479:118597

Piri T (1996) The spreading of the S type of *Heterobasidion annosum* from Norway spruce stumps to the subsequent tree stand. Eur J For Pathol 26:193–204

Piri T, Valkonen S (2013) Incidence and spread of *Heterobasidion* root rot in uneven-aged Norway spruce stands. Can J For Res 43:872–877

Piri T, Korhonen K, Sairanen A (1990) Occurrence of *Heterobasidion annosum* in pure and mixed spruce stands in southern Finland. Scand J For Res 5:113–125

Piri T, Selander A, Hantula J, Kuitunen P (2019) Juurikääpätuhojen tunnistaminen ja torjunta (Identification and control of root rot damage). http://urn.fi/URN:NBN:fi-fe2019091828606

Piri T, Silver T, Hantula J (2023) Preventing mycelial spread of *Heterobasidion annosum* in young scots pine stands using fungal and viral biocontrol agent. Biol Control 184:105623

Potterf M, Eyvindson K, Blattert C, Burgas D, Burner R, Stephan JG, Mönkkönen M (2022) Interpreting wind damage risk—how multifunctional forest management impacts standing timber at risk of wind felling. Eur J For Res 141:347–361

Pouttu A, Annila E (2011) Kirjanpainajalla kaksi sukupolvea kesällä 2010. Metsätieteen aika-kauskirja 4(2010):6951

Pukkala T, Laiho O, Lähde E (2016) Continuous cover management reduces wind damage. For Ecol Manag 372:120–127

Raffa KF, Grégoire J-C, Lindgren S (2015) Natural history and ecology of bark beetles. In: Vega FE, Hofstetter RW (eds) Bark beetles. Elsevier, New York

Rassi P, Karjalainen S, Clayhills T, Helve E, Hyvärinen E, Laurinharju E, Malmberg S, Mannerkoski I, Martikainen P, Mattila J, Muona J, Pentinsaari M, Rutanen I, Salokannel J, Siitonen J, Silfverberg H (2015) Provincial list of Finnish Coleoptera 2015. Sahlbergia 21(Supplement 1):1–164

Rautio P, Hyppönen M, Hallikainen V, Niemelä J, Välikangas P, Jalkanen R, Winsa H, Hiltunen A, Bergsten U (2014) Kangasmetsien uudistamisen ongelmat Lapissa—kasvatetaanko kanervaa vai mäntyä? (The challenges of regenerating dry forests in Lapland—is heather or pine promoted?). Metsäntutkimuslaitoksen työraportteja 321

Ravn HP (1985) Expansion of the populations of *Ips typographus* (L.) (Coleoptera, Scolytidae) and their local dispersal following gale disaster in Denmark. Z Angew Entomol 99:26–33

Read DJ (1968) Some aspects of the relationship between shade and fungal pathogenicity in an epidemic disease of pines. New Phytol 67:39–48

Roberge J-M, Laudon H, Björkman C, Ranius T, Sandström C, Felton A, Sténs A, Nordin A, Granström A, Widemo F, Bergh J, Sonesson J, Stenlid J, Lundmark T (2016) Socio-ecological implications of modifying rotation lengths in forestry. Ambio 45(Suppl. 2):S109–S123

Ruosteenoja K, Jylhä K, Kämäräinen M (2016) Climate projections for Finland under the RCP forcing scenarios. Geophysica 51:17–50

Samils B, Stenlid J (2022) A review of biology, epidemiology and management of *Cronartium pini* with emphasis on Northern Europe. Scand J For Res 37:153–171

Samils B, Kaitera J, Persson T, Stenlid J, Barklund P (2021) Relationship and genetic structure among autoecious and heteroecious populations of *Cronartium pini* in northern Fennoscandia. Fungal Ecol 50:101032

Schebeck M, Schopf A, Ragland GJ, Stauffer C, Biedermann PHW (2022) Evolutionary ecology of the bark beetles *Ips typographus* and *Pityogenes chalcographus*. B Entomol Res 113:1–10

Schiebe C, Unelius CR, Ganji S, Binyameen M, Birgersson G, Schlyter F (2019) Styrene,(+)-trans-(1 R, 4 S, 5 S)-4-thujanol and oxygenated monoterpenes related to host stress elicit strong electrophysiological responses in the bark beetle *Ips typographus*. J Chem Ecol 45:474–489

Seidl R, Rammer W, Jäger D, Lexer MJ (2008) Impact of bark beetle (*Ips typographus* L.) disturbance on timber production and carbon sequestration in different management strategies under climate change. For Ecol Manag 256:209–220

Seidl R, Schelhaas M-J, Lexer MJ (2011) Unraveling the drivers of intensifying forest disturbance regimes in Europe. Glob Chang Biol 17:2842–2852

Sniezko RA, Liu J-J (2022) Genetic resistance to white pine blister rust, restoration options, and potential use of biotechnology. For Ecol Manag 520:120168

Spitzer R, Coissac E, Felton AM, Fohringer C, Landman M, Singh NJ, Taberlet P, Widemo F, Cromsigt JPGM (2021) Small shrubs with large importance? Smaller deer may increase the moose-forestry conflict through feeding competition over *Vaccinium* shrubs in the field layer. For Ecol Manag 480:118768

Stephens PA, Sutherland WJ, Freckleton RP (1999) What is the Allee effect? Oikos 87:185–190

Suvanto S, Peltoniemi M, Tuominen S, Strandström M, Lehtonen A (2019) High-resolution mapping of forest vulnerability to wind for disturbance-aware forestry. For Ecol Manag 453:117619

Suvanto S, Lehtonen A, Nevalainen S, Lehtonen I, Viiri H, Strandström M, Peltoniemi M (2021) Mapping the probability of forest snow disturbances in Finland. PLoS One 16:e0254876

Thomsen IM (2009) Precipitation and temperature as factors in *Gremmeniella abietina* epidemics. For Pathol 39:56–72

Uotila A (1985) Siemenen siirron vaikutuksesta männyn versosyöpäalttiuteen Etelä- ja Keski-Suomessa (On the effect of seed transfer on the susceptibility of scots pine to *Ascocalyx abietina* in southern and Central Finland). Folia For 639:1–12

Uotila E (1994) Hyönteistuhot metsätaloudessa – taustaa tuhojen taloudelliselle analyysille (insect damage in forestry – background for the economic analysis of damages). Metsätieteen aikakauskirja 1(1994):69–78

Vainio EJ, Jurvansuu J, Hyder R, Kashif M, Piri T, Tuomivirta T, Poimala A, Xu P, Mäkelä S, Nitisa D, Hantula J (2018) *Heterobasidion* Partitivirus 13 mediates severe growth debilitation and major alterations in the gene expression of a fungal forest pathogen. J Virol 92:e01744-17

Venäläinen A, Lehtonen I, Laapas M, Ruosteenoja K, Tikkanen O-P, Viiri H, Ikonen V-P, Peltola H (2020) Climate change induces multiple risks to boreal forests and forestry in Finland: a literature review. Glob Chang Biol 26:4178–4196

Viiri H, Viitanen J, Mutanen A, Leppänen J (2019) Metsätuhot vaikuttavat Euroopan puumarkkinoihin—Suomessa vaikutukset toistaiseksi vähäisiä (Forest damages impact European timber markets—Effects in Finland remain minimal for now). Metsätieteen aikakauskirja 2019:10200. (In Finnish)

Wallentin C, Nilsson U (2014) Storm and snow damage in a Norway spruce thinning experiment in southern Sweden. Forestry 87:229–238

Wermelinger B (2004) Ecology and management of the spruce bark beetle *Ips typographus* — a review of recent research. For Ecol Manag 202:67–82

Witzell J, Karlman M (2000) Importance of site type and tree species on disease incidence of *Gremmeniella abietina* in areas with a harsh climate in northern Sweden. Scand J For Res 15:202–209

Wulff S, Roberge C (2021) Inventering av granbarkborreangrepp i Götaland och Svealand 2021. (Inventory of spruce bark beetle infestations in Götaland and Svealand 2021), Department of Forest Resource Management, Swedish University of Agricultural Sciences, Report 534, Umeå

Zeng H, Pukkala T, Peltola H (2007) The use of heuristic optimization in risk management of wind damage in forest planning. For Ecol Manag 241:189–199

Zeng H, Garcia-Gonzalo J, Peltola H, Kellomäki S (2010) The effects of forest structure on the risk of wind damage at a landscape level in a boreal forest ecosystem. Ann For Sci 67:111–111

Chapter 13
Carbon Exchange, Storage and Sequestration

Lars Högbom, Aleksi Lehtonen, Line Nybakken, Anna Repo, Sakari Sarkkola, and Monika Strömgren

Abstract

- Boreal forests sequester and store large amounts of carbon both above and below ground.
- Forest management could influence carbon storage.
- Differences between upland soils and peatlands are important. In peatlands, large amounts of carbon are stored in the peat, making them more susceptible to differences in forest management.
- On peatlands, carbon balance is mostly determined by groundwater levels.
- Carbon storage on both upland and peat soils depends on harvest intensity since most carbon losses, apart from harvested forest products, come from decomposition of roots and logging residues.
- Scales in both space and time are both important considerations when estimating the effect on carbon balances.

L. Högbom (✉)
The Forestry Research Institute of Sweden – Skogforsk, Uppsala Science Park, Uppsala, Sweden

A. Lehtonen · A. Repo · S. Sarkkola
Natural Resources Institute Finland (LUKE), Helsinki, Finland
e-mail: aleksi.lehtonen@luke.fi; anna.repo@luke.fi; sakari.sarkkola@luke.fi

L. Nybakken
Faculty of Environmental Sciences and Natural Resource Management, Norwegian University of Life Sciences (NMBU), Ås, Norway
e-mail: line.nybakken@nmbu.no

M. Strömgren
The Forestry Research Institute of Sweden – Skogforsk, Uppsala Science Park, Uppsala, Sweden

Department of Soil and Environment, Swedish University of Agricultural Sciences (SLU), Uppsala, Sweden
e-mail: monika.stromgren@skogforsk.se

© The Author(s) 2025   243
P. Rautio et al. (eds.), *Continuous Cover Forestry in Boreal Nordic Countries*,
Managing Forest Ecosystems 45, https://doi.org/10.1007/978-3-031-70484-0_13

Keywords Soil carbon · Upland soil · Peatlands · Carbon storage · Carbon sequestration

13.1   Forestry's Role in Mitigating Climate Change

Understanding how carbon (C) cycling in forest soils is affected by forest management has long been an important research topic (e.g. Hesselman 1926). Large quantities of C are exchanged annually between forests and the atmosphere. Established forests are estimated to offset about 30% of global fossil fuel emissions (Birdsey and Pan 2015). The offset could be increased by expanding forested areas, improving C management in existing forests, and using wood for products and energy that substitute for fossil fuel emissions (McKinley et al. 2011). These approaches are interconnected and influenced by forest-management practices. For instance, reducing harvest rates immediately increases C storage in the forest, but it also decreases the influx of new C into harvested wood products.

The boreal biomes account for approximately 17% of the worlds land area, but more than 30% of the total terrestrial C stock, with the majority found in the soil (Bradshaw and Warkentin 2015). Lal (2005) estimated that in the boreal zone around 90% of the ecosystem C is in soils. A large part of this C is stored in peatlands. Still, in life-cycle analyses of forest C balances, soils are typically neglected.

To provide various ecosystem services, it is essential to optimise synergies and possibly minimise related trade-offs. Thus, we need to better understand forest-based mitigation potential and identify areas of potential conflict among different ecosystem services.

13.1.1   Comparing Carbon Impacts of Forest-Management Methods: A Challenging Task

Productive Fennoscandian forests (production >1 m³/ha/yr) cover about 55 million ha. The climate varies enormously along the 2000 km latitudinal gradient, from low-arctic climate in the boreal north to the nemoral climate zone in southern Sweden. The west-east gradient changes from an oceanic climate and boreal rainforests on the Norwegian Atlantic coast, to the continental, dry forests of eastern Finland. Today's forest-production landscape is largely a twentieth century creation of active management and governance, aiming to optimise economic value and increase standing forest biomass. Today's forest landscape is a mosaic of different stand ages, dominant tree species, productivity classes, and patch sizes (< 1 to >100 ha), intermixed with semi-open and open land with low or no forest productivity (e.g. mires, bogs, and areas above the treeline and with thin soil). In addition, there is a large diversity among private forest owners' goals for their holdings. All these factors influence how forests are managed.

The dominant silvicultural method is rotation forestry (RF) or even-aged forestry. There is an ongoing debate regarding whether transitioning to continuous cover forestry (CCF) or uneven-aged forestry would enhance climate benefits. However, there are several management strategies that, depending on definitions, fall into one of these groups (see Chap. 2 for definitions). This variety of methods complicates the interpretation of available data. Further, contradictory results arise as the timeframe (present, 20, or 100 years), spatial scale (individual trees, stands, landscapes), and the choice of system boundaries (e.g., biological cycles within forests, industrial cycles, and the value of substitution for fossil C) vary among studies.

Additionally, several other factors should be considered when comparing the carbon balance of CCF and RF:

- What serves as the reference case for silviculture? What are we comparing CCF or RF against?
- How will forests and forestry develop in the future, considering legislation, forest certification and climate change?
- How quickly can fossil fuels be phased out and what is the future energy portfolio like?
- How will the value of substitution for fossil C change over time?

Considering these factors, comparing results from different studies is not generally straightforward. Differences in assumptions made in modelling and different studies could have led to sometimes contrasting results.

Currently, there are only a few comparisons of the C impacts of RF and CCF, and no clear conclusions can be drawn from these studies. Conclusions are affected by factors such as the examined C stocks, tree growth and decomposition models, and the baseline used in the comparison. Studies have examined the development of C stocks in trees and soil (e.g. Peura et al. 2018), while some have also included substitution calculations in the assessment (e.g. Pukkala et al. 2011; Pukkala 2014; Lundmark et al. 2016; Table 13.1). When comparing different studies, the examination period selected also has an impact on conclusions. In a Norwegian study by Nilsen and Strand (2013), the C stock of trees in RF was found to be three times higher than in CCF at the time of the measurement. However, as both CCF and RF can be implemented using a range of harvesting methods and intensities, making direct comparisons between the management methods is challenging. Hence highlighting the significance of the timing of field measurements for comparisons.

13.1.2 Carbon Cycling in Forest Ecosystems

In this chapter, we review the carbon cycle in forest ecosystems. For an in-depth look at the cycle and the processes involved, see, for example, Ågren and Andersson (2012) or Gower (2003).

The biogenic C cycle (Fig. 13.1) in forest ecosystems is dominated by two prominent fluxes: the uptake of carbon dioxide (CO_2) via photosynthesis (gross primary production), and the release of C through respiration. This can be autotrophic, originating from living biomass, like root exudates, or heterotrophic respiration, such as decomposition of organic matter including roots, stumps, and soil organic matter (SOM). Most assimilated C is lost via autotrophic respiration, leaving only a minor part of the assimilated C to accumulate in biomass and soil as net ecosystem production. Forest logging operations reduce the uptake via photosynthesis at least temporarily, regardless of management system, since both even- and uneven-aged forestry will lower standing biomass, although with substantial differences.

The north-south climatic gradient is also reflected in C storage. In Finland and Sweden, C storage in biomass and the soil are much higher in the south than the north (Stendahl 2017; Merilä et al. 2023).

Carbon cycling in boreal forests is closely linked with nitrogen (N) cycling. Nitrogen is the mineral nutrient that limits growth rates in boreal conditions. Detailed descriptions of how the nitrogen cycle interacts with decomposition are found in a long list of publications (e.g. Tamm 1990). Low N availability enhances decomposition while N addition via deposition or fertilisation slows decomposition (Mayer et al. 2020), especially on nutrient-poor soils, leading to an accumulation of soil organic matter. Stability of SOM is also controlled by soil pH. At high pH, the

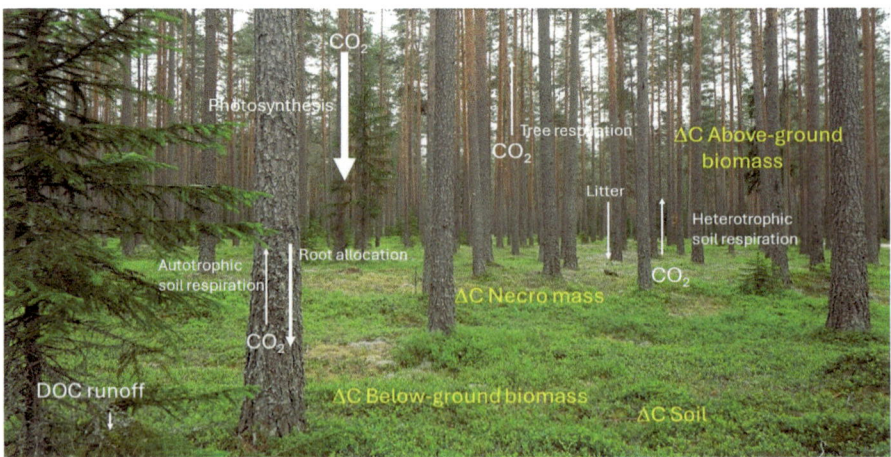

Fig. 13.1 Details of forest carbon cycling. Note that the length of the arrows does not reflect the size of the fluxes. Forest management can change the rate of these processes. Necro mass refers to dead organic material including woody debris. (Photo Lars Högbom)

solubility of the SOM increases, making more C available for soil microorganisms.

13.1.3 Soil Organic Matter (SOM)

Soil organic matter (SOM) is the largest C pool in boreal forest ecosystems. Carbon enters the SOM pool as dead organic litter from both above- and belowground. Over time, a significant portion of this is decomposed by fungi, soil fauna and bacteria and returned to the atmosphere. Forest fires also deplete SOM. The balance between C input and output determines whether there is a net increase or decrease in the soil's C stock.

Soil organic matter is highly variable and contains an array of organic substances with different degrees of recalcitrance and longevity. It is also important to differentiate between labile and stable soil C (Jandl et al. 2007). For technical reasons, fine roots (diameter < 2 mm) are usually included in the soil organic matter pool.

Quantifying changes in SOM and soil C content following various forest measures is notoriously difficult because of large within-stand variability and relatively small changes in very large soil C pools (Peltoniemi et al. 2004). Much of the C stored in SOM is highly stable, with a turnover time from hundreds to thousands of years. Only a minor fraction of soil C is actively cycled on monthly to yearly scales. Schmidt et al. (2011) estimated a mean SOM residence time of around 50 years. In addition to SOM, other C pools in the soil include stumps and coarse roots (> 2 mm).

Apart from the large CO_2 fluxes mentioned earlier, there are some other natural fluxes of greenhouse gases (GHG) in forested ecosystems. Leaching of dissolved organic carbon (DOC) constitutes a small fraction of the overall gross primary production. However, during heavy rain, the losses can be quite substantial. Methane (CH_4) fluxes, originating primarily from the soil surface, are minor under aerobic conditions. On the other hand, under anaerobic conditions in organic soils, CH_4 emissions can be substantial, for example from ditches (Rissanen et al. 2023). Since CH_4 is a potent GHG, it impacts climate change considerably. Another important GHG is nitrous oxide (N_2O) which can be produced in certain waterlogged conditions in nutrient-rich soils.

The litter derived from biomass, harvest residues and natural mortality constitutes an input into the soil C stock. Over time, a significant portion of this is decomposed by fungi, soil fauna and bacteria, and returns to the atmosphere. Additionally, soil C can be depleted when burnt by forest fires. This balance between C input and output determines whether soil C stocks grow or shrink.

Since the end of the last ice age, about 73 Mg C/ha has accumulated in the organic layer and the uppermost 50 cm of mineral soil of Swedish forests. Presently, 40–190 kg C/ha/yr accumulates within forests on mineral soils in Fennoscandia (Högberg et al. 2021). This increase is due to higher forest productivity and active suppression of forest fires. Notably, more C is stored in soil than in living tree biomass.

The carbon stored in peat soils (i.e. mire ecosystems) significantly surpasses that found in mineral soils, implying that peat soils have sequestered C faster since the last ice age. In contrast to most mineral soils, however, many organic soils are currently C sources (Jauhiainen et al. 2019). The total emissions of CO_2, N_2O and CH_4 from organic soils correspond to a loss of 1400 kg CO_2 eq. /ha/yr according to Swedish forestry-related climate reporting (Högberg et al. 2021). Drained peatlands emit the most, as aerobic conditions enhance peat decomposition, estimated to be around 5500 kg CO_2 eq./ha/yr.

Beyond C sequestered within biomass and soil, another important C stock is coarse dead wood, including standing or fallen dead trees and stumps following harvesting, although this stock is smaller compared to soils and live trees.

13.2 How Does Continuous Cover Forestry Affect Carbon Cycling in Forests?

13.2.1 Variability within Cycles and among Sites

Forest stands under RF and CCF differ considerably over time in C sequestration and stocks. Forests under RF begin their rotation period as a significant GHG source during clearcutting. Clearcuts on mineral soil lose around 16–20 Mg CO_2 ha/yr. (Grelle et al. 2012; Vestin et al. 2022; Grelle et al. 2023). For drained peat soils, this loss is larger. Tong et al. (2022a, 2022b) estimated the first-year loss in Sweden at 48 Mg CO_2/ha in a boreal stand and 26 Mg CO_2/ha in a hemi-boreal stand. In the second year, the losses had decreased to 26 and 7 Mg CO_2/ha. Korkiakoski et al. (2023) estimated the C loss at 30 and 22 Mg CO_2/ha for years one and two, respectively, in a nutrient-rich drained peatland in southern Finland.

As vegetation establishes and trees begin to grow, stands become C sinks, often within 10 years on mineral soils (Peichl et al. 2023; Grelle et al. 2023). The peak CO_2 uptake and most robust sink phase occur when the forest reaches about 40–50% of the normal length of a rotation period (e.g. Grelle et al. 2023; cf. Magnani et al. 2007). Subsequently, the sink slows as the forest ages (Odum 1969; Pregitzer and Euskirchen, 2004; Besnard et al. 2018; Repo et al. 2021). C stocks in dead wood and soil vary, influenced by factors like litter input and decomposition. Sequestration rates and stocks also vary among CCF methods, primarily triggered by harvesting. Since CCF forests retain standing trees, the variation is reduced. However, CCF forests require sufficient openness to facilitate regrowth, which lowers their sink strength and C stock in biomass compared to RF forests.

The variation in C sequestration and stocks depends on the specific CCF method used. For instance, CCF with gap cutting probably shows similar patterns to RF, particularly when evaluated at the scale of a gap.

Large broad-scale emissions at clearcutting are compensated by the high uptake of CO_2 in growing forest stands. Any annual variation due to active RF will be levelled out. However, a simulation study by Lehtonen et al. (2023) showed that at a national level, with equal harvesting levels, sinks are stronger when clearcuts are avoided on fertile drained peatlands. From a climate-mitigation standpoint, an important question is which forest-management approach stores most C in soil and biomass and reduces peat-soil-related emissions. The answer depends on underlying assumptions. For instance, biomass C stocks increase with extended RF rotation lengths.

How much continuous biomass storage can occur in CCF without impeding regrowth remains uncertain. Notably, ecosystem productivity emerges as a pivotal factor. A more productive management approach can sustain significant harvest potential and support litter production contributions to C stocks in woody debris and soil. Existing studies provide equivocal conclusions on productivity differences between RF and CCF, suggesting a small difference or large variation due to stand characteristics (see Chap. 4). However, the effect on peatlands, particularly nutrient-rich drained peatlands, differs from upland soils. On peatlands, climate benefits following CCF have been reported (Korkiakoski et al. 2023; Lehtonen et al. 2023). It is vital to note that ground vegetation and moss also contribute to overall ecosystem productivity, and that litter quality and local climate influence decomposition.

An alternative perspective is to begin with the present state of the forest (e.g., Pukkala et al. 2011) and to seek the optimal forest-management strategy for climate-change mitigation. In this perspective, the response may vary based on the forest's status, and therefore diverge among stands. Is it a 5-year old stand established after clearcutting? Does the analysed area encompass a forest with a well-distributed array of tree ages and sizes? Is it an old-growth spruce forest on fertile peat soil with elevated N_2O emissions? Is it an aging spruce forest where the sink strength has begun to wane? Can a present-day loss be offset by higher future uptake or vice versa? This final query underscores the importance of the timeframe considered.

13.2.2 Comparing CCF and RF

In this section, we discuss how the choice of RF or CCF may influence different elements of the forest C cycle. Direct comparisons between these two management systems are scarce in Fennoscandia and they differ in approaches, methods, and system boundaries. Effects of CCF and RF on growth and yield are discussed in Chap. 4.

Differences in forest soil C stocks between management regimes are the result of changes in organic material inputs to the soil and changes in decomposition rates.

CCF can create a more open stand structure, which may lead a to higher live-crown ratio in spruce forests (Bianchi et al. 2020; Kumpu et al. 2020). For Scots pine, 10-year-old clearcuts were shown to have lower above- and below-ground litter input than the partially cut stands (Roth et al. 2023). The highest litter input, however, occurs in uncut mature forest at the end phase of RF. Moreover, CCF has continuity of living roots and thus more below-ground litter input than RF (Prescott and Grayston 2023; Roth et al. 2023).

Decomposition is the processes of chemical and physical breakdown of organic material. It consists of leaching of organic compounds, fragmentation by animals and chemical alteration mainly by microbes, but light is also important. These processes are further regulated by the physical environment. Temperature, humidity, soil texture, soil disturbance, litter quality (chemistry, texture, etc.) and the decomposer community's composition and abundance all play roles.

The CO_2 emissions in the first stages after clearcutting in RF are probably the largest difference in stand-level C storage compared to CCF. During this period, an RF stand is subjected to highly variable sunlight, temperature, and humidity, compared to the more stable climate in a closed-canopy forest. At the same time, cutting large trees causes a flush of litter input and a total change in the decomposer community. Early-stage RF forests are dominated by saprotrophs, while mycorrhizal fungi that break down the most recalcitrant organic compounds largely disappear for some years after final felling (Wallander et al. 2010). There is also a transient period of litter inputs from a wider array of species, as the clearcut is first occupied by pioneer species.

During this stage, any differences in decomposition will depend heavily on the local climate, and also on decomposer communities. The question is whether the large-scale disturbance and successional setback changes full-rotation decomposition rates compared with a corresponding period in a forest with more frequent, but much-smaller-scale disturbances. Not surprisingly, we find no studies on this subject. There are also very few studies comparing mature, previously clearcut forests with CCF or other management regimes. However, in a comparison of Scots pine stands, Roth et al. (2023) found that fresh litter decomposed slower in a 10-year old clearcut than in a retention cut or an uncut stand. The clearcut had similar decomposition rates to a gap cut. They attributed the slower decomposition to a cooler microclimate and faster litter input in the retention and uncut stands. The clearcut and uncut mature stands represent the different extremes of RF, while gap and retention cuttings are two different types of CCF. Further, Purahong et al. (2015) compared decomposition of fresh leaf litter among even-aged spruce and beech forests, beech forests under CCF-like "near-to-nature" forest management, and an unmanaged beech forest in Germany. They found significantly faster general decomposition, lignin decomposition, and mineralisation of key elements in both RF forests compared with the CCF and natural forests.

Clearcutting has long-term effects on mycorrhizal-community composition (Kyaschenko et al. 2017; Hasby 2022) compared to never-clearcut forests. As some ectomycorrhizal fungi specialise on decomposing lignin and other recalcitrant litter components, such a change in the decomposer community might lead to differences in SOC. A recent meta-analysis by Latterini et al. (2023) concludes that a legacy of previous clearcutting decreases decomposition compared with unharvested controls, while retention-tree systems show increased decomposition. However, neither unharvested controls nor retention-tree systems are comparable with CCF and none of the included studies were performed in European boreal forests. In summary, existing knowledge suggests that potential long-term changes in the decomposer community might be the most important difference between CCF and RF, but it remains unclear to what degree this causes SOC differences.

13.3 Continuous Cover Forestry and Carbon Balance on Mineral Soil

Continuous cover forestry changes the timing of C fluxes compared to RF. After felling, RF forests are a C source until biomass growth and litter production exceeds the C released through the decomposition of soil C and logging residues (Kolari et al. 2004; Schulze et al. 2021). CCF may have C benefits since the source phase is shorter or non-existent compared to RF, leading to greater long-term average C sequestration. However, this might not be true if the C release in RF is later offset by larger uptake in RF than in CCF (see Chap. 4 for growth and yield comparisons). Nevertheless, for climate change mitigation, the timing of the emissions or uptake matters because of residence times in the atmosphere.

Only a few studies have compared the impact of RF and CCF on C sequestration in forests on mineral soils in Fennoscandia. There is, to our knowledge, only one published field study. This showed decreased C sequestration because of lower biomass production in CCF (Nilsen and Strand 2013). Studies based on models or simulations have found different results. Lagergren and Jönsson (2017) observed no difference in C sequestration between RF and CCF management in their ecosystem model analysis. Their model, however, involved a shift to more shade-tolerant tree species after CCF. A study by Peura et al. (2018) simulates a transition from RF to CCF forestry. The average annual C sequestration at a landscape level over a 100-year period was 0.68 Mg C/ha/yr, in CCF when harvested at 15-year intervals, reducing the basal area density to 10–12 m²/ha, while in RF, C sequestration was 0.23 Mg C/ha/yr. However, the forest landscape shifted to CCF during this study's simulation period, and the total harvest was roughly 15% smaller in CCF than in RF, which may partly explain the difference. Both C-sequestration values are of the same order of magnitude as is typical of forests on mineral soils. These ranged from 0.45–0.51 Mg C/ha/yr from 1990–2017 when estimated using the methods of Finland's national greenhouse-gas inventory (EU NIR 2019). In another simulation study, harvesting at 10–30-year intervals in CCF reduced tree basal area to 8–15 m²/ha (Shanin et al. 2016). In this study, net ecosystem production, excluding harvested biomass, ranged from −3 to +2 Mg C/ha/yr in various CCF methods. While CCF lacks a clearcut phase, this study shows that forest stands managed with CCF are not always C sinks but can be either a sink or source depending on harvesting intensity.

How will CCF affect soil C stocks in comparison to RF? In a Norwegian study based on field measurements, soil C stocks were higher in CCF than in even-aged forests, but the difference was not statistically significant (Nilsen and Strand 2013). A simulation study by Peura et al. (2018) found a somewhat higher average soil C stock in CCF than RF over a 100-year period. Since the harvested volume from the CCF simulation was lower than from the RF, and a decreased harvest led to higher soil C (cf. Mäkipää et al. 2023), it is difficult to distinguish whether the result is due to the management method or the decreased harvest. The hypothesis of colder microclimates and continuous litter input causing higher SOC stocks in CCF than on clearcuts was tested in a 10-year Finnish field experiment (Roth et al. 2023). No statistically significant differences were found in the SOC density or stocks between treatments despite the measured warmer microclimate and estimated lower litter inputs on the clearcut plots. Roth et al. 2023 view 10 years as too short of a period to detect measurable changes in stocks. It is well known that soils lose C the first year after clearcutting (e.g. Peichl et al. 2023; Grelle et al. 2023). However, the effect on soil C over the whole rotation period is still unclear. The microclimate, for example, can be less favourable for decomposition in a dense mature RF stand compared to a more open CCF stand (Roth et al. 2023). In addition, the large variation in CO_2 uptake within a rotation period makes the comparison more difficult. Some simulation studies, accounting for the whole rotation period, show that RF in comparison to CCF either harbours larger soil C stocks (the no biofuel alternative in Pukkala 2014), similar C stocks (Lagergren and Jönsson 2017) or smaller C stocks (Lundmark et al. 2016; Kellomäki et al. 2021; see Table 13.1).

However, a study by Roth et al. (2023) detected changes in the processes controlling organic matter accumulation and decomposition. In-situ decomposition was lower in retention cuts and in a mature uncut stand, where forest cover caused a cooler microclimate and higher litter input. Decomposition rates were equally high on clearcut sites and in canopy gaps of gap-cut stands, also indicating differences between CCF methods. In addition, the study found differences in litter quality between treatments. The study concludes that the accumulation of labile compounds in retention cuts together with decreased decomposition rates indicate a higher soil-C-accumulation potential in this CCF method.

Even small changes in soil C stocks can significantly impact boreal forest C budgets as two-thirds of the total C stock is below ground on mineral soils and even more on peatlands. Harvest timing and intensity determine the effects of CCF on soil-C stocks. Shanin et al. (2016) used the EFIMOD and ROMUL models to estimate that thinning from above every 10–20 years to lower tree basal area below 12 m^2/ha led to reduced soil C stocks compared to forests with an initially larger basal area. If harvesting was less intensive (basal area 16 m^2/ha after harvest), soil C stocks increased regardless of thinning frequency.

Soil preparation in RF is suggested to increase C losses from mineral soils. Consequently, the lack of soil preparation could be seen as a C benefit for CCF. However, a recent review by Mäkipää et al. (2023) concludes that the effect of post-clearcutting soil preparation on CO_2 emissions is minor. The potential C benefits depend on the effects on both soil and biomass. The simulation study of Kellomäki et al. (2021) concluded that in spruce forests of central Finland over a long period (401–1000 years), soil C stocks were significantly higher under CCF than RF, whereas C stocks in CCF trees were roughly twice those in even-aged forests. However, the ecosystem C benefit varies among studies (see Table 13.1). In

Table 13.1 Ecosystem-level C sequestration of continuous cover forestry (CCF) compared with rotation forestry (RF) on mineral soils. Parentheses () indicate a weak effect. HWP stands for harvested wood product and n.d. for no difference. Blank cells indicate that the topic was not covered by the study

Biomass C	Soil C	Ecosyst C	HWP C	Comment	Reference
−	n.d.	−	−	Field study, 81 years, no replicate	Nilsen and Strand (2013)
+	−	+		Model. RF low thinning no biofuel vs. CCF	Pukkala (2014)
−	−	−		Model. RF high thinning biofuel vs. CCF	Pukkala (2014)
n.d.	+	+	−	Model. Assumed same biomass production in CCF and RF	Lundmark et al. (2016)
−	+		−	Model. Assumed CCF had 80% of mean annual increment vs. RF	Lundmark et al. (2016)
n.d.	n.d.	n.d.	+	Model. Broadleaf fraction was 45% in RF and 13–20% in CCF	Lagergren and Jönsson (2017)
		(+)	−	Model. Decreased harvest in CCF	Peura et al. (2018)
−	+	−	+	Model. Increased biomass production in CCF	Kellomäki et al. (2021)

a simulation study comparing different management options in a changing climate with the LPJ-GUESS model, Lagergren and Jönsson (2017) found only minor differences in ecosystem C stocks and C sequestration between RF and CCF.

13.4 Continuous Cover Forestry and Greenhouse Gas Exchange on Peatlands

CCF is suggested as a feasible way to decrease C emissions and water runoff relative to even-aged management of drained nutrient-rich spruce peatland forests (Nieminen et al. 2018). CCF relies on continuously maintaining a tree stand with significant transpiration and interception capacity to moderate the water table, thereby avoiding large openings whose C sinks are temporarily weak due to having only surface vegetation.

Carbon emissions vary greatly between peatland sites and even within sites (e.g. Jauhiainen et al. 2019). Furthermore, there are significant uncertainties whether individual drainage locations are C sources or sinks. However, two key site characteristics increase peat decomposition and thus net C emissions: fertile soil and deep water tables (Minkkinen et al. 2020; Ojanen et al. 2013; Ojanen and Minkkinen 2019). Emissions particularly seem to increase if the water level drops below 30 cm (Ojanen et al. 2013). On the other hand, water levels approaching the ground surface (0–20 cm) after regeneration felling increase methane (CH_4) emissions (Ojanen et al. 2010; Korkiakoski et al. 2020).

Drained peatlands continue to lose C until the peat layer is fully decomposed. The amount of C sequestered in trees may not fully compensate for the loss from peat, and the end use of the harvested wood determines the ultimate contribution of forestry to climate change mitigation (Ojanen 2014). If CCF methods can maintain peatland water levels within a range that minimises C emissions, they may offer opportunities, at least in peatland areas, to control peat decomposition and reduce soil GHG emissions. The first findings of CCF impacts on C emissions from peatlands support this hypothesis. Studies conducted by Korkiakoski et al. (2020, 2023) found that a 13-ha test area subjected to felling based on CCF (selection harvesting, roughly 70% of total volume removed) was a minor source of CO_2, while net CO_2 emissions were five times higher in a clearcut. Correspondingly, the CCF site remained a CH_4 sink, whereas the clearcut became a minor source of CH_4 (Korkiakoski et al. 2020). Mäkiranta et al. (2010) and Korkiakoski et al. (2020, 2023) also measured N_2O emissions from clearcut sites that were many times higher than from otherwise-similar stocked peatlands. According to a simulation by Shanin et al. (2021), a nutrient-rich spruce-dominated peatland forest ecosystem managed by selection harvesting remained a C sink over the examined 240-year period, regardless of felling, provided the stand basal area remained above 6 m^2/ha (Fig. 13.2).

A modelling study found that prohibiting clearcuts on fertile drained peatlands would increase Finland's C sink by 1 Tg CO_2 eq./yr compared to business-as-usual management (Lehtonen et al. 2023). This simulation coupled the MELA simulator with the SpaFHy-peat hydrological model (Launiainen et al. 2019), the Yasso07 soil model (Tuomi et al. 2011) and empirical emissions models (Ojanen and Minkkinen 2019). In this work the major reasons for CCF's larger climate benefits were avoiding clearcuts and the period of seedling stands with marginal tree growth.

After clearcutting, emissions can be up to 30 Mg CO_2/ha/yr. In the 5 years following felling, a stand can lose an amount of C from peat equivalent to the timber harvested in the final felling (Korkiakoski et al. 2023). In the control area managed by partial harvesting (CCF), the net emissions were significantly smaller. The large emissions from the clearcut area seem to be due to the increase in the decomposition of surface peat, and especially the decrease in new C input as litterfall drops precipitously after cutting (Korkiakoski et al. 2023). After partial harvesting, the water table rises, reducing the fraction of the peat layer susceptible to rapid decomposition. However, on drained peatlands, even with CCF the water table may remain so low that the peat respiration rate will not necessarily decline significantly. There is therefore a risk that a change in management method alone will not be enough to reduce peat-soil respiration. However, more research is needed to verify this (Table 13.2).

Fig. 13.2 A spruce-dominated peatland forest after regeneration felling (upper photo, rotation forestry) and selection harvesting (lower photo, continuous cover forestry) in a drained, thick-peated site. Photos: Sakari Sarkkola

Table 13.2 Effects of CCF vs. RF on carbon sequestration in peatland forest ecosystems. HWP stands for harvested wood product and n.d. for no difference. An empty cell indicates that the topic was not addressed in the study

Biomass C	Soil C	Ecosyst C	HWP C	Comment	Reference
		+		Field study, partial harvest compared to clearcut	Korkiakoski (2023)
	n.d.			Field study, uncut compared to clearcut	Mäkiranta et al. (2010)
+	+	+		MELA simulations (Finland) with equal harvesting. CCF on fertile peat soils	Lehtonen et al. (2023)

13.5 Challenges and Future Research Directions

Several potential areas for further inquiry and research will help improve our understanding of CCF's carbon-sequestration potential. Carbon balance is closely linked with forest production and harvesting, so there is a need to incorporate prediction of biomass and soil-C development under CCF into simulation models like MELA, Heureka and others. MELA- and Heureka-type models should be coupled with soil models that can incorporate both mineral soils and drained peatlands in a way that allows users to optimise forestry for climate change mitigation and adaptation. The current MELA and Heureka models allow calculation of tree-biomass carbon, but soils are excluded. More data and information are needed about net ecosystem exchange after clearcutting on mineral soils and drained peatlands to estimate the full carbon impact of different forest practices. More studies are also needed on different harvesting regimes' effects on peatland groundwater levels.

References

Ågren GI, Andersson FO (2012) Terrestrial ecosystem ecology. Principles and applications. Cambridge University Press, New York. ISBN 987-1107011076

Besnard S, Carvalhais N, Arain MA, Black A, de Bruin S, Buchmann N, Cescatti A, Chen J, Clevers JGPW, Desai AR, Gough CM, Havrankova K, Herold M, Hörtnagl L, Jung M, Knohl A, Kruijt B, Krupova L, Law BE, Lindroth A, Noormets A, Roupsard O, Steinbrecher R, Varlagin A, Vincke C, Reichstein M (2018) Quantifying the effect of forest age in annual net forest carbon balance. Environ Res Lett 13:124018. https://doi.org/10.1088/1748-9326/aaeaeb

Bianchi S, Huuskonen S, Siipilehto J, Hynynen J (2020) Differences in tree growth of Norway spruce under rotation forestry and continuous cover forestry. For Ecol Manag 458:117689. https://doi.org/10.1016/j.foreco.2019.117689

Birdsey R, Pan Y (2015) Trends in management of the world's forests and impacts on carbon stocks. For Ecol Manag Arial Age 355:83–90. https://doi.org/10.1016/j.foreco.2015.04.031

Bradshaw CJA, Warkentin IG (2015) Global estimates of boreal forest carbon stocks and flux. Glob Planet Chang 128:24–30. https://doi.org/10.1016/j.gloplacha.2015.02.004

EU NIR (2019) European Union. 2019 National Inventory Report (NIR). Submitted to the UNFCCC. https://unfccc.int/documents/194921

Gower ST (2003) Patterns and mechanisms of the forest carbon cycle. Annu Rev Environ Resourc 28:169–204. https://doi.org/10.1146/annurev.energy.28.050302.105515

Grelle A, Strömgren M, Hyvönen R (2012) Carbon balance of a forest ecosystem after stump harvest. Scand J For Res 27(8):762–773. https://doi.org/10.1080/02827581.2012.726371

Grelle A, Hedwall P-O, Strömgren M, Håkansson C, Bergh J (2023) From source to sink—recovery of the carbon balance in young forests. Agric For Meteorol 330:109290. https://doi.org/10.1016/j.agrformet.2022.109290

Hasby F (2022) Impacts of clear-cutting on soil fungal communities and their activities in boreal forests—A metatranscriptomic approach. PhD Thesis No 2022:11 Faculty of Natural Resources and Agricultural Sciences. Swedish University of Agricultural Sciences. ISBN: 978-91-7760-897-4

Hesselman H (1926) Studier över barrskogens humustäcke, dess egenskaper och beroende av skogsvården. (Studies over the forest humus layer, properties and dependency of forest management.) Reports from the Swedish Institute of Experimental Forestry, no 22-1926. [In Swedish with German summary]. https://res.slu.se/id/publ/124773

Högberg P, Arnesson Ceder L, Astrup R, Binkley D, Bright R, Dalsgaard L, Egnell G, Filipchuk A, Genet H, Ilintsev A, Kurz WA, Laganière J, Lemprière T, Lundblad M, Lundmark T, Mäkipää R, Malysheva N, Mohr CW, Nordin A, Petersson H, Repo A, Schepaschenko D, Shvidenko A, Soegaard G, Kraxner F (2021) Sustainable boreal forest management—challenges and opportunities for climate change mitigation. Report from an Insight Process conducted by a team appointed by the International Boreal Forest Research Association (IBFRA). Report 2021/11, Swedish Forest Agency. https://pure.iiasa.ac.at/17778

Jandl R, Lindner M, Vesterdal L, Bauwens B, Baritz R, Hagedorn F, Johnson DW, Minkkinen K, Byrne KA (2007) How strongly can forest management influence soil carbon sequestration? Geoderma 137:253–268. https://doi.org/10.1016/j.geoderma.2006.09.003

Jauhiainen J, Alm J, Bjarnadottir B, Callesen I, Christiansen JR, Clarke N, Dalsgaard L, He H, Jordan S, Kazanavičiūtė V, Klemedtsson L, Lauren A, Lazdins A, Lehtonen A, Lohila A, Lupikis A, Mander Ü, Minkkinen K, Kasimir Å, Olsson M, Ojanen P, Óskarsson H, Sigurdsson BD, Søgaard G, Soosaar K, Vesterdal L, Laiho R (2019) Reviews and syntheses: greenhouse gas exchange data from drained organic forest soils—a review of current approaches and recommendations for future research. Biogeosciences 16:4687–4703. https://doi.org/10.5194/bg-16-4687-2019

Kellomäki S, Väisänen H, Kirschbaum MUF, Kirsikka-Aho S, Peltola H (2021) Effects of different management options of Norway spruce on radiative forcing through changes in carbon stocks and albedo. Forestry 94:588–597. https://doi.org/10.1093/forestry/cpab010

Kolari P, Pumpanen J, Rannik Ü, Ilvesniemi H, Hari P, Berninger F (2004) Carbon balance of different aged scots pine forests in southern Finland. Glob Chang Biol 10(7):1106–1119. https://doi.org/10.1111/j.1529-8817.2003.00797.x

Korkiakoski M, Ojanen P, Penttilä T, Minkkinen K, Sarkkola S, Rainne J, Laurila T, Lohila A (2020) Impact of partial harvest on CH_4 and N_2O balances of a drained boreal peatland forest. Agric For Meteorol 295:108168. https://doi.org/10.1016/j.agrformet.2020.108168

Korkiakoski M, Paavo Ojanen P, Tuovinen J-P, Minkkinen K, Nevalainen O, Penttilä T, Aurela M, Laurila T, Lohila A (2023) Partial cutting of a boreal nutrient-rich peatland forest causes radically less short-term on-site CO_2 emissions than clear-cutting. Agric For Meteorol 332:109361. https://doi.org/10.1016/j.agrformet.2023.109361

Kumpu A, Piispanen R, Berninger F, Saarinen J, Mäkelä A (2020) Biomass and structure of Norway spruce trees grown in uneven-aged stands in southern Finland. Scand J For Res 35:252–261. https://doi.org/10.1080/02827581.2020.17881

Kyaschenko J, Clemmensen K, Hagen A, Karltun E, Lindahl BD (2017) Shift in fungal communities and associated enzyme activities along an age gradient of managed *Pinus sylvestris* stands. ISME J 11:863–874. https://doi.org/10.1038/ismej.2026.184

Lagergren F, Jönsson AM (2017) Ecosystem model analysis of multi-use forestry in a changing climate. Ecosyst Serv 26:209–224. https://doi.org/10.1016/j.ecoser.2017.06.007

Lal R (2005) Forest soils and carbon sequestration. For Ecol Manag 220:242–258. https://doi.org/10.1016/j.foreco.2005.08.015

Latterini F, Dyderski M, Horodecki P, Picchio R, Venanzi R, Lapin K, Jagodzinski AM (2023) The effect of forest operations and silvicultural treatments on litter decomposition rate: a meta-analysis. Curr For Rep 9:276–290. https://doi.org/10.1007/s40725-023-00190-5

Launiainen S, Guan M, Salmivaara A, Kieloaho AJ (2019) Modeling boreal forest evapotranspiration and water balance at stand and catchment scales: a spatial approach. HESS 23:3457–3480. https://doi.org/10.5194/hess-23-3457-2019

Lehtonen A, Eyvindson K, Härkönen K, Leppä K, Salmivaara A, Peltoniemi M, Salminen O, Sarkkola S, Launiainen S, Ojanen P, Räty M (2023) Potential of continuous cover forestry on drained peatlands to increase the carbon sink in Finland. Sci Rep 13:15510. https://doi.org/10.1038/s41598-023-42315-7

Lundmark T, Berg J, Nordin A, Fahlvik N, Poudel BC (2016) Comparison of carbon balances between continuous-cover and clear-cut forestry in Sweden. Ambio 45:S203–S213. https://doi.org/10.1007/s13280-015-0756-3

Magnani F, Mencuccini M, Borghetti M, Berbigier P, Berninger F, Delzon S, Grelle A, Hari P, Jarvis PG, Kolari P, Kowalski AS, Lankreijer H, Law BE, Lindroth A, Loustau D, Manca G, Moncrieff JB, Rayment M, Tedeschi V, Valentini R, Grace J (2007) The human footprint in the carbon cycle of temperate and boreal forests. Nature 447:848–850. https://doi.org/10.1038/nature05847

Mäkipää R, Abramoff R, Adamczyk B, Baldy V, Charlotte Biryol C, Bosela M, Pere Casals P, Curiel Yuste J, Dondini M, Filipek S, Garcia-Pausas J, Gros R, Gömöryová E, Hashimoto S, Hassegawa M, Immonen P, Laiho R, Li H, Qian Li Q, Luyssaert S, Menival C, Mori T, Naudts K, Santonja M, Smolander A, Toriyama J, Tupek B, Xavi Ubeda X, Verkerk PJ, Lehtonen A (2023) How does management affect soil C sequestration and greenhouse gas fluxes in boreal and temperate forests?—a review. For Ecol Manag 529:120637. https://doi.org/10.1016/j.foreco.2022.120637

Mäkiranta P, Riutta T, Penttilä T, Minkkinen K (2010) Dynamics of net ecosystem CO_2 exchange and heterotrophic soil respiration following clearfelling in a drained peatland forest. Agric For Meteorol 15:1585–1596. https://doi.org/10.1016/j.agrformet.2010.08.010

Mayer M, Prescott CE, Abaker WEA, Augusto L, Cécillon L, Ferraira GWF, James J, Jandl R, Katzensteiner K, Laclau J-P, Laganière J, Nouvellon Y, Paré D, Stanturf JA, Vanguelova EI, Vesterdal L (2020) Tamm review: influence of forest management activities on soil organic carbon stocks: a knowledge synthesis. For Ecol Manag 466:118127. https://doi.org/10.1016/j.foreco.2020.118127

McKinley DC, Ryan MG, Birdsey RA, Giardina CP, Harmon ME, Heath LS, Houghton RA, Jackson RB, Morrison JF, Murray BC, Pataki DE, Skog KE (2011) A synthesis of current knowledge on forests and carbon storage in the United States. Ecol Appl 21:1902–1924. https://doi.org/10.1890/10-0697.1

Merilä P, Lindroos A-J, Helmisaari H-S, Hilli S, Nieminen TM, Nöjd P, Rautio P, Salemaa M, Tupek BM, Ukonmaanaho L (2023) Carbon stocks and transfers in coniferous boreal forests along a latitudinal gradient. Ecosystems 27(1):151–167. https://doi.org/10.1007/s10021-023-00879-5

Minkkinen K, Ojanen P, Koskinen M, Penttilä T (2020) Nitrous oxide emissions of undrained, forestry-drained, and rewetted boreal peatlands. For Ecol Manag 478:118494. https://doi.org/10.1016/j.foreco.2020.118494

Nieminen M, Hökkä H, Laiho R, Juutinen A, Ahtikoski A, Pearson M, Kojola S, Sarkkola S, Launiainen S, Valkonen S, Penttilä T, Lohila A, Saarinen M, Haahti K, Mäkipää R, Miettinen J, Ollikainen M (2018) Could continuous cover forestry be an economically and environmentally feasible management option on drained boreal peatlands? For Ecol Manag 424:78–84. https://doi.org/10.1016/j.foreco.2018.04.046

Nilsen P, Strand LT (2013) Carbon stores and fluxes in even- and uneven-aged Norway spruce stands. Silva Fenn 47:1–15. https://doi.org/10.14214/sf.1024

Odum EP (1969) The strategy of ecosystem development – an understanding of ecological succession provides a basis for resolving man's conflict with nature. Science 164:262–270. https://doi.org/10.1126/science.164.3877.262

Ojanen P (2014) Estimation of greenhouse gas balance for forestry-drained peatlands. https://doi.org/10.14214/df.176

Ojanen P, Minkkinen K (2019) The dependence of net soil CO_2 emissions on water table depth in boreal peatlands drained for forestry. Mires Peat 24:27. https://doi.org/10.19189/MaP.2019.OMB.Sta.1751

Ojanen P, Minkkinen K, Alm J, Penttilä T (2010) Soil–atmosphere CO_2, CH_4 and N_2O fluxes in boreal forestry-drained peatlands. For Ecol Manag 260:411–421. https://doi.org/10.1016/j.foreco.2010.04.036

Ojanen P, Minkkinen K, Penttilä T (2013) The current greenhouse gas impact of forestry-drained boreal peatlands. For Ecol Manag 289:201–208. https://doi.org/10.1016/j.foreco.2012.10.008

Peichl M, Martínez-García E, Fransson JES, Wallerman J, Laudon H, Lundmark T, Nilsson MB (2023) On the uncertainty in estimates of the carbon balance recovery time after forest clearcutting. Glob Chang Biol 29(15):16772. https://doi.org/10.1111/gcb.16772

Peltoniemi M, Mäkipää R, Liski J, Tamminen P (2004) Changes in soil carbon with stand age—an evaluation of a modelling method with empirical data. Glob Chang Biol 10(12):2078–2091. https://doi.org/10.1111/j.1365-2486.2004.00881.x

Peura M, Burgas D, Eyvindson K, Repo A, Mönkkönen M (2018) Continuous cover forestry is a cost-efficient tool to increase multifunctionality of boreal production forests in Fennoscandia. Biol Conserv 217:104–112. https://doi.org/10.1016/j.biocon.2017.10.018

Pregitzer KS, Euskirchen ES (2004) Carbon cycling in world forests: biome patterns related to forest age. Glob Chang Biol 10:2052–2077. https://doi.org/10.1111/j.1365-2486.2004.00866.x

Prescott CE, Grayston SJ (2023) Tamm review: continuous root forestry—living roots sustain the belowground ecosystem and soil carbon in managed forests. For Ecol Manag 532:120848. https://doi.org/10.1016/j.foreco.2023.120848

Pukkala T (2014) Does biofuel harvesting and continuous cover management increase carbon sequestration? Forest Policy Econ 43:41–50. https://doi.org/10.1016/j.forpol.2014.03.004

Pukkala T, Lähde E, Laiho O, Salo K, Hotanen J-P (2011) A multifunctional comparison of even-aged and uneven-aged forest management in a boreal region. Can J For Res 41:661–668. https://doi.org/10.1139/x11-009

Pukkala T, Pukkala T, Lähde E, Laiho O (2014) Optimizing any-aged management of mixed boreal forest under residual basal area constraints. J For Res 25:627–636. https://doi.org/10.1007/s11676-014-0501-y

Purahong W, Kapturska D, Pecyna MJ, Jariyavidyanont K, Kaunzner J, Juncheed K, Uengwetwanit T, Rudloff R, Schulz E, Hofrichter M, Schloter M, Krüger D, François Buscot F (2015) Effects of forest management practices in temperate beech forests on bacterial and fungal communities involved in leaf litter degradation. Microb Ecol 69:905–913. https://doi.org/10.1007/s00248-015-0585-8

Repo A, Rajala T, Henttonen HM, Lehtonen A, Peltoniemi M, Heikkinen J (2021) Age-dependence of stand biomass in managed boreal forests based on the Finnish National Forest Inventory data. For Ecol Manag 498:119507. https://doi.org/10.1016/j.foreco.2021.119507

Rissanen AJ, Ojanen P, Stenberg L, Larmola T, Anttila J, Tuominen S, Minkkinen K, Koskinen M, Mäkipää R (2023) Vegetation impacts ditch methane emissions from boreal forestry-drained peatlands—moss-free ditches have an order-of-magnitude higher emissions than moss-covered ditches. Front Environ Sci 11:1121969. https://doi.org/10.3389/fen-vs.2023.1121969

Roth E-M, Karhu K, Koivula M, Helmisaari H-S, Tyittola E-S (2023) How do harvesting methods applied in continuous-cover forest and rotation forest management impact soil carbon storage and degradability in boreal scots pine forest. For Ecol Manag 544:121144. https://doi.org/10.1016/j.foreco.2023.121144

Schmidt MWI, Margaret S, Torn MS, Abiven S, Dittmar T, Guggenberger G, Janssens IA, Kleber M, Kögel-Knabner I, Lehmann J, DAC M, Nannipieri P, Rasse DP, Weiner S, Trumbore SE

(2011) Persistence of soil organic matter as an ecosystem property. Nature 478:49–56. https://doi.org/10.1038/nature10386

Schulze E-D, Lloyd J, Kelliher FM, Wirth C, Rebmann C, Lühker B, Mund M, Knohl A, Milyukova IM, Schulze W, Ziegler W, Aβ V, Sogachev AF, Valentini R, Dore S, Grigoriev S, Kolle O, Panfyorov MI, Tchebakova N, Vygodskaya NN (2021) Productivity of forests in the Eurosiberian boreal region and their potential to act as a carbon sink – a synthesis. Glob Chang Biol 5:703–722. https://doi.org/10.1046/j.1365-2486.1999.00266.x

Shanin V, Valkonen S, Grabarnik P, Mäkipää R (2016) Using forest ecosystem simulation model EFIMOD in planning uneven-aged forest management. For Ecol Manag 378:193–205. https://doi.org/10.1016/j.foreco.2016.07.041

Shanin V, Juutinen A, Ahtikoski A, Frolov P, Chertov O, Rämö J, Lehtonen A, Laiho R, Mäkiranta P, Nieminen M, Laurén A, Sarkkola S, Penttilä T, Ťupek B, Mäkipää R (2021) Simulation modelling of greenhouse gas balance in continuous-cover forestry of Norway spruce stands on nutrient-rich drained peatlands. For Ecol Manag 496:119479. https://doi.org/10.1016/j.foreco.2021.119479

Stendahl J (2017) Tema: Skogsmarkens kolförråd. (Carbon storage in forest soils.) In skogdata 2017, p 15–23. Department of Forest Resource Management, Swedish University of Agricultural Sciences [in Swedish]. https://pub.epsilon.slu.se/14487/27/skogsdata_2017_170905.pdf

Tamm C-O (1990) Nitrogen in terrestrial ecosystems—questions of productivity vegetational change and ecosystem stability, Ecological Studies 81. Springer Verlag, Berlin, p 117. ISBN 0-387-51807-X

Tong CHM, Nilsson MB, Drott A, Peichl M (2022a) Drainage ditch cleaning has no impact on the carbon and greenhouse gas balances in a recent forest clearcut in boreal Sweden. Forest 13:842. https://doi.org/10.3390/f13060842

Tong CHM, Nilsson MB, Sikström U, Ring E, Drott A, Eklöf K, Futter MN, Peacock M, Segersten J, Peichl M (2022b) Initial effects of post-harvest ditch cleaning on greenhouse gas fluxes in a hemiboreal peatland forest. Geoderma 426:116055. https://doi.org/10.1016/j.geoderma.2022.116055

Tuomi M, Rasinmäki J, Repo A, Vanhala P, Liski J (2011) Soil carbon model Yasso07 graphical user interface. Environ Model Softw 26:1358–1362. https://doi.org/10.1016/j.envsoft.2011.05.009

Vestin P, Mölder M, Kljun N, Cai Z, Hasan A, Holst J, Klemedtsson L, Lindroth A (2022) Impacts of stump harvesting on carbon dioxide, methane and nitrous oxide fluxes. iForest 15:148–162. https://doi.org/10.3390/f13060842

Wallander H, Johansson U, Sterkenburg E, Brandström Durling M, Lindahl BD (2010) Production of ectomycorrhizal mycelium peaks during canopy closure in Norway spruce forests. New Phytol 187:1124–1134. https://doi.org/10.1111/j.1469-8137.2010.03324.x

Chapter 14
Water Quality

Sakari Sarkkola, Mika Nieminen, Hjalmar Laudon, Nicholas Clarke, and Eliza Maher Hasselquist

Abstract

- Conventional forest operations can exert significant impacts on the hydrology and water quality of downstream aquatic environments.
- Few research results have been published on the impacts of continuous cover forestry (CCF) on water quality.
- CCF could be useful for reducing nutrient, carbon, and suspended solid exports in waterways.
- CCF may be a better alternative to rotation forestry (RF) on mineral soils and drained peatlands.
- Further research is needed on the many processes controlling nutrient and carbon exports in CCF and RF.

Keywords Upland soil · Peatland · Water loading · Brownification · Forest hydrology

S. Sarkkola (✉) · M. Nieminen
Natural Resources Institute Finland (LUKE), Helsinki, Finland
e-mail: sakari.sarkkola@luke.fi; mika.nieminen@luke.fi

H. Laudon · E. M. Hasselquist
Department of Forest Ecology and Management, Swedish University of Agricultural Sciences (SLU), Umeå, Sweden
e-mail: hjalmar.laudon@slu.se; eliza.hasselquist@slu.se

N. Clarke
Division of Environment and Natural Resources, Norwegian Institute of Bioeconomy Research (NIBIO), Ås, Norway
e-mail: nicholas.clarke@nibio.no

P. Rautio et al. (eds.), *Continuous Cover Forestry in Boreal Nordic Countries*, Managing Forest Ecosystems 45, https://doi.org/10.1007/978-3-031-70484-0_14

14.1   How Does Forest Management Impact Water Quality?

Conventional forest operations in Fennoscandia typically follow a forest rotation management approach, involving clearcut harvesting, subsequent soil preparation treatments, planting, and one or more thinnings. These practices can exert significant impacts on the hydrology and water quality of downstream aquatic environments. Loadings, i.e., the waterborne exports of elements to water courses, can arise from the release of suspended solids and nutrients resulting from these forestry operations. In conventional rotation forestry (RF), the most substantial loads tend to occur following regeneration cuttings, particularly clearcuttings with soil preparation (Nieminen 2004; Kaila et al. 2015; Nieminen et al. 2015), ditch network maintenance, DNM (Nieminen et al. 2010, 2018a), and forest fertilisation activities (e.g., Piirainen et al. 2013). On mineral soil sites, the increase in nutrient exports caused by harvesting and soil preparation is due to the increase in erosion, the release of nutrients due to decomposition of harvesting residues, and the lack of nutrient uptake by trees and other vegetation (Kreutzweiser et al. 2008). The exports are commonly greatest in the first years after the treatment and decrease over time. In drained peatlands, the exports caused by felling are largely due to the rise in the water level, because the harvest results in reduced evapotranspiration. This will in turn change the release of elements because of altered chemical reactions under anaerobic conditions (redox reactions) in the peat (Nieminen et al. 2017a). Erosion can also be significant, especially in areas treated by ditch mounding.

14.2   Peatland Forestry and Water Quality

Forestry on drained peatlands has a great economic significance, particularly in Finland, where about 25% of the productive forest land is on this type of soil. In Sweden and Norway, the corresponding figures are 14% and 3%, respectively. Forestry on drained peatlands also causes significant impacts on water quality. Historically, the initial drainage of peatlands was regarded as one of the most significant forestry practices influencing the quality of surface waters. It was previously believed that post-drainage export levels would return to those of an undisturbed peatland within 10–20 years (Finér et al. 2010), but current understanding suggests that these exports will remain consistently higher than the background levels of pristine peatlands (Nieminen et al. 2017a; Finér et al. 2021).

This additional export resulting from drainage has been estimated to exceed by a considerable margin the combined exports resulting from DNM, fertilisation, and timber harvesting (Nieminen et al. 2017b; Finér et al. 2021). The mechanisms behind the additional export caused by drainage are poorly understood, but the decomposition rate of peat is known to strongly depend on the water table level, which determines the oxygenation state in the peat (Ojanen and Minkkinen 2019). As the trees within the stand grow after drainage, their evapotranspiration (ET) increases (Sarkkola et al. 2010), the water table level falls, and more oxygen flows into the deeper peat layers, increasing decomposition. The nutrients released from deep peat layers are primarily not usable by trees and other vegetation, so are more likely to leach into groundwater. Several recent studies show that the nutrient exports from drained peatlands can increase over time (Nieminen et al. 2017b, 2023; Räike et al. 2019).

Drainage significantly increases nutrient exports over time, by altering the hydrological dynamics within a catchment. In their pristine state, minerotrophic, nutrient-rich peatlands have the capacity to retain mineral nutrients carried by the water flowing from the catchment area above them. However, after drainage, these nutrients released from the catchment area can swiftly bypass the vegetation and peat of the minerotrophic peatlands and are directly transported into ditches and subsequently into water courses, without any interaction with or filtration by the peatland ecosystem (Sallantaus 1988).

The importance of afforestation and the increase of forest biomass as a possible source of organic carbon export and the brownification of water bodies has been highlighted (Finstad et al. 2016; Škerlep 2021; Nieminen et al. 2021). In recent decades, the forest stands have matured, and many forests have become spruce dominated. This 'sprucification' has increased the amount of recalcitrant organic needle and root litter fall in the forests, which in turn may contribute to the leaching of organic carbon and increased brownification (Kritzberg et al. 2020).

14.3 Effects of Standard Forestry Treatments—Peatlands in Focus

DNM has been estimated to produce more than 90% of the export of suspended solids caused by forestry treatments (Finér et al. 2010; Marttila and Kløve 2010). Particulate organic and inorganic phosphorus is also leached along with the solid matter, and is estimated to average about 0.1% of the leached suspended solids (Finér et al. 2010). The peak export of suspended solids is 1–2 years after the treatment (Nieminen et al. 2010).

Suspended sediments (mineral particles, organic matter) are widely recognised for their detrimental impact on downstream aquatic ecosystems, by increasing turbidity and brownification of water and filling up watercourses. They can harm aquatic habitats, suffocate spawning beds, lead to fish population declines, and significantly alter the abundance and diversity of aquatic invertebrates (Annala et al. 2014; Kjelland et al. 2015; Rajakallio et al. 2021). Their detrimental impacts can endure over many years; often it may take several decades before habitats can recover, assuming recovery occurs at all. Sediment erosion, transport, and deposition are identified as one of the most serious, yet most understudied, aspects of water quality in Sweden (Futter et al. 2016).

Even though DNM increases the export of inorganic nitrogen (nitrate, ammonium) to some extent, the total dissolved nitrogen leaching does not usually increase, because the leaching of dissolved organic nitrogen (DON) usually decreases (Laudon et al. 2023). Also, the export of dissolved organic carbon (DOC), which affects the water colour, will not increase, and may even decrease (Joensuu et al. 2002; Nieminen et al. 2018a). However, over the long term, the leaching of organic carbon is greater from drained peatlands than from pristine mires (Nieminen et al. 2021).

Clearcuttings on peatlands increase the leaching of all main nutrients; for example, nitrogen export is about five times greater from peatlands than from mineral soil sites (Nieminen 2004; Finér et al. 2010). In peatlands, the harvest of evapotranspiring tree stands quickly raises the water level, which significantly increases the risk of leaching of nutrients and organic carbon. When the water level rises, i.e., when the previously aerobic peat layer becomes anaerobic, redox reactions start in the peat soil, which can especially increase the leaching of phosphorus, iron, organic carbon, organic nitrogen, and ammonium nitrogen (Kaila et al. 2014; Koskinen et al. 2017). There are observations of large exports especially from nutrient-rich peatland forests after regeneration harvestings and restoration treatments, which both raise the water level (Nieminen et al. 2015, 2020).

The leaching caused by the rise of the water level is difficult to mitigate by water protection methods currently used in practical forestry, such as sediment pits and ponds, especially because they do not retain the carbon and nutrients transported in dissolved organic form (Haahti et al. 2018). The rise of the water level can be prevented through more efficient drainage, but this in turn would increase the export of suspended solids, particulate nutrients, and heavy metals (Joensuu et al. 2002; Nieminen et al. 2010). Effective drainage would also increase the decomposition of peat, thereby increasing carbon and nutrient exports to recipient water bodies (Nieminen et al. 2023) and atmospheric carbon emissions (Ojanen and Minkkinen 2019).

An alternative to drainage for controlling water level is to retain evapotranspiring trees in the forest stand. Remaining trees and even understorey vegetation may also capture the nutrients released from harvest residues and soil after harvesting. This could be attained through continuous cover forestry (CCF), in which large-scale clearcuttings are avoided, and selection cutting, strip cutting, shelterwood cutting and small gap cuttings are instead used to retain evapotranspiring vegetation.

14.4   Can CCF Decrease Water Loading?

In peatlands, the CCF concept operates on the principle that tree removal should not significantly impact the evapotranspiration (ET) capacity of the stand. In this way, the effects on water table levels are moderate, and the drainage would remain sufficient to avoid affecting tree growth (Nieminen et al. 2018b). The remaining trees within the stand also play a crucial role in reducing nutrient leaching, as they capture some of the nutrients released, particularly from the decomposition of harvest residues.

CCF is suggested to reduce leaching from peatlands because it would reduce the need for DNM, and because the water level in the peat would be somewhat higher on average than in stands managed with conventional RF. This would reduce the decomposition of peat and thereby the release of nutrients, especially from deep peat layers. CCF also has potential to limit another form of soil disturbance, namely site preparation. This, however, hinges on the success of natural regeneration in CCF forests, which departs from the conventional artificial regeneration approach commonly practiced in Nordic forestry. Traditional site preparation, often involving techniques like disc-trenching, has been associated with adverse impacts on water quality, such as increased DOC export (Schelker et al. 2012) and elevated levels of methylmercury (Eklöf et al. 2014). By reducing the need for site preparation, CCF not only benefits water quality but also promotes recreational value and biodiversity by minimising physical disruption to the site (Fig. 14.1).

There are no research results on the loading effects of CCF practised on mineral soils, and only a few results relating to peat soils. Load estimates based on thinning cuttings of RF cannot be presented, because they have not been empirically studied. However, bearing in mind that the exports caused by clearcutting on mineral soil sites have been estimated to be relatively minor (Kreutzweiser et al. 2008), the water quality effects of CCF can also be expected to be minor.

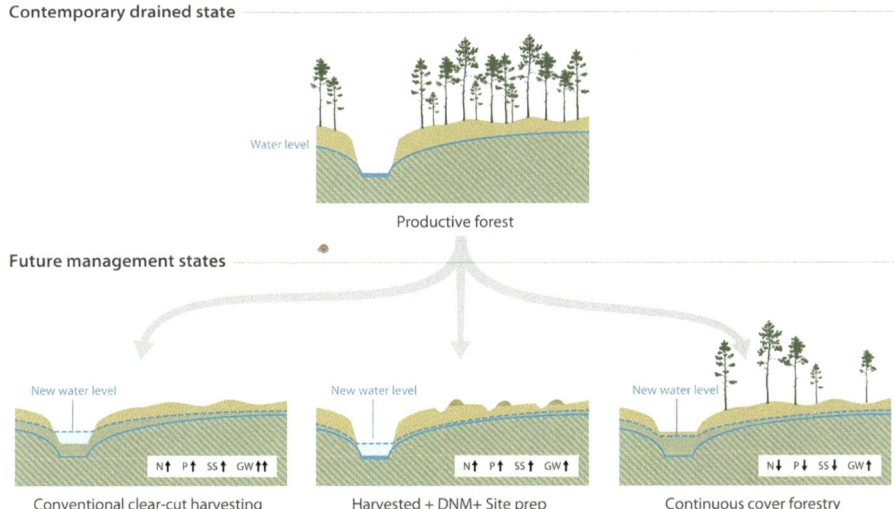

Fig. 14.1 Conceptual model of the effect of forest management methods on the ground water table level (GW) and the exports of nitrogen (N), phosphorus (P) and suspended sediments (SS) in drained peatlands. DNM = ditch network maintenance; Site Prep = site preparation (original figure: Kritzberg et al. 2020, CCBY 4.0, link to the license: creativecommons.org/licenses/by/4.0/)

In Alberta, Canada, partial cuts showed intermediate decreases in ammonium (NH_4^+) in the forest-floor soil between uncut and clearcut sites of both deciduous and coniferous stands, as well as for phosphate (PO_4^{3-}) in forest floor soil in the coniferous stands (Lindo and Visser 2003). In both deciduous and coniferous stands, nitrate (NO_3^-) in the forest floor soil was elevated in partial-cut corridors and clearcuts compared with partial-cut patch and uncut treatments (Lindo and Visser 2003). In Québec, partial harvesting (corresponding to one-third and two-thirds removal of tree basal area) in aspen stands resulted in lower potential mineralisation of nitrogen and a lower relative nitrification index (net [(NO_3^-): (NO_3^- + NH_4^+)] production ratio) in the forest floor soil, compared to clearcutting (Lapointe et al. 2005).

Finnish and Canadian studies examined the impact of different partial harvest treatments on water table level in boreal settings. Roy Proulx et al. (2021) found that a 40% basal area partial harvest in a Canadian black spruce stand did not affect water table level 1 year later. Leppä et al. (2020) showed that removing up to 70% of basal area in Finnish Norway spruce sites had minimal impact on water table level, thanks to rapid field layer vegetation growth offsetting ET loss by tree harvest (Leppä et al. 2020). Pothier et al. (2003) found that cutting 50–60% of basal area in red spruce-balsam fir sites promoted vegetation regeneration and water table recovery compared to clearcutting. Päivänen and Sarkkola (2000) found no significant water table impact from 30% stand removal, rendering DNM to maintain drainage conditions after harvesting unnecessary. Older studies had similar findings on partially harvested sites (e.g., Heikurainen 1966; Päivänen 1982).

Recent Finnish research suggested strip cuts could maintain lower water table levels than clearcuts, depending on soil conductivity and stand density (Stenberg et al. 2022). While not as effective as selection harvesting in maintaining sufficiently low water table levels for undisturbed tree growth without DNM, strip cuts could be used as a special type of CCF for shade-intolerant tree species that cannot be managed by selection harvestings (Saarinen et al. 2020). CCF appears effective in controlling water table levels in many contexts but requires further testing across various geographical and soil settings, along with improved modelling, before being adopted for drained forested peatlands on a larger scale.

In the above experiments, phosphorus exports after strip-cutting on nutrient-poor pine peatland sites were lower than the exports shown in earlier clearcutting experiments, but not clearly lower compared to all clearcutting experiments. Nieminen et al. (2023) suggest the amount of harvested volume per catchment area could be a factor that correlates with the variation in nutrient exports more strongly than the specific harvesting method, regardless of whether it involves clearcutting, strip-cutting, or single-tree harvesting. The amount of harvested volume is directly related to the amount of harvest residues, the rate of decrease in nutrient uptake, as well as the rate of soil-water level rise following harvesting.

Based on process-based modelling and nutrient-export coefficients for different forest operations, Nieminen et al. (2023) investigated the regional effects of CCF and RF on nitrogen and phosphorus exports from forested catchment areas in Finland. They employed the MONSU forest planning model (Pukkala 2011) to simulate forest development and management operations on approximately 15 million hectares of forests over a 50-year projection into the future. According to the results, the transition from RF to CCF would clearly reduce water loading, both when following the practical forest management guidelines of Finnish forestry and when harvesting volumes are kept at the current level. The exports from forestry on drained peatlands were significantly greater than from upland sites, but the transition to CCF would nevertheless decrease the export per hectare by approximately the same amount on uplands and peatlands over the next five decades. The largest load benefits from CCF would be reached in southern Finland, where the climatic conditions are more favourable and where harvesting volumes from clearcuttings are larger and water table levels in peatlands are lower than in northern Finland (Nieminen et al. 2023). Nevertheless, the study underscores the necessity for further examination of the impacts of forest management on water quality. Experimental studies would be needed on various harvesting and tillage methods, as well as greater understanding about the processes influencing the release of exports, particularly from drained peatland forests.

14.5 Risks Associated with CCF

So far, we have highlighted potential benefits of CCF on peatlands, but it is also important to consider associated risks and address unanswered questions before advocating for widespread CCF implementation in Fennoscandia. One major concern is the risk of damage caused by heavy off-road forestry machinery, with shorter intervals between use for harvesting compared with RF. Peatlands are highly vulnerable to rutting, leading to water channelisation, creating hotspots for methylation of mercury , and erosion (Eklöf et al. 2014). Their low bearing capacity makes them more sensitive than other boreal soils (Ågren et al. 2014). Traditionally, rutting is avoided by harvesting in frozen winter conditions, but with warming winters this strategy may become less effective. CCF may not allow the use of branches and stems from less profitable trees to create haul roads and protect against rutting, as these trees are typically saved for later selection harvestings (Andersson et al. 2016).

Tree species composition may be difficult to maintain with some CCF methods. Norway spruce, a shade-tolerant secondary tree species, contrasts with species like Scots pine and downy birch, which thrive in direct sunlight, typically regenerating after fire, storm felling, or clearcutting. Norway spruce could therefore become the dominant tree species, particularly in selectively harvested forests, potentially increasing water brownification compared to RF forests (Finstad et al. 2016; Škerlep 2021).

The key question in managing forests by CCF is, however, successful natural regeneration. If natural regeneration is not satisfactory, complementary planting or soil preparation of unregenerated patches is needed to establish a sufficiently dense tree stand. The studies published so far indicate promising economic performance for CCF compared to RF both in mineral soil forests and drained peatlands (see

Chap. 8), but those studies assumed satisfactory natural regeneration without any need for complementary activities.

14.6 Conclusions

CCF has great potential for reducing the effects of forestry on water quality, both in mineral soil and drained peatland catchments. In particular, CCF is a better alternative to RF because it avoids load-producing practices like DNM and ditch mounding. Although relatively little research has been published to support the performance of CCF from a water quality perspective, the theory behind controlling ground water levels by the ET of the tree stand is strong and the load benefit achieved by avoiding DNM and soil preparation is clear. Implementation of CCF in productive wet peatland forests to mitigate nutrient exports and GHG emissions is becoming widely accepted, and many forest enterprises and the Finnish State now manage those forests exclusively with CCF. The use of CCF as a tool to improve water quality will therefore soon be tested on a large scale.

References

Ågren AM, Buffam I, Cooper DM et al (2014) Can the heterogeneity in stream dissolved organic carbon be explained by contributing landscape elements? Biogeosciences 11:1199–1213. https://doi.org/10.5194/bg-11-1199-2014

Andersson E, Andersson M, Blomquist S et al (2016) Nya och Reviderade Målbilder för god Miljöhänsyn: Skogssektorns Gemensamma Målbilder för god Miljöhänsyn vid Skogsbruksåtgärder (New and revised environmental targets: Common targets for the forest sector for good environmental considerations in forestry). Skogsstyrelsen, Jönköping, Sweden

Annala M, Mykrä H, Tolkkinen M et al (2014) Are biological communities in naturally unproductive streams resistant to additional anthropogenic stressors? Ecol Appl 24:1887–1897. https://doi.org/10.1890/13-2267.1

Eklöf K, Schelker J, Sørensen R et al (2014) Impact of forestry on total and methyl-mercury in surface waters: distinguishing effects of logging and site preparation. Environ Sci Tech 48(9):4690–4698. https://doi.org/10.1021/es404879p. Epub 2014 Apr 8. PMID: 24666406

Finér L, Mattsson T, Joensuu S et al (2010) Metsäisten valuma-alueiden kuormituksen laskenta (A method for calculating nitrogen, phosphorus and sediment load from forest catchments). The Finnish environment 10/2010. Edita Prima Oy, Helsinki, 33 pp

Finér L, Lepistö A, Karlsson K et al (2021) Drainage for forestry increases N, P and TOC export to boreal surface waters. Sci Tot Environ 762:144098. https://doi.org/10.1016/j.scitotenv.2020.144098

Finstad A, Andersen T, Larsen S et al (2016) From greening to browning: catchment vegetation development and reduced S-deposition promote organic carbon load on decadal time scales in Nordic lakes. Sci Rep 6:31944. https://doi.org/10.1038/srep31944

Futter MN, Högbom L, Valinia S et al (2016) Conceptualizing and communicating management effects on forest water quality. Ambio 45:188–202. https://doi.org/10.1007/s13280-015-0753-6

Haahti K, Nieminen M, Finér L (2018) Model-based evaluation of sediment control in a drained peatland forest after ditch network maintenance. Can J For Res 48(2):130–140. https://doi.org/10.1139/cjfr-2017-0269

Heikurainen L (1966) Effect of cutting on the ground-water level on drained peat lands. International symposium on Forest hydrology. Pergamon Press, pp 345–354

Joensuu S, Ahti E, Vuollekoski M (2002) Effects of ditch network maintenance on the chemistry of runoff water from peatland forests. Scand J For Res 17:238–247. https://doi.org/10.1080/028275802753742909

Kaila A, Sarkkola S, Laurén A et al (2014) Phosphorus export from drained scots pine mires after clear-felling and bioenergy harvesting. For Ecol Manag 325:99–107. https://doi.org/10.1016/j.foreco.2014.03.025

Kaila A, Laurén A, Sarkkola S et al (2015) Effect of clear-felling and harvest residue removal on nitrogen and phosphorus export from drained Norway spruce mires in southern Finland. Bor Environ Res 20:693–706

Kjelland ME, Woodley CM, Swannack TM, Smith DL (2015) A review of the potential effects of suspended sediment on fishes: potential dredging-related physiological, behavioral, and transgenerational implications. Environ Syst Decis 35:334–350. https://doi.org/10.1007/s10669-015-9557-2

Koskinen M, Tahvanainen T, Sarkkola S et al (2017) Restoration of nutrient-rich forestry-drained peatlands poses a risk for high exports of dissolved organic carbon, nitrogen, and phosphorus. Sci Tot Environ 586:858–869. https://doi.org/10.1016/j.scitotenv.2017.02.065

Kreutzweiser DP, Hazlett PW, Gunn JM (2008) Logging impacts on the biogeochemistry of boreal forest soils and nutrient export to aquatic systems: a review. Environ Rev 16:157–179. https://doi.org/10.1139/A08-006

Kritzberg ES, Hasselquist EM, Škerlep M et al (2020) Browning of freshwaters: consequences to ecosystem services, underlying drivers, and potential mitigation measures. Ambio 49:375–390. https://doi.org/10.1007/s13280-019-01227-5

Lapointe B, Bradley RL, Shipley B (2005) Mineral nitrogen and microbial dynamics in the forest floor of clearcut or partially harvested successional boreal forest stands. Plant Soil 271:27–37. https://doi.org/10.1007/s11104-004-1830-y

Laudon H, Mosquera V, Eklöf K et al (2023) Consequences of rewetting and ditch cleaning on hydrology, water quality and greenhouse gas balance in a drained northern landscape. Sci Rep Nat 13:20218. https://doi.org/10.1038/s41598-023-47528-4

Leppä K, Hökkä H, Laiho R et al (2020) Selection cuttings as a tool to control water table level in boreal drained peatland forests. Front Earth Sci 8:576510. https://doi.org/10.3389/feart.2020.576510

Lindo Z, Visser S (2003) Microbial biomass, nitrogen and phosphorus mineralization, and mesofauna in boreal conifer and deciduous forest floors following partial and clear-cut harvesting. Can J For Res 33:1610–1620. https://doi.org/10.1139/X03-080

Marttila H, Kløve B (2010) Dynamics of suspended sediment transport and erosion in a drained peatland forestry catchment. J Hydrol 388:414–425. https://doi.org/10.1016/j.jhydrol.2010.05.026

Nieminen M (2004) Export of dissolved organic carbon, nitrogen and phosphorus following clearcutting of three Norway spruce forests growing on drained peatlands in southern Finland. Silva Fenn 38(2):123–132. https://doi.org/10.14214/sf.422

Nieminen M, Ahti E, Koivusalo H (2010) Export of suspended solids and dissolved elements from peatland areas after ditch network maintenance in south-Central Finland. Silva Fenn 44(1):161. https://doi.org/10.14214/sf.161

Nieminen M, Koskinen M, Sarkkola S et al (2015) Dissolved organic carbon export from harvested peatland forests with differing site characteristics. Water Air Soil Poll 226:181. https://doi.org/10.1007/s11270-015-2444-0

Nieminen M, Sarkkola S, Laurén A (2017a) Impacts of forest harvesting on nutrient, sediment and dissolved organic carbon exports from drained peatlands: a literature review, synthesis, and suggestions for the future. For Ecol Manag 392:13–20. https://doi.org/10.1016/j. foreco.2017.02.046

Nieminen M, Sallantaus T, Ukonmaanaho L et al (2017b) Nitrogen and phosphorus concentrations in discharge from drained peatland forests are increasing. Sci Tot Environ 609:974–981. https://doi.org/10.1016/j.scitotenv.2017.07.210

Nieminen M, Piirainen S, Sikström U et al (2018a) Ditch network maintenance in peat-dominated boreal forests: review and analysis of water quality management options. Ambio 47:535–545. https://doi.org/10.1007/s13280-018-1047-6

Nieminen M, Hökkä H, Laiho R et al (2018b) Could continuous cover forestry be an economically and environmentally feasible management option on drained boreal peatlands? For Ecol Manag 424:78–84. https://doi.org/10.1016/j.foreco.2018.04.046

Nieminen M, Sarkkola S, Tolvanen A et al (2020) Water quality management dilemma: increased nutrient, carbon, and heavy metal exports from forestry-drained peatlands restored for use as wetland buffer areas. For Ecol Manag 465:118089. https://doi.org/10.1016/j.foreco.2020.118089

Nieminen M, Sarkkola S, Sallantaus T et al (2021) Peatland drainage – a missing link behind increasing TOC concentrations in waters from high latitude forest catchments? Sci Tot Environ 774:145150. https://doi.org/10.1016/j.scitotenv.2021.145150

Nieminen M, Pukkala T, Stenberg L et al (2023) Jatkuvan kasvatuksen ja tasaikäismetsätalouden vaikutus metsäisten valuma-alueiden vesistökuormitukseen Suomessa (Effect of continuous cover forestry and even-aged forestry on the water loadings from forested headwater catchments in Finland). Metsätieteen aikakauskirja 2023 article 22001. https://doi.org/10.14214/ ma.22001

Ojanen P, Minkkinen K (2019) The dependence of net soil CO2 emissions on water table depth in boreal peatlands drained for forestry. Mires Peat 24(27):1–8

Päivänen J (1982) Hakkuun ja lannoituksen vaikutus vanhan metsäojitusalueen vesitalouteen. Summary: the effect of cutting and fertilisation on the hydrology of an old forest drainage area. Folia For 516: 1–19

Päivänen J, Sarkkola S (2000) The effect of thinning and ditch network maintenance on the water table level in a Scots pine stand on peat soil. Suo 51:131–138

Piirainen S, Domisch T, Moilanen M, Nieminen M (2013) Long-term effects of ash fertilization on runoff water quality from drained peatland forests. For Ecol Manag 287:53–66. https://doi. org/10.1016/j.foreco.2012.09.014

Pothier D, Prévost M, Auger I (2003) Using the shelterwood method to mitigate water table rise after forest harvesting. For Ecol Manag 179:573–583. https://doi.org/10.1016/ S0378-1127(02)00530-3

Pukkala T (2011) Optimising forest management in Finland with carbon subsidies and taxes. Forest Policy Econ 13:425–434. https://doi.org/10.1016/j.forpol.2011.06.004

Räike A, Taskinen A, Knuuttila S (2019) Nutrient export from Finnish rivers into the Baltic Sea has not decreased despite water protection measures. Ambio 49:460–474. https://doi.org/10.1007/ s13280-019-01217-7

Rajakallio M, Jyväsjärvi J, Muotka T, Aroviita J (2021) Blue consequences of the green bioeconomy: clear-cutting intensifies the harmful impacts of land drainage on stream invertebrate biodiversity. J Appl Ecol 58:1523–1532. https://doi.org/10.1111/1365-2664.13889

Roy Proulx S, Jutras S, Leduc A et al (2021) Partial harvest in Paludified black spruce stand: short-term effects on water table and variation in stem diameter. Forests 12:271. https://doi. org/10.3390/f12030271

Saarinen M, Valkonen S, Sarkkola S et al (2020) Jatkuvapeitteisen metsänkasvatuksen mahdollisuudet ojitetuilla turvemailla (Potential of continuous cover forestry in drained peatlands). Metsätieteen aikakauskirja 2020 article 10372. https://doi.org/10.14214/ma.10372

Sallantaus T (1988) Water quality of peatlands and man's influence on it. Suomen Akatemian julkaisuja 5(1988):80–98

Sarkkola S, Hökkä H, Koivusalo H et al (2010) Role of tree stand evapotranspiration in maintaining satisfactory drainage conditions in drained peatlands. Can J For Res 40:1485–1496. https://doi.org/10.1139/X10-084

Schelker J, Eklöf K, Bishop K, Laudon H (2012) Effects of forestry operations on dissolved organic carbon concentrations and export in boreal first-order streams. J Geophys Res 117:G01011. https://doi.org/10.1029/2011JG001827

Škerlep M (2021) Changing land cover as a driver of surface water browning. Academic Dissertation. Lund University. 65 p. https://lucris.lub.lu.se/ws/portalfiles/portal/101442357/Changing_land_cover_as_a_driver_of_surface_water_browning_Thesis_Kappa.pdf

Stenberg L, Leppä K, Launiainen S et al (2022) Measuring and modeling the effect of strip cutting on the water table in boreal drained peatland pine forests. Forests 13:1134. https://doi.org/10.3390/f13071134

Chapter 15
Forest Owners' and Forestry Stakeholders' Perceptions

Emmi Haltia, Louise Eriksson, Terhi Koskela, Per Kr. Rørstad, Ida Wallin, and Jasmine Zhang

Abstract

- In Finland, the new Forest Act in 2014 made continuous cover forestry (CCF) a possible forest management option. This triggered research on how forest owners and forestry professionals perceive CCF. In Sweden and Norway, the CCF method has been legal but not encouraged. Research on stakeholder views on CCF has only recently emerged, so only a few studies have been published on the topic.
- In Finland, according to surveys, less than 10% of forest owners have converted to CCF in all their forests and around 20–25% in part of their forests. About a fifth of forest owners expressed an interest in testing it.
- In the Finnish studies, CCF has been of particular interest to forest owners whose holdings are smaller than average and where they have recreational objectives.

E. Haltia (✉) · T. Koskela
Natural Resources Institute Finland (Luke), Helsinki, Finland
e-mail: emmi.haltia@luke.fi; terhi.koskela@luke.fi

L. Eriksson
Department of Geography, Umeå universitet, Umeå, Sweden
e-mail: louise.eriksson@umu.se

P. K. Rørstad
Faculty of Environmental Sciences and Natural Resource Management, Norwegian University of Life Sciences (NMBU), Ås, Norway
e-mail: per.kristian.rorstad@nmbu.no

I. Wallin
Southern Swedish Forest Research Centre, Swedish University of Agricultural Sciences (SLU), Alnarp, Sweden
e-mail: ida.wallin@slu.se

J. Zhang
Department of Urban and Rural Development, Swedish University of Agricultural Sciences (SLU), Uppsala, Sweden
e-mail: jasmine.zhang@slu.se

© The Author(s) 2025   273
P. Rautio et al. (eds.), *Continuous Cover Forestry in Boreal Nordic Countries*,
Managing Forest Ecosystems 45, https://doi.org/10.1007/978-3-031-70484-0_15

- The interest of forest owners clearly increased if they were compensated for converting to CCF.
- Forestry companies, as well as their associations in Sweden and Finland, promote CCF as an alternative forest management method, according to statements on their official websites. How well this supportive attitude will materialise in their actions remains an open question.
- The current forestry culture, the power of industrial networks, uncertainties concerning economic profitability and ecological outcomes, as well as current forestry education and technical knowledge, are the main barriers for CCF.

Keywords Forest owners · Forestry professionals · Forest industry · Perceptions · Barriers

15.1 Introduction

To understand the potential for implementation of continuous cover forestry (CCF) at a larger scale in the Nordic countries it is important to understand the actors influencing forest management decisions. While the forest owner has the legal right to make these decisions, other forestry actors and the institutional, operational, and social context in which these decisions are made are also highly influential (Andersson and Keskitalo 2018, Matilainen et al. 2023). Rather than considering CCF implementation to be the outcome of an individual owner decision (i.e. only determined by the owner's perceptions), other actors in the forest sector, including owner associations, the forest industry, and government actors are likely to play an important role, through formal and informal interactions with the owners.

The ownership structure differs somewhat between Norway, Sweden, and Finland, but non-industrial private forest (NIPF) owners (sometimes referred to as family forest owners or individual private forest owners) is an important ownership category in all countries. In this chapter we use the terms NIPF or forest owners when referring to this group. Finland has the highest number of individual NIPF forest owners (Fig. 15.1), and 11% of the population are forest owners. The corresponding figures in Sweden and Norway are 3% and 2%, respectively.

There are some other differences between the countries in terms of ownership structure (see Fig. 15.1). In Finland, the average size of holdings is 48 ha including shrub and idle land (Karppinen et al. 2020). In Sweden the corresponding figure is 45 ha (Skogsstyrelsen 2021a, 2021b) and in Norway 45 ha (Statistics Norway 2022). Forest owners are heterogeneous in their socioeconomic backgrounds, attitudes, values, and objectives for their forest ownership (Ficko et al. 2019). Ficko et al. also report that forest owner groups appearing in earlier forest owner surveys include multi-objective forest owners, recreationists, investors, and indifferent owners. Together with site-specific characteristics of forests, the varying objectives have led to various forest management choices. For example, forest owners in the

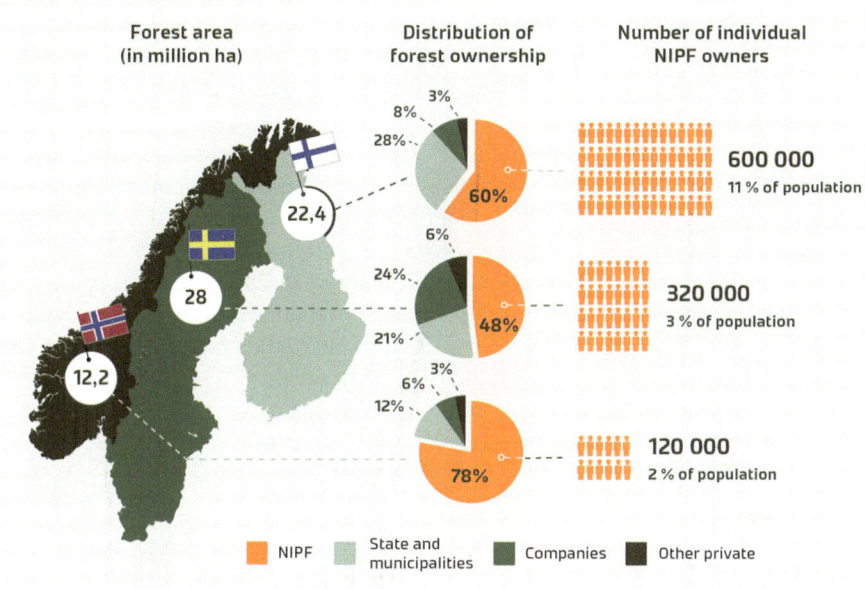

Fig. 15.1 Forest ownership in Fennoscandia

different countries use varying management strategies, and there also seem to be different drivers for their management choices (Westin et al. 2023).

Current perceptions regarding CCF among forest owners and professionals should be viewed in the light of how CCF has been used historically in the Nordic countries. In the nineteenth century and earlier, forest harvesting in Norway, Sweden and Finland was based on selective cutting of trees that exceeded a certain diameter, and this form of management had some CCF characteristics (detailed description in Chap. 2). It was, however, practised in such a way that the most valuable trees were harvested, and regeneration of the forest was not necessarily managed, resulting in a depletion of forest resources.

Awareness of this depletion brought about change in forest management practices, which were switched to rotation forestry (RF). After the 1950s and until recently, CCF management was rarely applied in the Nordic countries. In Finland, CCF was not even permitted under the provisions of the Forest Act until 2014, with the exception of some special cases. The Forest Act reform diversified forest management, removing previous constraints and allowing CCF to be adopted as a method of growing the forest. In Sweden, while there have been no formal barriers, and in recent years CCF has been increasingly supported by institutional factors (e.g., certification schemes), CCF is limited in certain forest contexts and few policy measures have promoted the method (Sténs et al. 2019). However, Espmark's (2017) report on the debate in Sweden shows that public interest in CCF has grown considerably since the 1990s, reflected in the number of news articles including keywords relating to CCF. The debate is characterised by conflicting arguments and opinions, representing different interests, and little is known about what forest owners and forestry professionals have specifically contributed to the debate and how they are influenced by it.

This chapter focuses on perceptions of CCF among stakeholders who have a direct or indirect influence over decisions regarding forest management. Emphasis is placed on examining how NIPF owners perceive and evaluate CCF, as well as their interest in adopting CCF practices. Given the importance of decision-support services in forest owners' decision making (Takala et al. 2023), the perceptions of forestry professionals—advisors at forest owner associations, timber buyers, etc.—are also reviewed. The limited number of empirical studies on NIPF owners and forestry stakeholders makes a comprehensive comparison of the Nordic countries difficult and prevents general conclusions regarding the perceptions of stakeholder groups. Nevertheless, the following review reveals some of the drivers and barriers relating to CCF adoption at larger scales in the Nordic countries.

15.2   Forest Owners and Forestry Professionals

15.2.1   Forest Owners in Finland

An evaluation of the revised Forest Law shows that CCF is slowly gaining popularity (Kniivilä et al. 2020). In the Finnish Forest Owner Survey conducted in 2019 (Hänninen et al. 2020), 8% of NIPF owners reported that they have converted to CCF in all their forests, and 18% in part of their forests. Approximately 20% of NIPF owners had not applied CCF forestry but were interested in doing so in the future (Horne et al. 2020). If compensation were paid for a transition to CCF on the grounds of climate benefits, 62% of NIPF owners expressed interest in implementing it in part of their forests (Koskela et al. 2021).

In another survey that asked NIPF owners to estimate the proportion of different management types in their forest holdings (Juutinen et al. 2020), RF averaged around 60% and CCF 19% of the reported areas, weighted by forest area. More than 60% of respondents intended to use two or more methods of managing their forest in the future (RF, short RF, long RF, CCF, or other management methods incorporating conservation). If an area had been naturally regenerated previously, there was a greater interest in CCF, but the use of seedlings, young stand management and clearcut in the 5-year period prior to the survey was associated with less interest in CCF.

According to the Finnish Forest Owner Survey, NIPF owners regarded CCF as being beneficial to certain species, so it was a more common forest management method among owners who prioritised recreational and natural values (Horne et al. 2020). Similar results have been reported in other studies. Nature and recreation goals, and smaller-than-average forest estates also correlated with the willingness of forest owners to adopt CCF in the future (Husa and Kosenius 2021). In a qualitative study that interviewed ten owners of different-sized forest estates in the Kuonanjoki river catchment area in southern Finland, the owners were particularly critical about the profitability of CCF, changes in timber value, and the risks associated with forest regeneration/ingrowth (Kietäväinen 2022). In contrast, landscape values, biodiversity, and the suitability of CCF for some forest stands and locations were assessed as positive features. The level of knowledge varied, and not all owners meant the same things even when using the same terms. Perceptions of CCF compared to RF varied: some forest owners thought that CCF was more profitable than RF, whereas others took the opposite view. Some forest owners thought that CCF required more management, improved the wood quality, or was better for soil, animals, and plants, while other forest owners felt the opposite.

Juutinen et al. (2020) examined which forest management types Finnish forest owners planned to implement in the future, and how their assessment of generally implemented forest management types impacted these plans. The area that owners planned to allocate to CCF was smaller if they were positive about current forest management in Finland in terms of timber production, biodiversity, and carbon sequestration. NIPF owners who described themselves as agricultural or forestry entrepreneurs allocated a smaller area of forest to CCF than forest owners of any other occupation. The other socioeconomic background factors—age, gender, education, living in a city or large town—had no effect on future plans for forest management.

The choice by Finnish NIPF owners between conventional forest management that prioritises wood production and alternative forest management that emphasises natural values has also been studied using the choice experiment method (Juutinen et al. 2021). In that study, forest management with an emphasis on natural values was described as being similar to CCF. The majority of owners were interested in making an agreement on management that prioritised natural values and was similar to CCF in return for financial compensation. Forest owners were also interested in choosing nature-oriented forestry without financial compensation if it was more profitable in economic terms than conventional RF.

The potential of CCF for reducing carbon dioxide emissions (Chap. 13) and nutrient leakage to water bodies (Chap. 14) from peatlands has entered the debate in recent years. According to the Finnish Forest Owner Survey, almost 40% of owners were willing to convert to CCF on peatlands without ditch maintenance if they received compensation (Koskela et al. 2021). About 30% of owners did not know where they stood on the issue. NIPF owners who were more often interested in CCF on peatlands lived in urban environments, were aged between 45 and 54, owned the forest through a joint ownership or estate of a deceased person, had multiple objectives for forestry, or emphasised recreational use. Other respondents reported economic objectives or had no specific goals for their forest ownership.

A recent survey (Viitala et al. 2023) examined the willingness of peatland forest owners to convert to CCF. Fifty-seven percent believed that clearcutting and ditch maintenance on peatlands should be avoided because of water quality issues, and

43% because of climate change mitigation. About half (52%) of respondents thought that the conversion to CCF on peatlands should be subsidised by the government. The same proportion felt they needed more information about CCF on peatlands. The owners were asked about what compensation they would need to convert to CCF in rich, spruce-dominated peatland forests. About 45% were willing to enter into an agreement on conversion to CCF, even if the compensation was lower than their self-estimated loss of income relating to the conversion. Thirty percent wanted their loss of income to be covered, and 11% required compensation that would exceed the estimated costs. Only 6% were not willing to enter into the proposed agreement for any financial compensation.

15.2.2 Forestry Professionals in Finland

Soon after the revision of the Finnish Forest Act allowing CCF, an educational seminar tour for forestry professionals was organised (Valkonen and Cheng 2014). Participants were asked to complete a questionnaire about their views on CCF; the 771 respondents corresponded to a response rate of 78%. Most had a neutral attitude towards CCF. As CCF had previously only been allowed in special cases, only a few respondents had any experience of it. Less than 20% of the respondents agreed with the statement that they found CCF appealing, and about 50% disagreed. However, a majority of respondents were willing to learn more about CCF, were interested in applying it, and were positive to the method being allowed in the new Forest Act. Forestry professionals were not eager to recommend CCF, but the majority were willing to apply it if forest owners requested it and if implementation would lead to successful timber production in relation to the characteristics of the forest. CCF was believed to be more suitable than RF for some specific sites and some forest owners. However, respondents felt that adoption of CCF would remain limited in the future, and that RF would retain its dominance because it was considered more profitable.

While not the main focus, some studies have touched upon forestry professionals' perceptions regarding CCF. Hujala et al. (2016) studied the development of decision-support services to protect biodiversity and diversify forest management. A survey was conducted of forestry professionals working in forest management associations and in a national company that offers forestry services (Otso Metsäpalvelut Ltd). Results showed that 30–50% of respondents had drawn up forest management plans that included CCF for their customers in the first year after the new Forest Act. The respondents submitted written comments about CCF, and their perceptions ranged from positive to very negative. Positive comments included that the change in the law made it possible to use CCF in a rational way rather than under other, misleading terms due to earlier legislative restrictions. The neutral respondents referred to the lack of knowledge about CCF and a current lack of interest among forest owners. In the negative comments, the forestry professionals were critical of CCF as a forest management method and perceived that both forestry professionals and forest owners were both sceptical. They also believed that forest industry companies would not be interested in buying wood from CCF sites.

Takala et al. (2023) investigated discourses on decision-support services that go beyond the traditional production-oriented focus and promote the concept of more sustainable forest use, including CCF. In the study, 12 forest owners and 12 forestry professionals were interviewed about their views on the need for development of decision-support services provided to forest owners. Four discourses related to the issue were identified.

1) The juggling discourse, which was produced exclusively by the forestry professionals (not NIPF owners), means that the purpose of decision-support services was to balance the different forest uses starting with the forest owners' objectives. In principle, the professionals were interested in providing new services and multi-objective forest management, but they also often saw restrictions regarding changing current systems and practices. The juggling discourse expressed a somewhat reserved attitude towards CCF since, among other reasons, forest owners may not necessarily understand all the difficulties, uncertainties, and risks associated with CCF.

2) The productivist discourse, produced by both forest owners and professionals, emphasised a pragmatic forestry based on forestry guidelines, which they judged to be a non-ideological approach. This discourse also expressed scepticism towards environmental perspectives. CCF was considered a challenge to the well-established practice of high-quality RF.

3) The loyal discourse, produced exclusively by forest owners, was satisfied with the mainstream decision-support services.

4) The critical discourse, produced primarily but not exclusively by forest owners, highlighted contentious issues pertaining to forests. Unlike the other three narratives, it showcased dissatisfaction with the mainstream, production-oriented decision-support services.

The findings demonstrated that the prevailing conditions for discourse may prevent notions that diverge from the dominant production-focused mindset in the mainstream decision-support services offered to forest owners in Finland.

15.2.3   Forest Owners and Forestry Professionals in Sweden and Norway

Studies from Sweden and Norway have compared perceptions and attitudes of different forestry stakeholder groups. In a study examining stakeholder preferences and willingness to pay[1] for forest management alternatives at landscape level, Nordén et al. (2017) showed that NIPF owners in Sweden had a slightly negative

---

[1] Willingness to pay reflects the value of product or service in monetary terms.

mean marginal willingness to pay for uneven-aged forests through CCF, which may be compared to the owners' positive mean marginal WTP for a mixed-tree composition. The owners' WTP for uneven-aged forests through CCF was comparable to public forest officials employed at the Swedish Forest Agency, whereas private forest officials displayed a lower WTP for CCF. The sample of owners in this study was considerable (also including a random sample of owners in Sweden), indicating that the study may provide an adequate overview of the large group of NIPF owners in Sweden.

In a small-scale study of Swedish NIPF owners recruited via forest owner associations, Eriksson and Klapwijk (2019) found that the owners displayed an overall neutral attitude towards CCF. The study also showed that the attitude towards CCF was less positive compared to evaluations of biodiversity considerations made as part of RF (e.g., leaving dead wood) but more positive compared to forest preservation. In both studies, the perceptions of CCF among NIPF owners were heterogeneous, suggesting that owners lacked a coherent view of CCF.

Axelsson and Angelstam (2011) examined how local forestry stakeholders, including NIPF owners and private and public forest professionals, perceived CCF in two case study areas in northern and central Sweden. They found a range of perceptions associated with CCF, including the historically based view of selective cutting and thinning as forms of forest exploitation, but also the more current perceptions regarding management for aims other than production. Results revealed concerns regarding the economic viability of CCF, but also perceived benefits regarding the social and ecological aspects of CCF. Another study showed that forestry stakeholders emphasising the importance of ecological and social ecosystem services, and problems associated with biodiversity loss, were more positive towards managing the forest for biodiversity benefits, including CCF (Eriksson and Klapwijk 2019).

Hertog et al. (2022) examined perceived barriers to endorsing CCF in a survey of Swedish stakeholders, i.e., selected institutional forest owners (e.g., municipal forest managers), one private forest owner, professionals working with CCF, and actors in mainstream forestry (e.g. forest owner associations and companies). Results showed that these actors had different perceptions of barriers to a large-scale uptake of CCF. The current forestry culture and the power of industrial networks were perceived to be more important barriers among the actors endorsing CCF than actors in mainstream forestry. Instead, the latter emphasised low economic profitability and ecological risks. Both groups considered current forestry education and technical knowledge about CCF as barriers to an increased uptake of CCF.

Changes in silviculture and harvesting strategies could be one way to adapt to a changing climate. Heltorp et al. (2018) used focus group interviews to investigate the views of Norwegian forest owners and forest professionals on forest management and adaptation to climate change. In general, climate change was viewed as an opportunity rather than a threat, with limited need to adapt. In a larger survey of members of owner associations, Norwegian NIPF owners were asked if they would consider CCF as a strategy to adapt to climate change (Heltorp 2019). Approximately half of the respondents were indifferent. These findings are supported in a study of forestry professionals by Roitsch et al. (2023). More research is needed to determine and understand the underlying factors behind these results. However, it is evident that CCF is currently attracting growing interest in the Norwegian forest sector, and courses are offered to NIPF owners and forest professionals such as harvester operators.

15.3 Forest Industry Companies

Positions taken by large forest companies on CCF are relevant to NIPF owners' management decisions, because the companies play a key role in the timber market, their employees are instructing the forest owners, and they are making decisions in the forest during harvesting. However, forest companies' perceptions of CCF have been examined in only a couple of Swedish studies so far. Companies regard CCF as inferior to conventional clearcut-based forestry (Heder Brandt et al. 2023; Hertog et al. 2022; Espmark 2017). CCF is only considered suitable for enhancing social values close to urban areas, for promoting reindeer husbandry, and for preserving certain species and ecological structures, as expressed for example by the Swedish Forest Industries Federation on their website:

> CCF (Hyggesfritt skogsbruk) has its place in achieving certain objectives and where special consideration needs to be given that cannot be met by clear-cutting. This may, for example, involve maintaining or enhancing recreational values in areas close to urban areas, preserving and promoting deciduous forests that are important for reindeer husbandry or for promoting fungi/mycorrhiza. (Skogsindustrierna 2023)

Similar statements can be found in information provided by the major forest companies Sveaskog, Holmen and Sydved. The practice of CCF is confined to areas managed for objectives other than production, such as Sveaskog's Ekoparks and SCA's areas of high conservation status. The Finnish forest industry companies, UPM and the Metsä Group, and Finnish-Swedish StoraEnso make similar statements on their websites, as does the Finnish Forest Industries Federation (Finnish Forest Industries Federation 2021). In Finland, CCF has been suggested for peatland forests to reduce carbon emissions. UPM has stated that they will convert to CCF in all their own nutrient-rich peatland forests dominated by Norway spruce (UPM 2021). UPM owns a total of 515,000 ha of forest in Finland (UPM 2023).

As arguments against an expansion of CCF, forest companies refer to a lack of technical knowledge, ecological uncertainties, and lower economic profitability (Hertog et al. 2022). Any larger shift towards alternative forest management methods must be preceded by a shift in company attitudes. According to the study by Heder Brandt et al. (2023), CCF was mentioned only three times in the magazines of three major forest companies in Sweden between 2019 and 2021. The EU Forest Strategy 2030 seems to have prompted a shift towards a stronger focus on environmental issues and led to more articles mentioning CCF during 2021–2022. The magazine Skogsnära, issued by Stora Enso Ltd., has opened the possibility of incorporating alternative forest management methods instead of clearcutting, but RF and clearcutting practices dominate the magazine content. The current RF model is described as positive for the environmental, economic, and social aspects of forest management, making it the superior option for sustainable forestry, at least according to Swedish forest companies. This shows that external pressures have not yet brought about changes in the attitudes of major forest companies (Heder Brandt et al. 2023).

15.4 Conclusions and Discussion

CCF is growing slowly in Norway, Sweden and Finland, but there are considerable differences between the countries regarding the extent of the CCF debate and interest in understanding how NIPF owners and forestry stakeholders perceive CCF. In Norway, CCF has not been used widely and only a few studies were found that touched on Norwegian forest owners' views on CCF, and no studies on forest industries and their policies were found. In Sweden, a comprehensive understanding of forest owners' and forestry professionals' views on CCF is still lacking. However, studies have addressed the policies of forest industry companies regarding CCF and how the companies promote their views to forest owners in their own publications. Barriers to CCF have been studied from the perspective of the Swedish forestry sector. In Finland, the revision of the Forest Act in 2014 allowing for CCF triggered many studies that examined the NIPF owners' willingness to adopt CCF. Forestry professionals' perceptions have also been studied in Finland, but research into how policies of forest industry companies and other institutional surroundings impact the adoption of CCF is still limited.

According to the studies carried out in Finland, NIPF owners are interested in testing CCF more in the future, at least in part of their forests. Interest for CCF is particularly common among those Finnish forest owners who emphasise nature and recreation objectives in their forestry. In Sweden, forestry practitioners' views have been similar, and CCF is considered a good management method for promoting forestry goals associated with ecosystem services other than timber production. Forestry practitioners both in Finland and in Sweden have concerns about the economic viability of CCF (Chap. 8), and RF is perceived to be more profitable.

Forest industry companies in Sweden and Finland communicate similar viewpoints, considering CCF as a possible management strategy if the forest owner wants to adopt it, but they also acknowledge the associated risks regarding profitability. The Swedish studies on forest industry companies show more clearly that CCF is considered inferior to RF. It is regarded as suitable for urban forests, for areas of reindeer husbandry, and other areas where certain recreational or ecological objectives are particularly important, but is not to be used if timber production goals are the priority. Since the forest industry companies have an important role in forest management, such as by providing decision support to forest owners, their views on CCF are important in understanding the viewpoints of owners and professionals. As Sténs et al. (2019) point out, public agencies, forestry professionals, and researchers still regard the economic uncertainty of CCF as an obstacle (see also Mason et al. 2021).

The reviewed studies suggest that the spread of CCF has been slowed by institutional factors. While there are currently no institutional barriers to CCF implementation, there is no direct support for CCF either. Cost-sharing or institutional support systems affect the economic profitability of forest management. In Norway and Sweden, forest owners do not generally get financial support for forest management operations. Although certification schemes support CCF in the Swedish context, this is mainly limited to certain forests (Stens et al. 2019). In Finland, the Government provides financial support for several forestry operations, such as tending of seedlings and young stands. The current Finnish system of support for private forestry favours RF, but the removal of the requirement for minimum area in the latest support scheme could enable financial support for group and gap cutting (Laturi et al. 2021; Viitala et al. 2023). Other studies have highlighted logistical barriers associated with mechanised harvesting, market mechanisms whereby forest industries prefer medium-sized rather than large logs, ecological barriers in the form of browsing pressure by ungulates, and ways to introduce CCF into even-aged forests (Mason et al. 2021).

The studies show that forest owners and forestry professionals lack knowledge and experience of CCF, and this is likely preventing greater uptake of CCF in the Nordic countries. The Finnish and Norwegian studies showed that many forest owners were unsure about CCF, and the uncertainties associated with CCF have also been highlighted by forestry professionals in Finland. Lack of knowledge and limited CCF skills among forestry professionals have been highlighted as a key barrier to wider adoption of CCF in Europe, but also a range of uncertainties and lack of scientific knowledge regarding the effect of CCF, for example, regarding climate change and the extent to which CCF can increase forest resilience (Mason et al. 2021).

The need to improve training of forest professionals and industrial workers in individual-based forest management was highlighted by Kruse et al. (2023). Forest owners frequently use advisory services to support their decision making, and advisory services and information to forest owners have an important role in the forest management solutions adopted (Hänninen et al. 2020; Pynnönen et al. 2021). Advisory services therefore play an important role in the spread of new forest management practices. With more experience of CCF, the forestry professionals' knowledge is likely to increase. However, it is still possible that current practices may restrict the adoption of CCF, given that the current RF is considered sustainable and the most efficient way to produce high-quality timber, as shown in the Finnish study of discourses in advisory services (Takala et al. 2023).

Even though forestry actors generally advocate CCF as an alternative management method for environmental, social and cultural ecosystem services, contrasting views are evident (Espmark 2017). Advocates of some CCF methods (Naturkultur with a focus on felling of individual trees providing the most value for the land, and the Lübeck model building on natural processes and endorsing cost minimisation rather than maximisation of production) generally emphasise how CCF is economically superior and the need to maximise revenues for forest owners. Less evident among forestry actors are the views that CCF should be used to combat climate change, which is generally endorsed by climate scientists, and a preference of CCF over RF because the latter is considered an environmental catastrophe, a view generally endorsed by environmental organisations.

If CCF is to be promoted on the basis of benefits such as biodiversity, climate change mitigation or water quality, the barriers or hindrances to its implementation need to be removed. The current RF has been in use for a long time, so a change in management regime requires thorough re-evaluation of the current course of action. It may also be necessary to develop specific forms of support for transitioning to CCF on sites where environmental benefits can be achieved by changing the forest management method. Research needs include the development of effective decision support and different kinds of incentive systems, and the impact these may have on forest owners' decision-making and choices.

References

Andersson E, Keskitalo C (2018) Adaptation to climate change? Why business-as-usual remains the logical choice in Swedish forestry. Glob Environ Chang 48:76–85. https://doi.org/10.1016/j.gloenvcha.2017.11.004

Axelsson R, Angelstam P (2011) Uneven-aged forest management in boreal Sweden: local forestry stakeholders' perceptions of different sustainability dimensions. Forestry 84(5):567–579. https://doi.org/10.1093/forestry/cpr034

Eriksson L, Klapwijk MJ (2019) Attitudes towards biodiversity conservation and carbon substitution in forestry: a study of stakeholders in Sweden. Forestry 92(2):219–229. https://doi.org/10.1093/forestry/cpz003

Espmark K (2017) Debatten om hyggesfritt skogsbruk i Sverige: En analys av begrepp och argument i svenskt pressmaterial 1994–2013 (The debate on continuous cover forestry in Sweden: An analysis of concepts and arguments in Swedish press material 1994–2013). Future forests Rapportserie 2017:2. Sveriges lantbruksuniversitet, Umeå

Ficko A, Lidestav G, Ní Dhubháin Á, Karppinen H, Zivojinovic I, Westin K (2019) European private forest owner typologies: A review of methods and use. Forest Policy Econ 99:21–31. https://doi.org/10.1016/j.forpol.2017.09.010

Finnish Forest Industries Federation (2021) Metsänhoidossa käytetään useita eri kasvatusmenetelmiä (Several different silvicultural methods are used in forest management). https://www.metsateollisuus.fi/uutishuone/metsanhoidossa-kaytetaan-useita-eri-kasvatusmenetelmia. Accessed 21 Aug 2023

Hänninen H, Valonen M, Haltia E (2020) Metsänomistajat palveluiden käyttäjinä: Metsänomistaja 2020-tutkimuksen tuloksia. Luonnonvara- ja biotalouden tutkimus 63/2020, Luonnonvarakeskus, Helsinki, p 63

Heder Brandt P, Olsson A, Dahlquist K, Inal T (2023) "Profitability is sustainability": framing of forest management practices by the Swedish forest industry. Scand J Forest Res 38:429–441. https://doi.org/10.1080/02827581.2023.2252740

Heltorp KMA (2019) Forestry and forest management in an uncertain environment – adaptation to climate change in Norwegian forestry. Philosophiae Doctor Thesis 2019:38, Norwegian University of Life Sciences. https://nmbu.brage.unit.no/nmbu-xmlui/handle/11250/2824156

Heltorp KMA, Kangas A, Hoen HF (2018) Do forest decision-makers in southeastern Norway adapt forest management to climate change? Scand J Forest Res 33(3):278–290. https://doi.org/10.1080/02827581.2017.1362463

Hertog IM, Brogaard S, Krause T (2022) Barriers to expanding continuous cover forestry in Sweden for delivering multiple ecosystem services. Ecosyst Serv 53:101392. https://doi.org/10.1016/j.ecoser.2021.101392

Horne P, Karppinen H, Korhonen O, Koskela T (2020) Metsien hoidon ja kasvatusmenetelmien hyväksyttävyys–Metsänomistaja 2020. PTT raportteja 266, Pellervon taloustutkimus PTT, Helsinki, p 81

Hujala T, Pynnönen S, Kurttila M, Arponen A, Kasurinen S, Tähtinen S, Primmer E, Paloniemi R (2016) Metsien monimuotoisuuden suojelu ja metsien käsittelyjä monipuolistavien neuvontapalvelujen kehittäminen: Maanomistajien ja metsäammattilaisten näkemyksiä (Protecting forest biodiversity and developing advisory services to diversify forest management: views of landowners and forest professionals). Luonnonvara- ja biotalouden tutkimus 69/2016. Luonnonvarakeskus, Helsinki, p 36

Husa M, Kosenius A-K (2021) Non-industrial private forest owners' willingness to man-age for climate change and biodiversity. Scand J For Res 36:614–625. https://doi.org/10.1080/0282758 1.2021.1981433

Juutinen A, Kurttila M, Pohjanmies T, Tolvanen A, Kuhlmey K, Skudnik M, Triplat M, Westin K, Mäkipää R (2021) Forest owners' preferences for contract-based management to enhance environmental values versus timber production. Forest Policy Econ 132:102587. https://doi.org/10.1016/j.forpol.2021.102587

Juutinen A, Tolvanen A, Koskela T (2020) Forest owners' future intentions for forest management. Forest Policy Econ 118:102220. https://doi.org/10.1016/j.forpol.2020.102220

Karppinen H, Hänninen H, Horne P (2020) Suomalainen metsänomistaja 2020 (Finnish Forest Owner 2020). Luonnonvara- ja biotalouden tutkimus 30/2020. Luonnonvarakeskus, Helsinki, p 73

Kietäväinen P (2022) Kuonanjoen valuma-alueen metsänomistajien näkemyksiä jatkuvapeitteisestä metsänkasvatuksesta. Lapin AMK, Opinnäytetyö, p 68

Kniivilä M, Hantula J, Hotanen JP, Hynynen J, Hänninen H, Korhonen KT, Leppänen J, Melin M, Mutanen A, Määttä K, Siitonen J, Viiri H, Viitala E-J, Viitanen J (2020) Metsälain ja metsätuholain muutosten arviointi. Luonnonvara- ja biotalouden tutkimus 3/2020, Luonnonvarakeskus, Helsinki, p 124

Koskela T, Horne P, Karppinen H, Korhonen O (2021) Metsien ekosysteemipalvelut ja jokamiehenoikeus metsänomistajan näkökulmasta–Metsänomistaja 2020. PTT raportteja 267, Pellervon taloustutkimus PTT, Helsinki, p 107

Kruse L, Erefur C, Westin J, Ersson BT, Pommerening A (2023) Towards a benchmark of national training requirements for continuous cover forestry (CCF) in Sweden. Trees For People 12:100391. https://doi.org/10.1016/j.tfp.2023.100391

Laturi J, Maidell M, Haltia E, Horne P, Määttä K, Uusivuori J (2021) Metsätalouden kannustinjärjestelmän evaluointi. Luonnonvara- ja biotalouden tutkimus 15/2021, Luonnonvarakeskus, Helsinki, p 80

Mason WL, Diaci J, Carvalho J, Valkonen S (2021) Continuous cover forestry in Europe: usage and the knowledge gaps and challenges to wider adoption. Forestry 95(1):1–12. https://doi.org/10.1093/forestry/cpab038

Matilainen A, Andersson E, Lähdesmäki M, Lidestav G, Kurki S (2023) Services for What and for whom? A literature review of private Forest owners' decision-making in relation to Forest-based services. Small-scale For 22:511–535. https://doi.org/10.1007/s11842-023-09541-3

Nordén A, Coria J, Jönsson AM, Lagergren F, Lehsten V (2017) Divergence in stakeholders' preferences: evidence from a choice experiment on forest landscapes preferences in Sweden. Ecol Econ 132:179–195. https://doi.org/10.1016/j.ecolecon.2016.09.032

Pynnönen S, Haltia E, Hujala T (2021) Digital forest information platform as service innovation: Finnish Metsaan. fi service use, users and utilisation. Forest Policy Econ 125:102404. https://doi.org/10.1016/j.forpol.2021.102404

Roitsch D, Abruscato S, Lovrić M, Lindner M, Orazio C, Winkel G (2023) Close-to-nature forestry and intensive forestry—two response patterns of forestry professionals towards climate change adaptation. Forest Policy Econ 154:103035. https://doi.org/10.1016/j.forpol.2023.103035

Skogsindustrierna (2023) Frågor och svar om hyggen i skogen (Questions and answers on forest harvesting). https://www.skogsindustrierna.se/om-skogsindustrin/vad-gor-skogsindustrin/skogsbruk/faq/om-hyggen-i-skogen/. Accessed 5 Jul 2023

Skogsstyrelsen (2021a) Hyggesfritt skogsbruk (Continuous cover forestry). https://www.skogsstyrelsen.se/bruka-skog/olika-satt-att-skota-din-skog/hyggesfritt-skogsbruk/. Accessed 5 Jul 2023

Skogsstyrelsen (2021b) Fastighets- och ägarstruktur i skogsbruk 2020 (Property and ownership structure in forestry in 2020). Sveriges officiella statistik. Statistiska meddelanden. JO1405 SM 2001. Skogsstyrelsen

Statistics Norway (2022) https://www.ssb.no/en/statbank/table/10613/tableViewLayout1/ Accessed 22 Aug 2023

Sténs A, Roberge JM, Löfmarck E, Beland Lindahl K, Felton A, Widmark C, Rist L, Johansson J, Nordin A, Nilsson U, Laudon H, Ranius T (2019) From ecological knowledge to conservation policy: a case study on green tree retention and continuous-cover forestry in Sweden. Biodivers Conserv 28:3547–3574. https://doi.org/10.1007/s10531-019-01836-2

Takala T, Tanskanen M, Brockhaus M, Kanniainen T, Tikkanen J, Lehtinen A, Hujala T, Toppinen A (2023) Is a sustainability transition possible within the decision-support services provided to Finnish forest owners? Forest Policy Econ 150:102940. https://doi.org/10.1016/j.forpol.2023.102940

UPM (2021) UPM korvaa avohakkuut korpikohteilla eri-ikäisrakenteisella kasvatuksella (UPM replaces clear-cutting in spruce dominating nutrient-rich peatlands with continuous cover forestry). https://www.upm.com/fi/tietoa-meista/medialle/tiedotteet/2021/10/upm-korvaa-avohakkuut-korpikohteilla-eri-ikaisrakenteisella-kasvatuksella/. Accessed 21 Aug 2023

UPM (2023) UPM Metsä (UPM Forest). https://www.upm.com/fi/liiketoiminnot/puunhankinta-ja-metsatalous/. Accessed 21 Aug 2023

Valkonen S, Cheng Z (2014) Metsäammattilaisten suhtautuminen metsän erirakenteiskasvatukseen (Forest professionals' attitudes to continuous cover forestry). Metsätieteen aikakauskirja 2/2014

Viitala E-J, Ahtikoski A, Haltia E, Hökkä H, Mäkipää R, Nieminen M, Saarinen M, Sarkkola S, Tolvanen A, Valkonen S (2023) Tehokkaat ohjauskeinot jatkuvapeitteisen metsänkäsittelyn edistämiseksi runsasravinteisilla turvemailla (Effective policy measures to promote continuous forest management in nutrient-rich peatlands). Luonnonvara- ja biotalouden tutkimus 100/2023. Luonnonvarakeskus, Helsinki, p 74

Westin K, Bolte A, Haeler E, Haltia E, Jandl R, Juutinen A, Kuhlmey K, Lidestav G, Mäkipää R, Rosenkranz L, Triplat M, Skudnik M, Vilhar U, Schueler S (2023) Forest values and application of different management activities among small-scale forest owners in five EU countries. Forest Policy Econ 146:102881. https://doi.org/10.1016/j.forpol.2022.102881

Concluding Remarks from the Editors

Christian Kuehne, Emma Holmström, Johanna Routa, Saija Huuskonen, Jonas Cedergren and Pasi Rautio

Continuous cover forestry (CCF) refers to forest management methods that exclude clearcutting to preserve forest-like conditions and promote natural regeneration. This book compiles and synthesises research results and experiences of CCF in Fennoscandia. It aspires to be a source of inspiration, a reference for further reading, and a guide for future research, but also a potential aid for decision making.

Because of differences in forest types, forest history, industrial development, ecological conditions, and land ownership among others, CCF (as defined in Chap. 2) has not been practised on a large scale in Fennoscandia (see Chap. 1 for the history of CCF use in Fennoscandia). In other parts of Europe, CCF has become

C. Kuehne
Swedish University of Agricultural Sciences, S. S. Forest Research Centre, Inst för sydsvensk skogsvetenskap, Lomma, Sweden
e-mail: christian.kuehne@nibio.no

E. Holmström
Norwegian Institute of Bioeconomy Research, Ås, Norway
e-mail: emma.holmstrom@slu.se

J. Routa
Natural Resources Institute Finland, Production Systems unit, Joensuu, Finland
e-mail: johanna.routa@luke.fi

S. Huuskonen
Natural Resources Institute Finland, Natural Resources unit, Helsinki, Finland
e-mail: saija.huuskonen@luke.fi

J. Cedergren
Skogforsk, Uppsala Science Park, Uppsala, Sweden
e-mail: jonas.cedergren@skogforsk.se

P. Rautio
Natural Resources Institute Finland, Natural Resources unit, Rovaniemi, Finland
e-mail: pasi.rautio@luke.fi

© The Author(s) 2025 287
P. Rautio et al. (eds.), *Continuous Cover Forestry in Boreal Nordic Countries*, Managing Forest Ecosystems 45, https://doi.org/10.1007/978-3-031-70484-0

increasingly common, mostly on public land, since the 1990s (e.g. Bauhus et al. 2013).

As defined for this book in Chap. 2, CCF encompasses different silvicultural systems, including some considered to be rotation forestry (RF). Harvest interventions in these systems vary in magnitude and periodicity, creating a variety of post-harvest conditions and stand dynamics suited for different forest types and management situations. The outcomes of these silvicultural systems can be diverse, despite all being considered as CCF. As a result, the different systems should be studied and evaluated separately, as they may be more or less suitable under different management settings. This is especially true for the two primary soil types found in the region: mineral and organic soils (e.g. Chaps. 8 and 13).

Many chapters of this book point out knowledge gaps and outline research needs, in particular how CCF impacts regeneration dynamics (Chap. 3), growth and yield (Chap. 4), genetic diversity (Chap. 7), susceptibility to biotic and abiotic disturbance agents (Chap. 12), and multiple use (Chap. 10). In general, comparisons between RF and CCF do not always lead to straightforward, consistent findings, as it depends on the subject studied. Contrasting findings can be observed even within the same discipline, e.g. forest damage (Chap. 12) and multiple use (Chap. 10). The outcome of a shift from RF to CCF also appears to vary among the different silvicultural systems that qualify as CCF (Chap. 2), corroborating that the different systems need to be studied separately. In addition, available studies are often short term and only evaluate the impact of single cuts rather than effects and dynamics over entire rotations (RF) or as a result of the multiple harvest cycles in CCF (e.g. Chap. 11). The long-term monitoring of well-designed experimental trials with permanent plots in CCF-managed forests thus deserves more attention in the near future.

Findings from beyond Fennoscandia on how CCF influences stand dynamics and the provision of forest goods and services cannot be transferred unconditionally to the Nordic region. Nevertheless, they can provide a first insight into the potential effects of CCF under Fennoscandian conditions. Fundamentals and basic theory can provide a foundation for inferences within a limited number of research fields covered in this book (e.g. Chap. 15). A summary of the most urgent research needs identified in each chapter is provided in Fig. A.1.

CCF is not likely to solve all problems currently attributed to RF in the region (e.g. Mönkkönen et al. 2018). Extending the fundamental principles of CCF with additional management actions and activities like deadwood retention and habitat tree preservation is a way toward better incorporating other management goals than timber production (e.g. Gustafsson et al. 2020a). This is especially true for promoting and conserving biodiversity (Chap. 11, see also e.g. Larsen et al. 2022), but some of these actions and activities are also relevant and suitable for RF (Gustafsson et al. 2020b). To what extent these additional management measures provide any benefit regarding CCF under Fennoscandian conditions remains largely unstudied.

CCF is not suitable or advisable in all locations. Successful and economically feasible implementation requires trained personnel and a permanent forest road infrastructure (e.g. Chap. 6). Whether CCF is a suitable approach depends primarily on the owner's management goals (Nyland 2016). It is plausible that CCF works

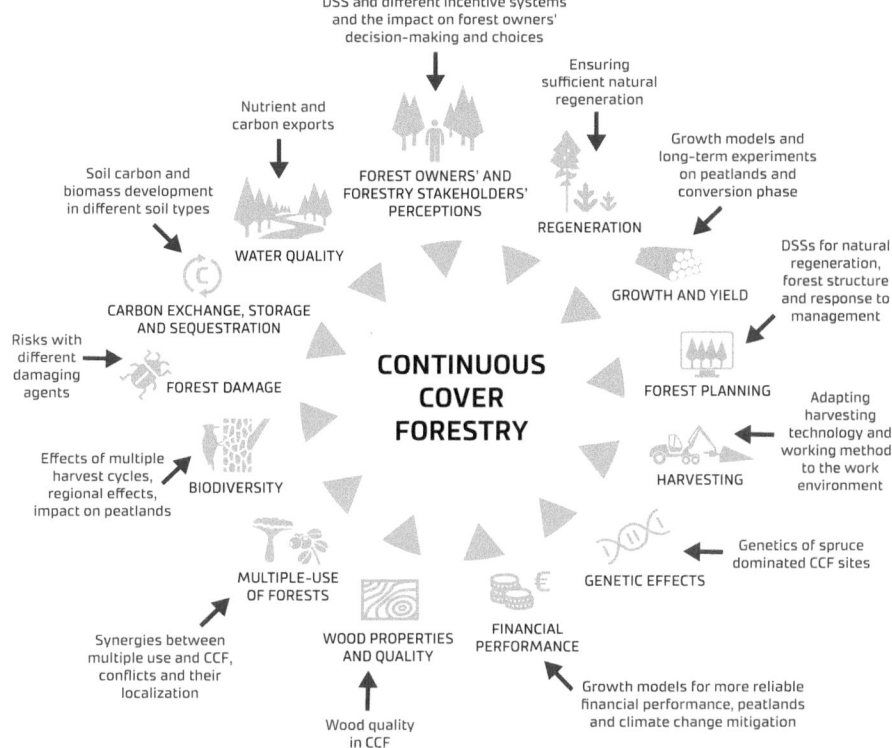

Fig. A.1 Major knowledge gaps and future research needs for CCF in the Nordic region

better than RF under specific conditions and for certain suites of management goals (see e.g. Chap. 10). This is likely why local forest administrations aim to or have already begun to shift toward CCF in forests close to larger settlements in Fennoscandia, such as in the Oslo community forest. However, a shift to CCF will often require conversion of stands established, and so far managed, under the RF paradigm. Best practices for the conversion of even-aged stands under Fennoscandian conditions are also lacking, and appropriate research is needed (Chap. 2).

Whether CCF is a suitable approach for adapting Fennoscandian forests to climate change remains to be seen (Felton et al. 2024). Proper species selection and the right choice of a CCF-conforming silvicultural system are crucial (Chap. 2). In addition, other forest management measures such as mixed-species forestry (Felton et al. 2016, Huuskonen et al. 2021), and suitable thinning regimes (Moreau et al. 2022) likely offer better prospects in the near term (see also Triviño et al. 2023b). However, these elements can be implemented both in CCF and RF.

Forest ecosystems in Fennoscandia are currently expected to meet many different management goals, such as raw material production, carbon sequestration, biodiversity conservation, and adaptation to climate change. Some of these goals may conflict with each other (Högbom et al. 2021). A wide range of forest management approaches and methods is likely needed to meet these contrasting goals (Eyvindson et al. 2021). In addition, stand- vs. landscape-level and mid- vs. long-term considerations are crucial in forest management and need to be taken into account. Defining the best targets and opportunities of each management approach and combining different treatment methods appropriately can provide greater well-being, income, biodiversity, carbon sequestration, and recreational opportunities than any single approach or method alone (e.g. Duflot et al. 2022, Triviño et al. 2023a). While CCF is likely to gain a wider application throughout Fennoscandia in upcoming years, this might happen in parallel to resumed larger-scale RF operations (Eyvindson et al. 2021). Research is needed on the effects of CCF management, to assure a better, more balanced, and targeted provision of goods and services from Nordic forests as a result of this development.

The multiple CCF-related knowledge gaps revealed in this book must be addressed to enable appropriate advice to forest owners and stakeholders on when (and where) to apply CCF or RF. The research needs highlighted here will help direct CCF-related research in Fennoscandia over the coming years. Without solid scientific evidence, knowledge-based decisions about best forest management practices cannot be made. This could lead to underuse of Fennoscandian forests' potential in achieving the UN Sustainable Development Goals, such as biodiversity protection, climate change mitigation, and sustainable economic growth.

References

Bauhus J, Puettmann KJ, Kühne C (2013) Close-to-nature forest management in Europe: does it support complexity and adaptability of forest ecosystems? In: Messier C, Puettmann KJ, Coates KD (eds) Managing forests as complex adaptive systems: building resilience to the challenge of global change. Routledge, The Earthscan Forest Library, pp 187–213

Duflot R, Fahrig L, Mönkkönen M (2022) Management diversity begets biodiversity in production forest landscapes. Biol Conserv 268:109514

Eyvindson K, Duflot R, Triviño M et al (2021) High boreal forest multifunctionality requires continuous cover forestry as a dominant management. Land Use Policy 100:104918

Felton A, Nilsson U, Sonesson J et al (2016) Replacing monocultures with mixed-species stands: ecosystem service implications of two production forest alternatives in Sweden. Ambio 45:124–139. https://doi.org/10.1007/s13280-015-0749-2

Felton A, Belyazid S, Eggers J et al (2024) Climate change adaptation and mitigation strategies for production forests: trade-offs, synergies, and uncertainties in biodiversity and ecosystem services delivery in Northern Europe. Ambio 53:1–16

Gustafsson L, Bauhus J, Asbeck T et al (2020a) Retention as an integrated biodiversity conservation approach for continuous-cover forestry in Europe. Ambio 49:85–97

Gustafsson L, Hannerz M, Koivula M et al (2020b) Research on retention forestry in Northern Europe. Ecol Process 9:3

Högbom L, Abbas D, Armolaitis K et al (2021) Trilemma of nordic–baltic forestry—how to implement UN sustainable development goals. Sustain For 13:5643

Huuskonen S, Domisch T, Finér L et al (2021) What is the potential for replacing monocultures with mixed-species stands to enhance ecosystem services in boreal forests in Fennoscandia? For Ecol Manag 479:118558

Larsen JB, Angelstam P, Bauhus J et al (2022) Closer-to-Nature Forest Management. From science to policy 12. European Forest Institute

Mönkkönen M, Burgas D, Eyvindson K et al (2018) Solving conflicts among conservation, economic, and social objectives in boreal production forest landscapes: Fennoscandian perspectives. In: Perera A et al (eds) Ecosystem services from Forest landscapes: Broadscale considerations. Springer, pp 169–219

Moreau G, Chagnon C, Achim A et al (2022) Opportunities and limitations of thinning to increase resistance and resilience of trees and forests to global change. Forestry 95(5):595–615

Nyland RD (2016) Silviculture: concepts and applications. Waveland Press

Triviño M, Morán-Ordoñez A, Eyvindson K et al (2023a) Future supply of boreal forest eco-
    system services is driven by management rather than by climate change. Glob Change Biol
    29(6):1484–1500

Triviño M, Potterf M, Tijerín J et al (2023b) Enhancing resilience of boreal forests through man-
    agement under global change: a review. Curr Landsc Ecol Rep 8:103–118